WA 1305392 2

D0228704

STANDARD LOAN

Renew Books on PHONE-it: 01443 654456
Help Desk: 01443 482625
Media Services Reception: 01443 482610

Books are to be returned on or before the last date below

INTELLIGENT FAULT DIAGNOSIS AND PROGNOSIS FOR ENGINEERING SYSTEMS

INTELLIGENT FAULT DIAGNOSIS AND PROGNOSIS FOR ENGINEERING SYSTEMS

GEORGE VACHTSEVANOS
FRANK LEWIS
MICHAEL ROEMER
ANDREW HESS
BIQING WU

WILEY

JOHN WILEY & SONS, INC.

Learning Resources
Centre

13053922

This book is printed on acid-free paper. ∞

For general information about our other products and services, please contact our Customer Care Department within the United States at (800) 762-2974, outside the United States at (317) 572-3993 or fax (317) 572-4002.

Wiley also publishes its books in a variety of electronic formats. Some content that appears in print may not be available in electronic books. For more information about Wiley products, visit our web site at www.wiley.com.

Library of Congress Cataloging-in-Publication Data:
Intelligent fault diagnosis and prognosis for engineering systems / George
Vachtsevanos . . . [et al.].
      p. cm.
   Includes bibliographical references and index.
   ISBN-13: 978-0-0471-72999-0 (cloth)
   ISBN-10: 0-471-72999-X (cloth)
  1. Fault location (Engineering) 2. System analysis. I. Title: Intelligent
fault diagnosis and prognosis for engineering systems. II. Vachtsevanos,
George J.
     TA169.6.I68 2006
     620'.0044—dc22

                           2006005574

Printed in the United States of America

10 9 8 7 6 5 4 3 2 1

To my wife Athena.
G.V.

To Roma, my son.
F.L.L.

To my wife Sharon, son Michael, and daughter Aubrey.
M.J.R.

To my parents, Yanying Chen and Qingliang Wu.
B.W.

CONTENTS

PREFACE

Condition-based maintenance (CBM) and prognostics and health management (PHM) have emerged over recent years as significant technologies that are making an impact on both military and commercial maintenance practices. We are indeed witnessing a true paradigm shift in the way complex dynamic systems (aircraft and spacecraft, shipboard systems, industrial and manufacturing processes, etc.) are designed, monitored, and maintained. Fault diagnosis and prognosis of the failing component's remaining useful life, as well as logistics support activities required to maintain, repair, or overhaul such critical systems, require active contribution from multiple disciplines. CBM and PHM technologies are characterized by the merging and strong coupling of interdisciplinary trends from the engineering sciences, computer science, reliability engineering, communications, management, etc. Writing a book that is not merely an edited collection of technical papers in this area but presents systematically and as thoroughly as possible the diverse subject matter from sensors and systems studies to logistics therefore is a challenging and arduous task that requires the contributions of many experts and the compilation of source and reference material across disciplines.

CBM and PHM technologies are evolving rapidly, their potential application domains are expanding, and the customer base for these technologies is increasing at a phenomenal rate. Although they targeted initially the equipment maintainer as the end user, substantial benefits also may accrue from their implementation for the operator, the process manager or field commander, and the equipment designer. The latter can take advantage of CBM/PHM studies in order to design or redesign critical systems or processes so that they exhibit improved attributes of fault tolerance and high confidence. It is therefore difficult, if not impossible, to capture in a single volume all aspects of such a broad, evolving, and multidisciplinary field of knowledge. We tried to provide the reader with those essential elements of these emerging technologies with the understanding that some topics are either omitted or abbreviated because of page limitations. Finally, it is certain that by the time the book reaches the reader's desk, new contributions, algorithms, case studies, etc. will reach the public domain.

The organization and basic subject matter for this book parallel a successful four-day intensive short course entitled "Fault Diagnostics/Prognostics for Equipment Reliability and Health Maintenance" offered by Dr. George Vachtsevanos and his colleagues at Georgia Tech and on location at various industries and government agencies over the past years.

ACKNOWLEDGMENTS

The integration task needed to pull together a cohesive and up-to-date presentation was enormous and would not have been possible without the unselfish and enthusiastic support we received from our colleagues and coworkers. The research work of faculty and students at the Intelligent Control Systems Laboratory of the Georgia Institute of Technology forms the backbone of many topics covered in this book. The authors are grateful to Marcos Orchard, Romano Patrick, Bhaskar Saha, Abhinav Saxena, Taimoor Khawaja, Nicholas Propes, Irtaza Barlas, Guangfan Zhang, and Johan Reimann, among others, for their valuable contributions. Their research was supported by DARPA, the U.S. Army's Research Office, ONR, AFRL, and industry. We are gratefully acknowledging the support we received over the past years from all our sponsors.

Some of the diagnostic signal-processing methods and the ideas on wireless sensor nets for CBM were developed at the University of Texas at Arlington's Automation & Robotics Research Institute.

We would like to acknowledge the contributions of Carl S. Byington and Gregory J. Kacprzynski of Impact Technologies with respect to the use cases on mechanical component diagnostics and prognostics.

Thanks to Rick Lally and his staff at Oceana Sensor for input to the sensors chapter. Our special thanks to Drs. Thomas and Jennifer Michaels of the School of Electrical and Computer Engineering at Georgia Tech for the section on ultrasonic sensors and to Mr. Gary O'Neill of the Georgia Tech Research Institute for contributing the chapter on logistics. Our gratitude to Kai Goebel and his GE colleagues for their contributions, Jonathan Keller of the U.S. Army, ARDEC, and Itzhak Green of Georgia Tech's School of Mechanical Engineering. It is impossible to acknowledge by name many researchers and CBM/PHM practitioners whose work we reference in this volume.

Finally, the prologue was written by Andy Hess, the recognized father and guru of PHM. It is his initiatives, encouragement, and support that provided the impetus for our work in this new and exciting area.

GEORGE VACHTSEVANOS
Georgia Institute of Technology

FRANK LEWIS
University of Texas, Arlington

MICHAEL ROEMER
Impact Technologies, LLC

ANDREW HESS
Joint Strike Fighter Program Office

BIQING WU
Georgia Institute of Technology

PROLOGUE

The desire and need for accurate diagnostic and real predictive prognostic capabilities have been around for as long as human beings have operated complex and expensive machinery. This has been true for both mechanical and electronic systems. There has been a long history of trying to develop and implement various degrees of diagnostic and prognostic capabilities. Recently, stringent advanced diagnostic, prognostics, and health management (PHM) capability requirements have begun to be placed on some of the more sophisticated new applications. A major motivation for specifying more advanced diagnostic and prognostic requirements is the realization that they are needed to fully enable and reap the benefits of new and revolutionary logistic support concepts. These logistic support concepts are called by many names and include *condition-based maintenance* (CBM), *performance-based logistics* (PBL), and *autonomic logistics*—all of which include PHM capabilities as a key enabler.

The area of intelligent maintenance and diagnostic and prognostic–enabled CBM of machinery is a vital one for today's complex systems in industry, aerospace vehicles, military and merchant ships, the automotive industry, and elsewhere. The industrial and military communities are concerned about critical system and component reliability and availability. The goals are both to maximize equipment up time and to minimize maintenance and operating costs. As manning levels are reduced and equipment becomes more complex, intelligent maintenance schemes must replace the old prescheduled and labor-intensive planned maintenance systems to ensure that equipment continues to function. Increased demands on machinery place growing importance on keeping all equipment in service to accommodate mission-critical usage.

While fault detection and fault isolation effectiveness with very low false-alarm rates continue to improve on these new applications, prognosis requirements are even more ambitious and present very significant challenges to system design teams. These prognostic challenges have been addressed aggressively for mechanical systems for some time but are only recently being explored fully for electronics systems.

A significant paradigm shift is clearly happening in the world of complex systems maintenance and support. Figure P.1 attempts to show the analogy of this shift between what is being enabled by intelligent machine fault diagnosis and prognosis in the current world of complex equipment maintenance and how the old-time coal miners used canaries to monitor the health of their mines. The "old" approach was to put a canary in a mine, watch it periodically, and if it died, you knew that the air in the mine was going bad. The "new" approach would be to use current available technologies and PHM-type capabilities to continuously monitor the health of the canary and get a much earlier indication that the mine was starting to go bad. This new approach provides the predictive early indication and the prognosis of useful mine life remaining, enables significantly better mine maintenance and health management, and also lets you reuse the canary.

Prognosis is one of the more challenging aspects of the modern prognostic and health management (PHM) system. It also has the potential to be the most beneficial in terms of both reduced operational and support (O&S) cost and life-cycle total ownership cost (TOC) and improved safety of many types of machinery and complex systems. The evolution of diagnostic monitoring systems for aircraft and other complex systems has led to the recognition that predictive prognosis is both desired and technically possible.

Aircraft diagnostic technologies have been around for well over 40 years in some form or another. With current advances in computing resources, the

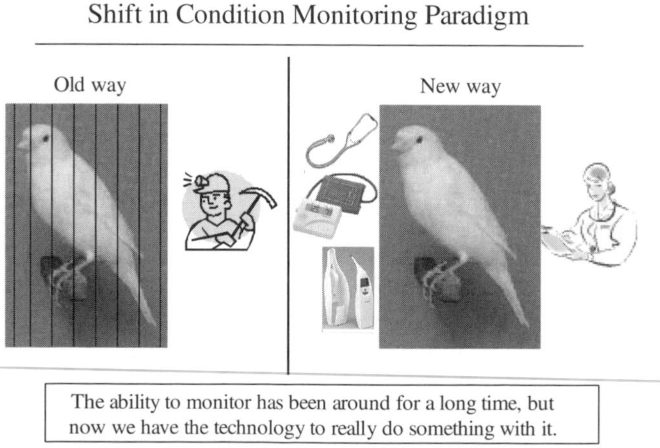

Shift in Condition Monitoring Paradigm

The ability to monitor has been around for a long time, but now we have the technology to really do something with it.

Figure P.1   The "old" canary paradigm and its "new" equivalent.

ability to perform a greatly enhanced diagnostic capability is available to the user fleet. This capability arises by allowing much of the processing of data to occur on-board the aircraft, thus eliminating the potential for lost data, data dropout, and incorrectly processed data at the ground station. Furthermore, today's highly digital aircraft have an enormous quantity of information available on Mux buses and the like, so it is possible to exploit these data and use them for diagnostic practices. By exploiting these data and turning them into information, many features relating to the health of the aircraft can be deduced in ways never before imagined. This information then can be turned into knowledge and better decisions on how to manage the health of the system. This same evolution is happening in other industries and with other complex systems—machines are getting more intelligent and providing larger, richer data sources.

The increase in this diagnostic capability naturally has evolved into something more: the desire for prognosis. Designers reasoned that if it were possible to use existing data and data sources to diagnose failed components, why wouldn't it be possible to detect and monitor the onset of failure, thus catching failures before they actually hamper the ability of the air vehicle to perform its functions. By doing this, mission reliability would be increased greatly, maintenance actions would be scheduled better to reduce aircraft down time, and a dramatic decrease in life-cycle costs could be realized. It is with this mind-set that many of today's "diagnostic systems" are being developed with an eye toward prognosis.

P.1 SOME BACKGROUND, NEEDS, AND IMPACTS

If you were to ask a fleet user or maintainer what his or her most pressing needs were, you most probably would get a mixture of many answers. He or she would identify many needs, including some of the following responses: maximize sortie-generation rates and aircraft or system availability, very low "can not duplicate" (CND) and "returned tested OK" (RTOK) rates, minimum or no periodic inspections, low number of spares required and small logistics footprint, quick turn-around time, maximized life usage and accurate parts life tracking, no false alarms, etc. Figure P.2 depicts some of these and other fleet user needs. What the maintainer actually may want most of all is no surprises. He or she would like to be able to have the ability to predict future health status accurately and to anticipate problems and required maintenance actions ahead of being surprised by "hard downing" events. Capabilities that would let the maintainer see ahead of both unplanned and even some necessary maintenance events would enable a very aggressive and beneficial opportunistic maintenance strategy.

These combined abilities both to anticipate future maintenance problems and required maintenance actions and to predict future health status are key

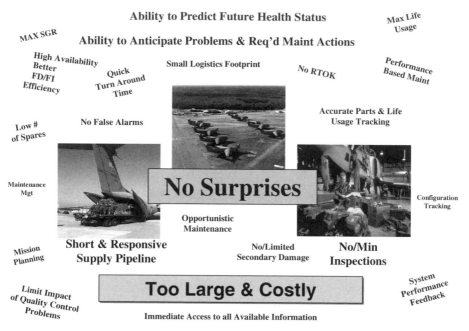

Figure P.2 Fleet user requirements.

enablers and attributes to any condition-based maintenance (CBM) or new logistics support concept and to "enlightened" opportunistic maintenance decisions. These predictive abilities and prognostic capabilities are also key enablers and very necessary attributes to both the evolving performance-based logistics (PBL) concepts and paradigm-changing new business case approaches.

Substantial improvements in both reliability and maintainability over legacy aircraft or system are essential if the logistics footprint, sortie generation rate, mission reliability, and anticipated O&S cost reductions are to be achieved. A modern and comprehensive prognosis and health management system needs to interface with the logistic information system to trigger (1) the activities of the supply-chain management system promptly to provide required replacement parts, (2) planning needed to perform the required maintenance, and (3) training of the maintainer. Typical turn-around time (TAT) to perform any required depot-level maintenance action is reduced significantly over that of legacy aircraft as a result of the advanced notification of maintenance requirements and readily available tools and replacement parts. Within security limitations, failure data should be available to all levels, including sustaining engineering personnel, OEMs, major command staff, and unit-level war fighter personnel. Some of the benefits of such a PHM system are listed below.

Change of Maintenance Philosophy

- On condition
- Opportunistic
- Not "on failure" nor "per schedule"
- Less interruption of mission schedule

Reduction in Test Equipment (TE)

- Less intermediate and flight-line TE
- 35 percent less peculiar support equipment during system design and development (SDD)
- Eliminated O-level TE

Benefits to the Maintainer

- Unprecedented insight into vehicle/squadron/fleet health
- Less time spent on inspections
- Better ability to plan maintenance
- Simplified training
- Improved fault detection

Historically, such predictions would have been more of a black art than a science. Essentially, prognosis provides the predictive part of a comprehensive health management system and so complements the diagnostic capabilities that detect, isolate, and quantify the fault, and prognosis, in turn, depends on the quality of the diagnostic system.

From an operator's, maintainer's, or logistician's—the user's—point of view, what distinguishes prognosis from diagnosis is the provision of a lead time or warning time to the useful life or failure and that this time window is far enough ahead for the appropriate action to be taken. Naturally, what constitutes an appropriate time window and action depends on the particular user of the information and the overall system of systems design one is trying to optimize. Thus prognosis is that part of the overall PHM capability that provides a prediction of the lead time to a failure event in sufficient time for it to be acted on.

What, then, constitutes a prognostic method—essentially any method that provides a sufficient answer (time window) to meet the user's needs. This is an inclusive approach and one that is seen as necessary to get the system of systems-wide coverage and benefits of prognosis. Some would argue that the only true prognostic methods are those based on the physics of failure and one's ability to model the progression of failure. While this proposition has scientific merit, such an unnecessarily exclusive approach is inappropriate given the current state of the art of prognostic technologies and the most likely outcome of insufficient coverage to provide system-wide prognosis.

Thus prognostic methods can range from the very simple to the very complex. Examples of the many approaches are simple performance degradation trending, more traditional life usage counting (cycle counting with damage assessment models and useful life remaining prediction models), the physics of failure-based, sometime sensor driven and probabilistic enhanced, incipient fault (crack, etc) propagation models, and others.

Modern and comprehensive PHM systems include many types of functional capabilities. Table P.1 lists most of these constituent functions, capabilities, and associated processes. Many of these functional capabilities, although much advanced and more accurate, are an evolution of capabilities developed by legacy diagnostic systems. Through the philosophy and concept of health management, the modern PHM system uses these functional capabilities to complement each other and provide a much larger impact and a broader set of maintenance-oriented benefits than any single function by itself.

Although most of these functional capabilities will be developed together and can complement each other during PHM system development for a particular platform, real predictive prognosis represents the new, hard, and often "risky" technology capability.

To understand the role of predictive prognosis, one has to understand the relationship between diagnosis and prognosis capabilities. Envisioning an initial fault to failure progression timeline is one way of exploring this relationship. Figure P.3 represents such a failure progression timeline. This timeline starts with a new component in proper working order, indicates a

TABLE P.1 Some PHM System Functional Capabilities

Fault detection
Fault isolation
Advanced diagnostic technologies
Predictive prognostic technologies
Useful life remaining, time-to-failure predictions
Component life usage tracking
Performance degradation trending
False-alarm mitigation
Warrant guarantee tracking, tools, and information for enabling new business
    practices
Health reporting
    Only tells operator what *needs* to be known immediately
    Informs maintenance of the rest
Aids in decision making and resource management
Fault accommodation
Information fusion and reasoners
Information management
    Right information to right people at right time

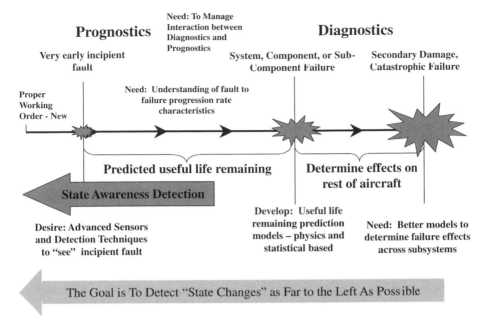

Figure P.3 Failure progression timeline.

time when an early incipient fault develops, and depicts how, under continuing usage, the component reaches a component or system failure state and eventually, if under further operation, reaching states of secondary system damage and complete catastrophic failure.

Diagnostic capabilities traditionally have been applied at or between the initial detection of a system, component, or subcomponent failure and complete system catastrophic failure. More recent diagnostic technologies are enabling detections to be made at much earlier incipient fault stages. In order to maximize the benefits of continued operational life of a system or subsystem component, maintenance often will be delayed until the early incipient fault progresses to a more severe state but before an actual failure event. This area between very early detection of incipient faults and progression to actual system or component failure states is the realm of prognostic technologies.

If an operator has the will to continue to operate a system and/or component with a known, detected incipient fault present, he or she will want to ensure that this can be done safely and will want to know how much useful life remains at any point along this particular failure progression timeline. This is the specific domain of real predictive prognosis, being able to accurately predict useful life remaining along a specific failure progression timeline for a particular system or component.

To actually accomplish these accurate useful life remaining prediction capabilities requires many tools in your prognostic tool kit. Sometimes available sensors currently used for diagnosis provide adequate prognostic state awareness inputs, and sometimes advanced sensors or additional incipient fault detection techniques are required. This brings up the questions: How early or small of an incipient fault do you want to detect, and how small of a material state change do you want or need to "see"? Other needed tools in your prognostic tool kit include a model or set of models that represents the understanding of a particular fault to failure progression rate, material physics of failure models, statistical and/or probabilistic-based models, models to represent failure effects across interconnected subsystems, and models to account for and address future operational mission usage.

In summary, the needs and benefits of intelligent machine and system fault diagnosis and prognosis are apparent. The positive impacts of these diagnostic and prognostic capabilities on reducing operation and support (O&S) cost and total ownership cost (TOC) are significant. These diagnostic and prognostic capabilities are the key enablers of the new and revolutionary maintenance concept being implemented today. All the technology elements and tool-kit capabilities are available and coming together to enable this vision of accurate and real intelligent machine and complex system diagnosis and prognosis and health management. And finally, all these diagnostic and prognostic technology elements, techniques, and capabilities must be applied and implemented wisely to obtain the maximum benefit impacts.

ANDREW HESS
Joint Strike Fighter Program Office

CHAPTER 1

INTRODUCTION

1.1 HISTORICAL PERSPECTIVE

Over the last several decades, there has been a wide range of approaches and implementation strategies for performing manual, semiautomated, or fully automated fault diagnosis and prognosis (i.e., health management) on critical systems in commercial and defense markets. Associated with these diagnostic and prognostic system designs are an equally diverse number of philosophies and associated architectures used to implement them for particular applications. For example, early-generation aircraft relied on manual detection and isolation of problems on the ground to meet their health management needs. These systems typically were analog and independent of one another, with only a schematic and voltmeter readings available to troubleshoot the problems. As aircraft systems became more complicated and integrated, built-in test (BIT) equipment, simple alarms, and trending analysis were implemented to warn the operators of safety-critical situations. However, the system maintainers still did not fully use this capability and often continued to rely on the voltmeter, schematics, and operator reports to solve problems or improve designs.

Continuing with aircraft diagnostic system evolution as an example (although other equipment has similar parallels), original equipment manufac-

1

turers (OEMs) and vehicle integrators began to realize that the output of the fault detection alarms or BIT equipment could be made available to support system troubleshooting. With this concept, fault indications were made available on the front of a line-replaceable unit (LRU) that indicated that a fault had been detected—they were originally mechanical but later were replaced with small light-emitting diodes (LEDs). These types of indicators progressed over the 1950s and 1960s. In many cases the LRU front panel contained a test switch to command the LRU to test itself, in a manner similar to how ground-support equipment could test the LRU. This capability became known as *built-in test equipment* (BITE). This capability began to decrease the need for some of the ground-support equipment previously used to test critical equipment. Depending on the system, the LRU fault indicators effectively could point the mechanic in the right direction, but schematics and voltmeters still were used in many conditions. However, the BITE of this era often was confusing, not reliable, and difficult to use. Mechanics often distrusted it. Many systems on airplanes such as the Boeing 707, 727, and early 737/747, and the McDonnell Douglas DC-8, DC-9, and DC-10 employed this type of diagnostic and maintenance practice.

In the 1970s, as computer systems began to be introduced, some of the increasingly complex systems began to use computers to perform their calculations. This was called *digital BITE*. With these computers came the ability to display fault detection and isolation information in digital form, normally via numeric codes, on the front panel of the LRU. The digital logic could produce codes that could better isolate the cause of the fault. The digital display offered the capability to display many different codes to identify each type of fault that was detected. These codes often pointed to some description in a manual that could be used to isolate and correct the fault. Many systems on the Boeing 757/767, Airbus A300/310, McDonnell Douglas DC-10, and Lockheed L-1011 employ this approach.

As the number of systems grew, use of separate front-panel displays to maintain the systems became less effective, particularly since each LRU often used a different technique to display its fault data. In addition, some of the systems had become increasingly integrated with each other. Digital data buses, such as ARINC 429, began to be used during this time period. Autopilot systems, for example, became among the first to use these digital data buses and depend on sensor data provided by many other systems. The success of these systems became a driving force in the definition of more sophisticated maintenance systems. The more sophisticated monitoring was necessary to meet the integrity and certification requirements of its automatic landing function. For example, the 767 maintenance control and display panel brought together the maintenance functions of many related systems. As the next step, ARINC 604, defined in 1986, provided a central fault display system (CFDS) that brought to one display the maintenance indications for all the systems on the airplane. This approach enabled more consistent access to maintenance data across the aircraft systems and a larger display than each of the systems could contain individually and saved the cost of implementing front-panel displays on many of the associated system LRUs. In this approach, the CFDS were used to select the system for which maintenance data were desired and then routed the maintenance text from that system to the display. This approach was applied to many of the systems on later Boeing 737s and most systems on the Airbus A320/330/340 and the McDonnell Douglas MD11.

Systems continued to become more complex and integrated. A single fault on an airplane could cause fault indications for many systems, even when displayed using systems such as the CFDS. In many cases, the mechanic had little help in determining which indication identified the source fault and which was merely an effect. To solve this and related issues, the ARINC 624 was developed in the early 1990s. It defines a more integrated maintenance system that can consolidate the fault indications from multiple systems and provide additional functionality to support condition-based maintenance. With such an architecture, minimal ground-support equipment is needed to test airplane systems because most of this capability is included in the maintenance system. For example, most factory functional tests of airplane systems on the Boeing 747-400 and 777 airplanes consist of little more than execution of selected tests, monitoring fault displays, and monitoring certain bus data using the integrated maintenance system.

The evolution of diagnostic and prognostic reasoning systems thus has strived to better identify the single LRU or component that is the source of the fault. This allows the maintainer to remove the failed component and correct the fault condition confidently. Although in many cases this is possible, there are many situations where it is not possible without the addition of sensors or wiring. Addition of these sensors increases the number of components that can fail and thus sometimes can worsen the maintenance effort. In addition, they add cost and weight to the airplane. There are clearly cases where the addition of such hardware can be beneficial, but the benefits of improved fault isolation must be weighed against the potential reduced reliability and increased cost and weight of the additional components.

1.2 DIAGNOSTIC AND PROGNOSTIC SYSTEM REQUIREMENTS

A critical part of developing and implementing effective diagnostic and prognostic technologies is based on the ability to detect faults in early enough stages to do something useful with the information. Fault isolation and diagnosis uses the detection events as the start of the process for classifying the fault within the system being monitored. Condition and/or failure prognosis then forecasts the remaining useful life (the operating time between detection and an unacceptable level of degradation). If the identified fault affects the life of a critical component, then the failure prognosis models also must reflect this diagnosis. Specific requirements in terms of confidence and severity levels must be identified for diagnosis and prognosis of critical failure modes. In general, the fault diagnosis detection level and accuracy should be specified separately from prognostic accuracy.

As a minimum, the following probabilities should be used to specify fault detection and diagnostic accuracy:

1. The probability of anomaly detection, including false-alarm rate and real fault probability statistics.
2. The probability of specific fault diagnosis classifications using specific confidence bounds and severity predictions.

To specify prognostic accuracy requirements, the developer/end user must first define

1. The level of condition degradation beyond which operation is considered unsatisfactory or undesirable to the mission at hand.
2. A minimum amount of warning time to provide the operator and maintainer required information that can be acted on before the failure or condition is encountered.
3. A minimum probability level that remaining useful life will be equal to or greater than the minimum warning level.

1.3 DESIGNING IN FAULT DIAGNOSTIC AND PROGNOSTIC SYSTEMS

Following the evolution of diagnostic systems in the aircraft industry, prognostic initiatives started to be introduced in order to try to take advantage of the maintenance planning and logistics benefits. However, the early prognostic initiatives often were driven by in-field failures that resulted in critical safety or high-cost failures, and thus retrofitted technology was hard to implement and costly to develop. Hence diagnostic and prognostic system developers found the need to analyze and describe the benefits associated with reducing in-field failures and their positive impact on safety, reliability, and overall life-cycle-cost reduction. This has lead to many cost-benefit analyses and ensuing discussions and presentations to engineering management about why the diagnostic and prognostic technologies need to be included in the design process of the system and not simply an afterthought once field failures occur. This had lead us to the point where many complex vehicle/system designs such as the Joint Strike Fighter (F-35), DD(X), Expeditionary Fighting Vehicle (EFV), and various future combat system (FCS) vehicles are now developing "designed in" health management technologies that can be implemented within an integrated maintenance and logistics system that supports the equipment throughout its life time. This "designed in" approach to health management is performed with the hardware design itself and also acts as the process for system validation and managing inevitable changes from in-field experiences and evaluating system design tradeoffs, as shown in Fig. 1.1.

Realizing such an approach will involve synergistic deployments of component health monitoring technologies, as well as integrated reasoning capabilities for the interpretation of fault-detect outputs. Further, it will involve the introduction of learning technologies to support the continuous improvement of the knowledge enabling these reasoning capabilities. Finally, it will involve organizing these elements into a maintenance and logistics architecture that governs integration and interoperation within the system, between its on-board elements and their ground-based support functions, and between the health management system and external maintenance and operations functions. In this book we present and discuss the required functions of health management technologies that, if applied correctly, can directly affect the operations and maintenance of the equipment and positively affect the life-cycle costs.

1.4 DIAGNOSTIC AND PROGNOSTIC FUNCTIONAL LAYERS

A comprehensive health management system philosophy integrates the results from the monitoring sensors all the way through to the reasoning software that provides decision support for optimal use of maintenance resources. A core component of this strategy is based on the ability to (1) accurately predict the onset of impending faults/failures or remaining useful life of critical components and (2) quickly and efficiently isolate the root cause of failures once failure effects have been observed. In this sense, if fault/failure predictions can be made, the allocation of replacement parts or refurbishment actions can

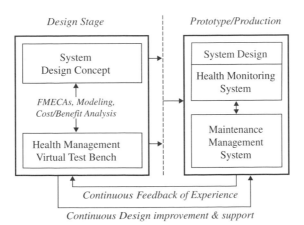

Figure 1.1 The "designed in" approach to health management.

be scheduled in an optimal fashion to reduce the overall operational and maintenance logistic footprints. From the fault isolation perspective, maximizing system availability and minimizing downtime through more efficient troubleshooting efforts is the primary objective.

In addition, the diagnostic and prognostic technologies being developed under a number of new acquisition programs implementing an autonomous logistics system concept require an integrated maturation environment for assessing and validating prognostics and health management (PHM) system accuracy at all levels in the system hierarchy. Developing and maintaining such an environment will allow for inaccuracies to be quantified at every level in the system hierarchy and then be assessed automatically up through the health management system architecture. The final results reported from the system-level reasoners and decision support is a direct result of the individual results reported from these various levels when propagated through. Hence an approach for assessing the overall PHM system accuracy is to quantify the associated uncertainties at each of the individual levels, as illustrated in Fig. 1.2, and build up the accumulated inaccuracies as information is passed up the vehicle architecture.

This type of hierarchical verification and validation (V&V) and maturation process will be able to provide the capability to assess diagnostic and prognostic technologies in terms of their ability to detect subsystem faults, diagnose the root cause of the faults, predict the remaining useful life of the faulty component, and assess the decision-support reasoner algorithms. Specific metrics to be discussed in this book will include accuracy, false-alarm rates, reliability, sensitivity, stability, economic cost/benefit, and robustness, just to name a few. The technical performance and accuracy of the diagnostic and prognostic algorithms also will need to be evaluated with performance met-

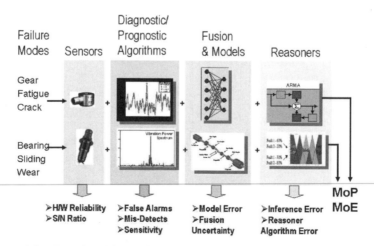

Figure 1.2   Functional layers in the health management system architecture.

rics, whereas system-level capabilities in terms of achieving the overall operational goals, economic cost/benefit, and assessment of the business case will be evaluated with effectiveness measures (also discussed in detail in Chap. 7).

Cost-effective implementation of a diagnostic or prognostic system will vary depending on the design maturity and operational/logistics environment of the monitored equipment. However, one common element to successful implementation is feedback. As components or LRUs are removed from service, disassembly inspections must be performed to assess the accuracy of the diagnostic and prognostic system decisions. Based on this feedback, system software and warning/alarm limits should be optimized until desired system accuracy and warning intervals are achieved. In addition, selected examples of degraded component parts should be retained for testing that can better define failure progression intervals.

1.5 PREFACE TO BOOK CHAPTERS

This book details the technologies in condition-based maintenance (CBM) and prognosis and health management (PHM) that have been introduced over the recent past by researchers and practitioners and are making significant inroads in such application domains as mechanical, thermal, electromechanical, and more recently, electrical and electronics systems. Thus our target applications are exclusively hardware-related. It is well recognized, though, that modern dynamical systems are a tightly coupled composite of both hardware and software. Software reliability is undoubtedly a serious concern and a challenge. Although we recognize its importance within the general area of reliability and maintainability, we are not addressing issues of software reliability in this book. The interested reader is referred to research publications on this topic stemming from the computer science community.

This book is structured as follows: Chapter 2 introduces those fundamental system concepts that set the stage for the effective design of fault diagnostic and prognostic technologies. We review systems-based methodologies that have a direct and significant impact on the design of CBM/PHM systems. The CBM/PHM designer must be thoroughly familiar with the physics of failure mechanisms and must posses an understanding of methods for the optimal selection of monitoring strategies, algorithms to detect and isolate faults and predict their time evolution, and systems approaches to design of experiments and testing protocols, performances metrics, and means to verify and validate the effectiveness and performance of selected models. A formal framework is established to conduct trade studies aimed at comparing alternative options and assisting in the selection of the "best" technologies that meet customer requirements. Failure modes and effects criticality analysis forms the foundation for good CBM/PHM design. We detail how it assists in deciding on the severity of failure modes, their frequency of occurrence, and their testability. It considers fault symptoms and the required sensor suite to monitor their behavioral patterns. It also may list the candidate diagnostic and prognostic algorithms that are best suited to address specific failure modes.

Chapter 3 discusses sensor systems employed for monitoring and interrogating critical system components/subsystems. Our focus here is not only on conventional sensors used primarily for monitoring and control purposes but also on sensor systems and sensing strategies that are specifically designed to respond to CBM/PHM design requirements. Advances in the latter category over the recent past have made available to the CBM/PHM community new sensors that are promising to improve substantially the reliability, cost-effectiveness, coverage, and sensitivity to fault signatures of monitoring and fault tracking devices. If CBM/PHM systems are to penetrate the reliability/maintainability domain for critical military and industrial assets successfully, sensor technology must provide the assurance that measurements are accurate and reliable. Hardware and software sensor systems must perform "better" than the components/subsystems they are intended to monitor. We discuss in this chapter basic transducing principles and signal conditioning issues. We review sensor systems used to monitor fundamental physical processes of heat transfer, pressure, speed, vibration, etc. Of special concern are devices that respond to structural damage and anomalies encountered in rotating equipment, electrical systems, etc. Thus we elaborate on properties and discuss ultrasonic systems, proximity devices, strain gauges, acoustic emission sensors, etc. Recent advances in wireless sensing, Micro-Electro-Mechanical Systems (MEMS), and smart sensors are introduced. An important consideration in sensing strategies refers to the optimal placement of sensors so that metrics of fault detectability and distinguishability are maximized. Wireless communication protocols, sensor interfaces, and standards for sensor systems are surfacing lately and are establishing a systematic framework for advanced CBM/PHM system design.

Chapter 4 addresses key issues regarding data processing and database management methodologies. Whereas the preceding chapter introduced the hardware technologies and sensing strategies required to acquire CBM/PHM-relevant data, this chapter focuses on and provides answers to such important questions as: How do we relate fault (failure) mechanisms to the fundamental "physics" of complex dynamic systems? Where is the information? How do we extract it from massive raw databases? What is an optimal database configuration and database management schema? Answers to these questions set the groundwork and constitute the underpinnings for fault diagnosis and prognosis. Thus we begin by discussing essential signal-processing tools for virtual sensing, data validation, and data analysis. The signal-processing toolbox addresses a variety of time-domain and frequency-domain methods for data analysis and fault feature extraction. The sections here are aimed specifically at familiarizing the reader with tools employed for the analysis of signals encountered in CBM/PHM applications. Examples from vibration monitoring are used to illustrate the basic concepts. Case studies from aircraft engine and gearbox fault signal analysis are presented. Feature selection and extraction techniques constitute the cornerstone for accurate and reliable fault diagnosis. We present a systematic approach and metrics of performance for feature selection and extraction. Tracking system usage patterns and factoring their potential impact into fault initiation and propagation studies are of great concern when designing CBM/PHM algorithms for such complex systems as aircraft, among many others. We conclude this chapter with a brief introduction to database management schemes. The preceding three chapters have layed the foundation for the main objective of this book: fault diagnosis and prognosis.

Chapter 5 is an extensive treatment of fault diagnostic technologies. We begin with a general framework for fault diagnosis, and then we introduce relevant definitions and a brief review of fault diagnosis requirements and performance metrics. A more detailed discussion of performance metrics is deferred to Chapter 7. Diagnostic methods historically have been divided into two major categories: methods that rely primarily on data and employ data-driven algorithms or tools and those that exploit physics-based models and measurements to classify fault conditions. Data-driven methods emphasize historical data diagnostic tools, statistical legacy data classification, and clustering algorithms based on fuzzy logic and neural network constructs. We introduce in this category reasoning and decision-making methods—fuzzy logic. Dempster-Shafer evidential theory is an effective means to combine fault evidence, resolve conflicting information, and address incomplete data sets, novel wavelet neural networks, and multimodel filter classifiers—whose primary objective is to detect and isolate incipient failures accurately and reliably while minimizing false alarms. We present next a concise summary of basic dynamic system models and system identification methods because they form the theoretical foundation for model-based diagnostic routines and assist in characterizing fault behavioral patterns. We continue in this section with a discussion of virtual (soft or analytical) sensors—software models that augment typically scarce fault databases—and filtering/estimation methodologies borrowed from the classic engineering disciplines—Kalman filtering, recursive least-squares estimation, and particle filtering—that take advantage of both state models that describe the fault or fatigue initiation and evolution and sensor measurements. We call on a variety of techniques to address the fault diagnosis problem: fatigue and crack-propagation models, finite-element analysis, and other modeling tools that are finding increased popularity because they provide valuable insight into the physics of failure mechanisms. Examples and case studies are used to illustrate the algorithmic developments throughout this chapter.

Failure prognosis is the Achilles' heel of CBM/PHM systems. Once an incipient failure or fault is detected and isolated, as discussed in the preceding chapter, the task of the prognostic module is to estimate as accurately and precisely as possible the remaining useful life of the failing component/subsystem. Long-term prediction entails large-grain uncertainty that must be represented faithfully and managed appropriately so that reliable, timely, and useful information can be provided to the user. Chapter 6 introduces recent advances in the development of prognostic algorithms. Such developments are still in their early stages, although a number of techniques have been proposed whose validity has not yet been confirmed via ground-truth data. The research community is approaching the prognosis problem not only recognizing the need for robust and viable algorithms but also understanding that sufficient and complete ground-truth failure data from seeded fault testing or actual operating conditions are lacking. This is a major impediment to the training and validation of the algorithmic developments. The failure prognosis problem basically has been addressed via two fundamental approaches. The first one builds on model-based techniques, where physics-based, statistical probabilistic, and Bayesian estimation methods are used to design fatigue or fault growth models. The second approach relies primarily on the availability of failure data and draws on techniques from the area of computational intelligence, where neural-network, neuro-fuzzy, and other similar constructs are employed to map measurements into fault growth parameters. We begin this chapter by suggesting the challenge faced by the designer, and we present a notional framework for prognosis. In the model-based category, we detail contributions in physics-based fatigue models. We introduce concepts from the dynamical model identification area that form the basis for such well-known statistical modeling techniques as auto regressive moving average (ARMA) and ARMA with exogenous input (ARMAX). The treatment here is intended to introduce the reader to fundamental concepts in prediction. Bayesian probability theory is discussed as the foundational framework for estimation and prediction, and two estimation methods—Kalman filtering and particle filtering—are detailed as two methods that are finding utility in failure prognosis. The methods are illustrated with examples from the area of aircraft failure mode prediction. We conclude this chapter with a discussion of data-driven prediction techniques, highlighting the utility of neural-network constructs employed as the mapping tools. Learning aspects are emphasized as essential ingredients in the improvement process, where the prediction algorithms continue to "learn" and adapt their parameters as new data become available. Tools for managing uncertainty are also suggested, aimed at reducing the uncertainty bounds that are unavoidably growing with the prediction horizon. Recent results from seeded fault testing on aircraft components complement the algorithmic modules. Probabilistic and particle-filtering methods seem to hold the greatest promise because they combine physics-based failure models with available measurements acquired online in real time. An adaptive framework allows for continuous adjustment of model parameters for improved prediction accuracy and precision. Research currently is under way to develop tools for material microstructural characterization that lead to meta-models capable of capturing more accurately the dynamics of failure mechanisms. It is anticipated that these developments, coupled with the availability of adequate seeded fault data, will drive the design and implementation of reliable and robust prognostic technologies. Performance metrics for prognostic algorithms are discussed in the next chapter.

Chapter 7 addresses important issues relating to CBM/PHM system requirements and performance/effectiveness metrics. Assessing the technical and economic feasibility of CBM/PHM systems constitutes an essential component of the overall design process. Researchers and CBM/PHM practitioners have been attempting to define such metrics for specific application domains without reaching as yet a consensus. We introduce in this chapter a set of basic metrics developed over the past years by researchers in this field with the objective of bringing to the reader an initial attempt at this difficult topic and to motivate further developments. Recent mandates to implement CBM/PHM technologies on critical military and commercial assets are dictating the need for the definition and development of fault diagnosis and prognosis requirements. Defining a candidate list of requirements is predicated on the availability of performance metrics for major modules of the CBM/PHM system. We begin by introducing fault feature evaluation metrics. Since features or condition indicators are the foundational elements of diagnosis, it is important that the selected feature vector meet certain criteria that will enable the realization of accurate and robust diagnostic routines. Information, distance, and dependence measures form the theoretical basis for feature evaluation. They lead to a two-sample z-test that assesses whether the means of two groups or feature distributions are statistically different from each other. Performance metrics for fault diagnostic systems are discussed next. Among them, such typical metrics as false positives and false negatives are defined. The receiver operating characteristic offers the designer a means to evaluate how well a fault is diagnosed and to trade off between false positives and false negatives. An architecture suitable for diagnostic metrics assessment is suggested, and examples are used to illustrate its utility. Defining metrics for prognostic systems presents unique challenges. The probabilistic nature of the prediction model and the inherent uncertainty dictate that measures of performance must be defined and assessed statistically. Accuracy, precision, and confidence are notions closely associated with our ability to predict the occurrence of an event such as the time to failure of a particular component. They are used to define a set of performance metrics for prognostic systems. Diagnostic and prognostic effectiveness metrics are discussed next, and cost-benefit analysis concepts are reviewed briefly as applied to CBM/PHM systems. We conclude this chapter by defining a candidate list of CBM/PHM system requirements based on specific performance metrics.

The eighth and final chapter answers the question: Who is the customer, and how does he or she benefit from CBM/PHM practices? The potential CBM/PHM customer list started initially with the maintainer. The maintainer is the recipient of the diagnostic and prognostic modules output; he or she exploits it to schedule maintenance so that equipment uptime is maximized; resources for repair, maintenance, and overhaul are optimized; and major failure events are avoided. The list, though, is growing, and it is generally recognized that additional benefits may be derived through CBM/PHM impacting the system designer's objectives, the operations manager, and the military field commander, among others. This chapter focuses on logistics support and enumerates potential benefits and enabling technologies needed to integrate CBM/PHM into the logistics culture. We describe how accurate prediction of the system's conditions affects the operational planning and scheduling tasks, the procurement of spare parts and material needed to restore the system to normal operation, financial performance, personnel training and system redesign considerations given the evidence provided by the CBM/PHM on the health of system components, the performance of sensors or lack thereof, and required sensing strategies. We introduce possible methodologies for logistics support and optimal scheduling strategies and present case studies to illustrate the concepts and techniques introduced in this chapter.

1.6   REFERENCES

G. Bloor et al., "An Evolvable Tri-Reasoner IVHM System," Proceedings of *IEEE Aerospace Conference,* Big Sky, MT, 2000.

J. Gray and G. Bloor, "Application Integration," Boeing's 3rd Annual Data Exchange Conference, Mesa, AZ, 2000.

G. Karsai and J. Gray, "Component Generation Technology for Semantic Tool Integration," *Proceedings of the IEEE Aerospace,* Big Sky, MT, 2000.

CHAPTER 2

SYSTEMS APPROACH TO CBM/PHM

2.1 INTRODUCTION

Condition-based maintenance (CBM) is the use of machinery run-time data to determine the machinery condition and hence its current fault/failure condition, which can be used to schedule required repair and maintenance prior to breakdown. *Prognostics and health management* (PHM) refers specifically to the phase involved with predicting future behavior, including remaining useful life (RUL), in terms of current operating state and the scheduling of required maintenance actions to maintain system health. Detecting a component fault or incipient failure for a critical dynamic system (aircraft, gas turbine, pump, etc.) and predicting its remaining useful life necessitate a series of studies that are intended to familiarize the CBM/PHM designer with the *physics of failure mechanisms* associated with the particular system/component. Moreover, the designer must have a thorough understanding of methods for optimal selection of monitoring strategies, tools, and algorithms needed to detect, isolate, and predict the time evolution of the fault, as well as systems approaches for design of experiments and testing protocols, performance metrics, and means to verify and validate the effectiveness and performance of the selected models. New systems approaches, as well as established ones from other disciplines, may be brought to the CBM/PHM domain with suitable modifications.

In this chapter we will review a number of systems-based methodologies that have a direct impact on the design of CBM/PHM systems: a formal framework to conduct *trade studies* that are intended to compare alternative options for the selection of components, sensors, and algorithms and to assist in the selection of "best" alternative technologies according to specified requirements. *Failure modes and effects criticality analysis* (FMECA) forms the foundation for good CBM/PHM design. In concert with *reliability-centered maintenance* (RCM), a FMECA study decides on the severity of candidate failure modes, their frequency of occurrence, and their testability. For each failure mode, it considers fault symptoms and the required sensor suite to monitor their behavioral patterns. In advanced versions, FMECA studies also may list the candidate diagnostic and prognostic algorithms that are best suited to address the identified failure modes. New fault data may be required in most cases that are essential for training and validating diagnostic and prognostic routines if historical data collected through on-system monitoring or testbench testing are not sufficient or nonexistent. Means, therefore, must be sought to devise and design a test plan, execute it, and assess the statistical significance of the collected data. Technical and economic performance metrics must be defined to guide the design process and evaluate the effectiveness and performance of the overall CBM/PHM system and its individual modules.

We will devote a separate chapter to performance assessment because it is an important and integral component of the CBM/PHM design. Here we will highlight the major concerns associated with a technoeconomic analysis and outline a possible approach to such studies. Finally, we will introduce the verification and validation (V&V) problem and suggest an approach to conduct V&V studies. Figure 2.1 depicts the main modules of an integrated approach to CBM/PHM system design with the systems-based components of the architecture described in this chapter highlighted. The schematic indicates feedback loops that are intended to optimize the approach and complete the data collection and analysis steps that are essential inputs to the development of the fault diagnostic and prognostic algorithms.

In Fig. 2.2 one can identify a preliminary offline phase and an online implementation phase of CBM/PHM. The online phase includes obtaining machinery data from sensors, signal preprocessing, extracting the features that are the most useful for determining the current status or fault condition of the machinery, fault detection and classification, prediction of fault evolution, and scheduling of required maintenance. These components are discussed in detail in Chaps. 5 and 6.

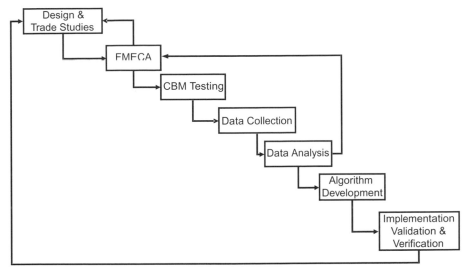

Figure 2.1   An integrated approach to CBM/PHM design.

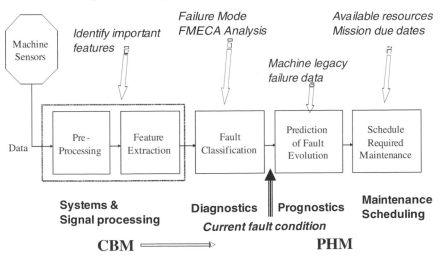

Figure 2.2   The CBM/PHM cycle.

The offline phase consists of the required background studies that must be performed offline prior to implementation of the online CBM phase and include determination of which features are the most important for machine condition assessment, the FMECA, the collection of machine legacy fault data to allow useful prediction of fault evolution, and the specification of available personnel resources to perform maintenance actions.

2.2   TRADE STUDIES

The major scope of a trade study is to arrive at the "best" or most balanced solution to the diagnosis and prognosis of critical component/subsystem failure modes leading to optimal CBM/PHM practices. Specific objectives of a trade study include

- Support the decision needs of the system engineering process.
- Evaluate alternatives (requirements, functions, configurations).
- Integrate and balance all considerations.
- Recommend "best" solution.
- Develop and refine a system concept.
- Determine if additional analysis, synthesis, or tradeoff studies are required to make a selection.

A formal framework for trade studies follows a methodology called *integrated product and process design* (IPPD) (Schrage et al., 2002). The IPPD methodology attempts to

- Establish the need.
- Define the problem.
- Establish value objectives.
- Generate feasible alternatives.
- Make decisions.

U.S. Department of Defense–mandated IPPD studies have been focusing on technology forecasting and technology identification, evaluation, and selection (TIES). Fault diagnosis and prognosis–related studies belong to the TIES category. An IPPD study addresses the following work tasks:

- *Define the problem.* Use the quality function deployment (QFD) tool to capture "customer" requirements, define design tradeoffs, and identify component functions and capabilities, monitoring and testing requirements, etc.

• *Establish value.* Establish value objectives and feasibility constraints. Employ the Pugh concept selection matrix to evaluate alternative component configurations, monitoring systems, etc.
• *Generate feasible alternatives.* Use the morphologic matrix tool to generate feasible alternatives through technology forecasting and synthesis. Of particular interest is the morphological matrix, which is used to identify the necessary technologies that are employed in the design of sensors, modeling tools, etc. and to recommend the selected system design architecture among those identified and evaluated by the Pugh concept.
• *Establish alternatives.* Use the Pugh concept selection matrix to evaluate feasible alternatives against the value objectives/criteria established in task 2.
• *Recommend a decision.* The recommended component design/monitoring enabling technologies are based on the results of the evaluation process.

The essential tools and their interaction for implementing the IPPD approach are shown in Fig. 2.3. In the figure, the *quality function deployment* (QFD), or *house of quality* (Sullivan, 1986) as it is sometimes known, captures the customer's requirements and defines design tradeoffs. It prioritizes the requirements, lists the characteristics of the components under study, and ranks capabilities versus the states requirements. It is used to translate the "voice of the customer" to identify the key "product and process characteristics/attributes." The *morphologic matrix* (MM) is a technology forecasting tool that can be used for initial synthesis of viable concepts that then can be evaluated, initially qualitatively, in a *Pugh concept evaluation*

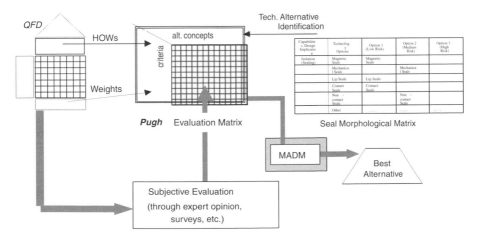

Figure 2.3 Tools and their interaction for implementing the IPPD approach.

matrix (PCEM) (Myers et al., 1989). The weighted criteria (or figures of merit) come from the QFD process. The MM identifies technology and subsystem solutions through brainstorming and engineering expertise and experience and should include "out of the box" and breakthrough technologies during the initial iteration. While the first evaluation can be purely subjective, using plus or minus assessments, use of *multiattribute decision making* (MADM) *model* allows a more quantitative selection approach that incorporates the weighted criteria evaluated in the concept(s) selection.

Refer to Fig. 2.3 during this discussion. The roof of the QFD shows the interrelationships between the defined capabilities. Typical conventions for ranking these capabilities employ a scale from 0 to 10, with 10 the most significant, or a qualitative metric, such as strong, medium, weak, for the interrelationships occupying the roof of the QFD. The relative importance of the capabilities in meeting the customer requirements is finally computed as the percent ranking. Of major concern from a CBM/PHM point of view are requirements addressing size, cost, materials, durability, maintainability/replaceability, etc.

As an example to illustrate the approach, Fig. 2.4 depicts a QFD for a trade study aimed at selecting the best seal configuration for the drive train of an advanced vehicle. Seal capabilities include performance/effectiveness, monitoring/sensing, suitability for fault diagnosis/prognosis, affordability, and physical model capability. The QFD of Fig. 2.4 suggests that suitability for fault diagnosis and prognosis is the most critical capability required to meet the customer requirements. A typical morphologic matrix for the seal example is shown in Table 2.1.

The capabilities identified in the QFD are next evaluated against the technology options. The matrix lists a number of potential sealing configurations, such as mechanical seals, lip seals, etc., and considers only three options for comparison purposes—best, average, and poor. Here, expert opinions may be solicited, or simulation studies, if models are available, can be performed to arrive at numerical rankings. Multiple morphologic matrices are generated depicting each capability versus the technology options. Comparative results of sealing effectiveness are summarized in a table format. Table 2.2 summarizes the comparative results with the technology options compared with typical seal failure modes. The convention used in Table 2.2 is explained in the template of Table 2.3, where the operating conditions, that is, temperature, pressure, etc., are ranked as high (A), medium (B), or low (C). A multiattribute decision making (MADM) approach is employed to select the best option.

2.3   FAILURE MODES AND EFFECTS CRITICALITY ANALYSIS (FMECA)

Understanding the physics of failure mechanisms constitutes the cornerstone of good CBM/PHM system design. FMECA studies are aiming to fulfill this

Roof;		Matrix;		
Strong	**S**	Strong	10	Θ
Positive	**P**	Medium	5	**O**
Weak	**W**	Weak	1	Δ

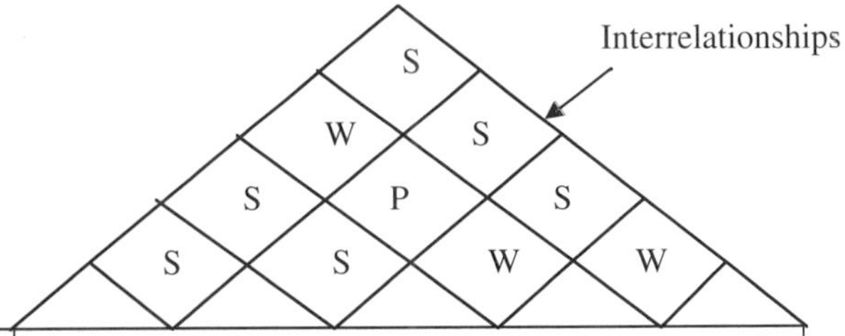

Interrelationships
S
W
S
S
P
S
S
S
W
W

Customer Requirements \ Design Requirements		SEAL CAPABILITIES/CHARACTERISTIC (HOW)				
SEAL REQUIREMENTS (WHAT) (AAA mission → Drivetrain mission → Seal requirements	Priority/Importance	Performance / Effectiveness (Tribological / Rotor Dynamics / Materials)	Monitoring/Sensing (Fault Detectability Identifiability)	Suitability for Fault Diagnostics/Prognostics	Affordability	Physical Model Capability for Fault Diagnostics/Prognostics
Size (precision, dimensioning, balance ratio etc.)	8	O	O	O	Δ	Δ
Cost	6	O	Δ	Δ	θ	Δ
Materials (graphite, ceramics, magnetic forces, etc.)	7	Δ	θ	θ	O	Δ
Operating Conditions (tribological, sealing, etc.)	9	θ	θ	θ	O	θ
Durability	9	O	O	O	O	Δ
Access for Monitoring/Networking	8	Δ	θ	θ	Δ	O
Immunity to shock, vibrations, noise, etc.	8	θ	O	θ	Δ	O
Maintainability/Replaceability	10	Δ	θ	θ	Δ	θ
Acceptable Impact on Program	8	θ	O	θ	Δ	O
Monitoring (temperature, pressure, speed range, environmental conditions)	9	Δ	θ	θ	Δ	O
Absolute Importance		399	601	681	636	385
Relative Importance		17.3%	26.01%	29.6%	10.3%	16.9%

Figure 2.4 Quality function deployment (QFD).

TABLE 2.1   Seal Morphologic Matrix

Capabilities/Design Implication	Technology Options	Option 1 "Best" (Low Risk)	Option 2 "Average" (Medium Risk)	Option 3 "Poor" (High Risk)
	Mechanical seals			
Performance	a. Contacting	x		
Containment/sealing/ mating	b. Contacting magnetic	x		
Rotor dynamic performance	c. Noncontacting	x		
Dynamics of shaft/seals/ bearings	Lip seals		x	
Tolerance to vibrations	Labyrinth			x
Effectiveness for liquid (water, mud), sand, solids, etc./ temperatures, speed, pressure, harsh environmental conditions	Rubber (O-rings, etc.)			x

essential objective and to provide the designer with tools and procedures that will lead to a systematic and thorough framework for design. An FMECA study basically attempts to relate failure events to their root causes. Toward this goal, it addresses issues of identifying failure modes, their severity, frequency of occurrence, and testability; fault symptoms that are suggestive of the system's behavior under fault conditions; and the sensors/monitoring apparatus required to monitor and track the system's fault-symptomatic behaviors. Furthermore, advanced FMECA studies may recommend algorithms to extract optimal fault features or condition indicators, detect and isolate incipient failures, and predict the remaining useful life of critical components. Such studies generate the template for diagnostic algorithms. The FMECA framework may be integrated into existing supervisory control and data acquisition (SCADA) or other appropriate data management and control centers to provide the operator with convenient access to information regarding failure events and their root causes.

Why and how do components/subsystems fail? To answer this question, consider as an example the case of motor failures. Bearing failures, which constitute the majority of motor failures, may be due to improper or inadequate lubrication, shaft voltages, excessive loadings, vibrations, etc. Electrical and mechanical problems may cause motor burnout, such as higher than normal current, overheating, load fluctuations, etc. Specific motor types exhibit phenomena that require special attention. For example, failure modes of

TABLE 2.2 Comparative Results

| | | | | Failure Modes | | | | | Thermal | |
Technology Option	Blistering	Cracking	Scoring/Wear	Dynamics/Vibration	Misalignment	Heat Generation	Corrosion	Aging[a]	Shock	Cavitation
Mechanical seal										
a. Contacting	C1, C3	B3, B4	A1, A3	B1, B3	B2, B3	B1, B2, B3	A1		A1, C1	A3, C2
b. Contacting magnetic	A2, B1, B3	B3, B4	A1, A3	B1, B3	B2, B3	B1, B2, B3	A1		A1, C1	A3, C2
c. Noncontacting	Startup/shutdown			A3	B3	A3	A1		A1, C1	A3, C2
Lip seal	C1, C2, C3, C4	A4, B2, C3	A1, A2, A3, A4			A3	B1	A1, B1, C1		B3, C2
Rubber (O-ring or packing)	Static condition	A4, B2, C3	A1, A2, A3, A4			A3	B1	A1, B1, C1		

[a] Aging primarily includes hardening, but here it implies also creep, compression set, stress relaxation, cracking, swell, and chemical degradation.

TABLE 2.3   Operating Conditions Template

	High, A	Medium, B	Low, C
1. Temperature	A1	B1	C1
2. Pressure	A2	B2	C2
3. Speed	A3	B3	C3
4. Contamination	A4	B4	C4

inverter-fed motors may be caused by pinhole insulation breakdown, reflected-wave overvoltages, heating due to current harmonics, voltage buildup between the rotor and stator by high-frequency switching, etc. Domain knowledge therefore is absolutely essential, as is a better understanding of the physics of failure mechanisms, to design a good FMECA and, eventually, an integrated CBM/PHM architecture.

FMECA studies require the contribution of domain experts, reliability engineers, monitoring and instrumentation specialists, and designers charged with the responsibility of developing diagnostic and prognostic reasoners. Enabling technologies for FMECA design begin with simple spreadsheet-type tables accompanied by explanation modules and move to more sophisticated tools such as rule-based expert systems, decision trees, and Petri nets, among others.

The design of an FMECA must identify failure modes and rank them according to their severity, frequency of occurrence, and testability. *Severity* categorizes the failure mode according to its ultimate consequence. For example, a possible class breakdown may be

Category 1: Catastrophic, a failure that results in death, significant injury, or total loss of equipment.

Category 2: Critical, a failure that may cause severe injury, equipment damage, and termination.

Category 3: Marginal, a failure that may cause minor injury, equipment damage, or degradation of system performance.

Category 4: Minor, a failure that does not cause injury or equipment damage but may result in equipment failure if left unattended, downtime, or unscheduled maintenance/repair.

For *frequency of occurrence,* four possible classifications are

Likely
Probable
Occasional
Unlikely

The four classes are distinguished on the basis of *mean time between failure* (MTBF) ranges. As an example, for a particular failure mode, based on an MTBF of 10,000 hours, the four categories may encompass the following ranges:

Category 1: Likely greater than 1000
Category 2: Probable from 100 to 1000
Category 3: Occasional from 10 to 100
Category 4: Unlikely less than 10

The probability of a fault occurrence may be based on a classification category number from 1 to 4 (or possibly more divisions), with 4 being the lowest probability to occur. Separation of the four classes is determined on a log power scale. The classification number is derived based on failure occurrence for the particular event standardized to a specific time period and broken down into likely, probable, occasional, and unlikely.

The term *testability* is intended to indicate whether symptoms or indicators of a particular failure mode can be monitored via conventional or PHM sensors. Thus it excludes from the candidate failure mode set those modes that cannot be measured. It may indicate any additional sensing instrumentation that will alleviate this problem, particularly if the failure mode is critical and frequent. Testability considerations are incorporated in the FMECA as comments.

Example 2.1 FMECA Study for a Navy Shipboard Chiller. Industrial chillers support electronics, communications, etc. on a navy ship, computing and communications in commercial enterprises, and refrigeration and other functions in food processing. A large number of failure modes are observed on such equipment, ranging from vibration-induced faults to electrical failures and a multitude of process-related failure events. Most chillers are well instrumented, monitoring vibrations, temperature, fluid pressure, and flow. Figure 2.5 shows a U.S. Navy centrifugal chiller and its possible fault modes.

An FMECA study was conducted for a shipboard chiller under sponsorship by the Office of Naval Research (Propes et al., 2002). The relevant study information regarding one particular fault condition is summarized below:

Fault	Refrigerant charge high
Occurrence	Probable
Severity	Critical
Testability	Can be monitored
Description	Due to overcharge during maintenance

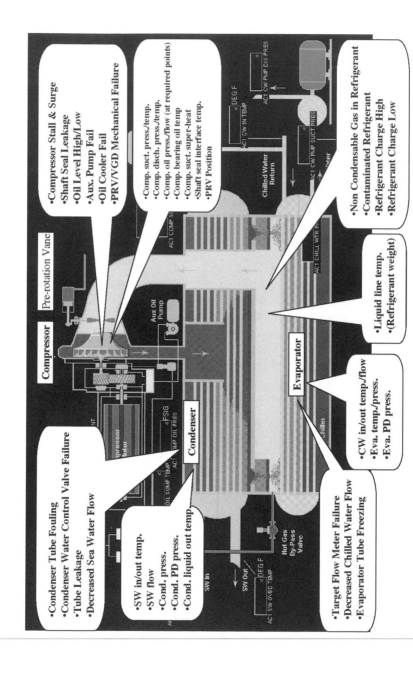

Figure 2.5 U.S. Navy centrifugal chiller and its fault modes.

The refrigerant is stored in the evaporator and under full load conditions should barely cover the tops of the cooler tubes. When refrigerant levels are high, the tubes are covered with refrigerant, and less refrigerant is boiled off to the compressors. The overall effect is decreased efficiency, which may result in loss of cooling. In addition, a very high charge level may result in the compressor sucking up liquid refrigerant droplets (instead of pure vapor), which can quickly erode the impeller.

Symptoms

Refrigerant level very high
Increased full-load ΔT across chill water
Low compressor discharge temperature
High compressor suction pressure
High compressor discharge pressure
High compressor motor amps
or
Compressor suction superheat less than 0°F

Comments: Some type of level gauge would be optimal for monitoring refrigerant charge. However, this could require modifications to the evaporator shell, which would be impractical. Currently, there is a site glass to view the level, but this is not known to be a very good indicator of charge owing to discrepancies in load conditions and chiller-tube/site-glass placement. Refrigerant levels should be monitored only during normal full-load operating conditions. (Since the boiling action within the cooler is much slower at partial loads than at full loads, the system will hold more refrigerant at partial load than at full load.)

Sensors

Some type of level gauge/sensor
Compressor suction pressure (10 inHg to 20 psig)
Compressor discharge pressure (0–60 psig)
Compressor discharge temp (30–220°F)
Chilled-water outlet temp (20–60°F)
Chilled-water inlet temp (20–60°F)
Pseudo–compressor suction superheat sensor
Prerotation vane position
Motor current sensing transformer

Example 2.2 Pump System FMECA Study. Pump systems are found in a vast array of application domains. Failure modes are associated with seals and bearings, the impeller and shaft, inlet strainer, valving, etc. Monitoring requirements and a motor-pump arrangement, including process and SCADA modules, are shown in Fig. 2.6.

An FMECA study ranking fault modes according to their severity, frequency of occurrence, testability, and component replaceability may follow the outline suggested in this section. As an example, for frequency of occurrence, the ranking is scaled from one to four as a function of how often the failure occurs, that is,

1 = less than one in two years
2 = 1 to 3 every two years
3 = 2 to 6 per year
4 = more than 6 per year

Figure 2.7 depicts one frame of the FMECA results as an illustration. For the pump, it suggests the fault candidate components/units, failure modes, primary causes, and symptoms; it evaluates the frequency of occurrence, severity, testability, and replaceability criteria as well as a composite criterion computed as the product of the constituent components. Furthermore, it recommends the most appropriate maintenance practice and the action required to mitigate the fault or failure conditions.

2.4 SYSTEM CBM TEST-PLAN DESIGN

The objective of a CBM test plan is to design the required instrumentation, data-collection apparatus, and testing procedures and operate the system/subsystem under controlled conditions on a test cell or under real operational

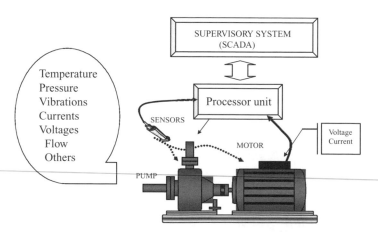

Figure 2.6 A motor-pump arrangement.

System	Components	Failure mode	Primary Cause	Symptoms	Evaluation					MF	Recommended action
					F	S	T	R	Q		
	Nonreturn valve	No pump delivery	Nonreturn valve blocked open	Noise and overheat	2	3	4	1	24	SM	Regularly open and clean valve
Pump	Bearing	Blistering									
		Cracks	Wear and tear Lack of lubrication Overheat Misalignment	Noise and vibration High temperature	3	2	2	1	18	CBM	Replacement
		Wear									
	Seal	Cracks	Overheat No flow	Noise	2	3	2	1	12	CBM	Replacement
		Wear	Wear and tear	Leaked							
	Impeller	Pump blocked	Solid parts Corrosion		1	4	2	1	8	BM	Pump replacement
		Axle loose	Corrosion	Noise No flow							

Figure 2.7 A sample of the pump system FMECA.

regimes in order to acquire baseline and fault data that will be used, eventually, to train and validate diagnostic and prognostic algorithms. Of particular importance are procedures to seed safely fault modes and acquire fault progression data, as well as the design of experiments that will guarantee a statistically sufficient and complete database. A systematic approach to test-plan design proceeds according to the following steps:

- Determine system operation modes.
- Decide on a set of fault modes that can be seeded safely; decision based on related control variables and Taguchi methods.
- Determine sensor suite and data acquisition apparatus needed for data collection.
- Use Taguchi matrix based on orthogonal array for multilevel experiments.
- Determine number and type of test runs for both baseline and fault data, personnel requirements, budget, etc.
- Make sure that sensors are calibrated!

For multilevel experiments, control variables such as load, temperature, speed, etc. must be identified, and a Taguchi matrix based on orthogonal array must be developed so that, for each test run, the dynamic range of the control variables is accounted for (Montgomery, 1999). Table 2.4 shows a generic approach to the design of a typical Taguchi matrix. The final number of test runs is, of course, dictated by additional considerations, such as cost, time, personnel availability, etc.

Example 2.3   CBM Test Plan for Navy Centrifugal Chiller. Consider the design again of a CBM/PHM system for a shipboard chiller. A critical concern in the design of a test plan is the ability to seed safely fault modes and monitor their time progression. Table 2.5 summarizes those chiller fault modes to be evaluated during beta testing that can be seeded without causing a catastrophic event.

Monitoring requirements are identified, and Table 2.6 presents a partial list of temperature measurements to be carried out during testing. Similar listings refer to all other measurements for pressure, flow rate, electrical quantities, valve positioning, etc.

Table 2.7 suggests a Taguchi matrix for baseline data, while Table 2.8 is a representative matrix listing control variables and a recommended range of values to be inserted during testing.

2.5 PERFORMANCE ASSESSMENT

CBM/PHM systems are designed to meet multiple objectives by providing useful information to a number of potential end users: the maintainer, the

TABLE 2.4 Taguchi Matrix Based on L9 Orthogonal Array for Multilevel Experiments

Control Variables

Run No.	A (−1: lower setting; 0: normal setting; +1: upper setting)	B (−1: lower setting; 0: normal setting; +1: upper setting)	C (−1: lower setting; 0: normal setting; +1: upper setting)	D (−1: lower setting; 0: normal setting; +1: upper setting)	Result (1: fault; 0: no fault)
1	−1	−1	−1	−1	1 or 0
2	−1	0	0	0	1 or 0
3	−1	+1	+1	+1	1 or 0
4	0	−1	0	+1	1 or 0
5	0	0	+1	−1	1 or 0
6	0	+1	−1	0	1 or 0
7	+1	−1	+1	0	1 or 0
8	+1	0	−1	+1	1 or 0
9	+1	+1	0	−1	1 or 0

TABLE 2.5   Summary of Chiller Fault Modes to Be Evaluated during Beta Testing

Component	Fault Mode	Can Be Seeded
Compressor	Compressor stall and/or surge	Yes
Condenser	Condenser tube fouling (simulated by decreasing water flow)	Yes
	Noncondensable gases	Yes
Evaporator	Evaporator tube fouling (simulated by decreasing water flow)	Yes
Refrigerant	Refrigerant charge low	Yes
Controls	PRV/VGD linkage/drive mechanism problems	Yes

operator, the process manager or field commander, and the system designer. Specifically, a CBM/PHM system assists the maintainer to determine the "optimum" time to perform maintenance given a host of constraints (e.g., spare parts, personnel availability, etc.). It provides the operator, the process manager, or field commander with confidence bounds on the availability of critical processes/assets to meet production schedules or mission requirements. Finally, the implementation means and findings of CBM/PHM practice may assist the system designer in improving the operational capabilities of legacy systems or in guiding the design of new systems so that they are more fault tolerant, robust, and reliable.

Performance-assessment studies are conducted to evaluate the technical and economic feasibility of various diagnostic and prognostic technologies. Technical performance metrics are defined for all algorithms in the CBM/PHM framework from sensing and feature extraction to diagnosis and prognosis. This subject is treated thoroughly in Chap. 7 because it constitutes a major component in the design, evaluation, and implementation of CBM/PHM systems. Cost-benefit analysis studies are essential in establishing not only the

TABLE 2.6   Temperature Measurements

No.	Description	No.	Description
1	Evaporator water inlet temperature	11	Condenser liquid temperature −2
2	Evaporator water outlet temperature	12	Compressor suction temperature
3	Condenser water inlet temperature	13	Compressor discharge temperature
4	Condenser water outlet temperature	14	Oil sump temperature
5	Oil cooler water inlet temperature	15	Oil temperature in seal cavity
6	Oil cooler water outlet temperature	16	
7	Evaporator vapor temperature	17	
8	Evaporator liquid temperature	18	
9	Condenser vapor temperature	19	
10	Condenser liquid temperature −1	20	

TABLE 2.7 Baseline Data Collection

			Variables		
	A Load (−1: 363 tons; 0: 250 tons; +1: 125 tons)	B SW in Temp (−1: 95°F; 0: 88°F; +1: 75°F)	C Ref. Pounds (−1: 1000 pounds; 0: 1200 pounds; +1: 1400 pounds)	D CW Flow (−1: 1200 gpm; 0: 1400 gpm; +1: 1600 gpm)	Result (1: fault; 0: no fault)
Run No.					
1	−1	−1	−1	−1	
2	−1	0	0	0	
3	−1	+1	+1	+1	
4	0	−1	0	+1	
5	0	0	+1	−1	
6	0	+1	−1	0	
7	+1	−1	+1	0	
8	+1	0	−1	+1	
9	+1	+1	0	−1	
10	0	0	0	0	

TABLE 2.8 Compressor Stall and/or Surge and Actuator Failure Data Collection

Run No.	A Load (tons) (−1: 363; 0: 250; +1: 125)	B VGD Position (−1: plus 6%; 0: optimum; +1: minus 6%)	C Ref. Pounds (−1: 1000; 0: 1200; +1: 1400)	D CW Flow (gpm) (−1: 1200; 0: 1400; +1: 1600)	B SW in Temp (°F) (−1: 95; 0: 88; +1: 75)	D PRV Position (−1: plus 6%; 0: optimum; +1: minus 6%)	Result (1: fault; 0: no fault)
1	−1	−1	−1	−1	−1	0	0 or 1
2	−1	+1	0	0	0	0	0 or 1
3	−1	0	+1	+1	+1	+1	0 or 1
4	0	−1	0	+1	−1	0	0 or 1
5	0	+1	+1	−1	0	0	0 or 1
6	0	0	−1	0	+1	−1	0 or 1
7	+1	−1	+1	0	−1	0	0 or 1
8	+1	+1	−1	+1	0	0	0 or 1
9	+1	0	0	−1	+1	+1	0 or 1

technical merits of these new technologies but also their economic feasibility. Unfortunately, sufficient performance data are not yet available from currently installed systems to estimate the true financial benefit, that is, return on investment, of such systems. A few studies that became public are based on "cost avoidance"; that is, if CBM/PHM practices are followed, what is the estimated cost that will be avoided in comparison with current breakdown or scheduled maintenance? A cost-benefit analysis, therefore, may proceed as follows:

- Establish baseline condition. If breakdown maintenance or time-based planned preventive maintenance is followed, estimate the cost of such maintenance practices and their impact on operations; frequency of maintenance, downtime for maintenance, dollar cost, etc. should be available from maintenance logs. Provide qualitative estimates of impact on other operations by interviewing key personnel.
- Condition-based maintenance will assist in reducing breakdown maintenance (BM) costs. A good percentage of BM costs may be counted as CBM benefits.
- If time-based planned maintenance is practiced, estimate how many of these maintenance events may be avoided by thresholding failure probabilities of critical components and computing approximate times to failure. The cost of such avoided maintenance events is counted as benefit of CBM.
- Intangible benefits: The impact of breakdown maintenance or extensive breakdown on system operations can be assessed only by interviewing affected personnel and creating an index assigned to the impact (severe, moderate, etc.).
- Estimate roughly the projected cost of CBM, that is, the dollar cost of instrumentation, computing, etc. and the cost of personnel training and installation and maintenance of the monitoring and analysis equipment.
- Aggregate life-cycle costs and benefits from the information just detailed.

Technical and economic measures, finally, are weighted appropriately and aggregated in a matrix form to provide management with a view of the potential benefits of the CBM system over its life cycle.

2.5.1 Verification and Validation of CBM/PHM Systems

Verification and validation (V&V) techniques for CBM/PHM technologies serve to ensure that delivered capabilities meet system design requirements. System performance metrics provide the foundation for V&V and are useful for system evolutionary design. Few formal or universally accepted techniques and methodologies exist. In the following subsections we discuss strategies and metrics useful for PHM system V&V. System verification and validation activities are in support of system development activities to achieve system accreditation. Successful system accreditation occurs when an affirmative answer is obtained to the question, "Do I trust that the system will meet the system performance specifications within stated system's constraints?" It is instructive to begin with a formal definition of the terms *verification* and *validation:*

- *Verification* answers the question, "Have I built the system right?" (i.e., Does the system as built meet the performance specifications as stated?).
- *Validation* answers the question, "Have I built the right system?" (i.e., Is the system model close enough to the physical system, and are the performance specifications and system constraints correct?).

Hughes (1997) provides a definition for model verification: "Model verification is the process of determining that a model implementation accurately represents the developer's conceptual description and specifications."

The V&V framework for CBM/PHM systems may entail the module of architecture understanding—understanding the architecture of a deployed CBM or PHM system presents major challenges owing to the inherent system complexity and constitutes a significant hurdle toward applying effective V&V methodologies. A possible approach is outlined below:

- Decompose system into a system of systems, and define constraints for interaction between the components; that is, partition the problem into well-understood subproblems.
- Interactions between components are captured using the Object Management Group's Interface Definition Language (IDL).
- To capture system behaviors for the system architecture and component descriptions, use the Universal Modeling Language (UML).

The offline process of system architecture understanding, component modeling, and system validation therefore may proceed as follows:

- Build a set of CBM (diagnostic, prognostic, etc.) architectures, and partition the tasks into functional modules.
- Assign modules to a computational structure.
- Choose a set of quality attributes with which to assess the architectures (pick success criteria).
- Choose a set of concrete tasks that test the desired quality attributes.
- Evaluate the degree to which each architecture provides support for each task.

2.5.2 Performance Metrics

Metrics for fault diagnosis and prognosis functional elements form the core performance assessment of a deployed CBM/PHM system. Simulation results and real fault data are employed to project the fielded performance of the system. Sound machinery condition monitoring and statistical signal-detection techniques can form the basis for development of performance metrics. Fault diagnosis and prognosis performance is discussed thoroughly in Chap. 7.

Prognosis. Once a changing condition is detected and isolated, prognosis provides the final output of a CBM/PHM system. A prognostic vector typically indicates time to failure and an associated confidence in that prediction. However, the definition and implementation of prognostic performance metrics must consider multiple viewpoints and objectives. Two such viewpoints follow:

- *The maintainer's viewpoint.* When should I perform maintenance? What is the most appropriate time (given such constraints as availability of spares, etc.) to maintain critical equipment?
- *The field commander's viewpoint.* Are my assets available to perform the mission? What is the confidence level that critical assets will be available for the duration of the mission? Or given a certain confidence level, what is the expected time to failure of a particular asset?

Answering the maintainer's question requires an estimate of the remaining useful lifetime of a failing component and assignment of uncertainty bounds to the trending curve that will provide him or her with the earliest and the latest (with increasing risk) time to perform maintenance and the associated risk factor when maintenance action is delayed.

Answering the commander's question requires an estimate of the confidence level assigned to completion of a mission considering mission duration, expected operating and environmental conditions, other contingencies, etc. or, given a desired confidence level, means to estimate the component's time to failure. Thus the prognostic curve (the output of prognostic routines) must be modulated by a time-dependent distribution (statistical/Weibull or possibilistic/fuzzy) that will provide the uncertainty bounds. An appropriate integral of such a distribution at each point in time will result in the confidence level committed to completing the mission or, given a desired confidence level, the time to failure of a failing component.

Chapter 7 presents several prognostic performance metrics. Relevant metrics for V&V studies include accuracy and precision of the predicted time-to-failure probability density function. We refer the reader to Chap. 7 for a detailed discussion of these metrics.

2.5.3 V&V Methods

CBM/PHM systems are composed of hardware, software, and models (algorithms) that work synergistically to analyze sensor data, detect and isolate an incipient failure, and predict the remaining useful life of failing components. The system's performance may be verified by collecting actual or simulated data and asserting whether the designed/deployed system meets the specifications. Model validation may follow more formal methods to ascertain whether the correct models have been designed. We will expand first on model V&V techniques, to be followed by a set of tools for system verification and validation.

The heart of the verification process is to ensure that the conditions implied by the CBM/PHM code meet the conditions and constraints that have been placed on the overall algorithm determined by the offline validation process. In general, this means solving for a Boolean expression that must evaluate to tautology. The satisfaction relationship is that the code or solutions must imply the specification; in other words, every state that the code allows to occur also must satisfy the description of the overall state contained in the specification. Thus the validation process involves extracting the Boolean expressions that describe the actions of the control code produced by the CBM algorithms and checking to ensure that those expressions satisfy the specifications. The control algorithms produced are executed in an environment that produces symbolic expressions in terms of sensor readings. These expressions are effectively the Boolean expressions that are compared against the specifications.

Model checking is a method for formally verifying finite-state concurrent systems. Specifications about the system are expressed as temporal logic formulas, and efficient symbolic algorithms are used to traverse the model defined by the system and check to see if the specification holds or not. A model checker is a utility that can be used in either of the V&V processes. If a model is characterized as a finite-state machine, a model checker will exercise all available permutations of each transition in the model and their interrelationships. As a validation tool, a conceptual model can be modeled as a state machine and be examined for the purpose of proof of concept. Available model checkers include SVC, SMV, SPIN, Cadence, etc. (Clarke and Wing, 1996). The Standard Validity Checker (SVC) is a program that checks to see if a Boolean expression is a tautology (evaluates to true) or, if not, produces a counterexample. By validating early in the development cycle, subsequent testing and debugging can be reduced significantly, leading to fault diagnosis systems that are more reliable and easier to develop.

Validity checking needs to specify the system in a formal language, such as SMV or PROMELA. Furthermore, formal specifications that indicate desirable properties of the system validated must be stated, and the model checker then determines whether the properties hold under all possible execution traces. Essentially, model checking exhaustively (but intelligently) considers the complete execution tree. SMV, developed by Carnegie-Mellon University researchers (Burch et al., 1994; McMillian 1993), is one of the most popular symbolic model checkers. An SMV specification uses variables with finite type grouped into a hierarchy of module declarations. Each module states its local variables, their initial value, and how they change from one state to the next. Properties to be verified can be added to any module. The properties are expressed in CTL (computation tree logic). CTL is a branching-time temporal logic, which means that it supports reasoning over both the breadth and the depth of the tree of possible executions. SMV checks whether the formulas are consistent with the model and reports a trace of counterexample when an inconsistency is found.

Tools for Verification and Validation. Robust tools for V&V can provide useful insights applicable to system development as well as ongoing validation and improvement. The capability to warehouse raw data, simulation models, and system configuration, along with an interface that permits easy exercise of developed detection, isolation, and prognostic algorithms, would add significant value to many acquisition programs that seek to employ PHM technology. Such a tool could incorporate real data, simulation models, and noise augmentation techniques to analyze effectiveness and overall robustness of existing and emerging technology. Impact Technologies, in cooperation with the Georgia Institute of Technology, developed a Web-based software application that provides Joint Strike Fighter (JSF) (F-35) system suppliers with a comprehensive set of PHM V&V resources. The application includes standards and definitions, V&V metrics for detection/diagnosis/prognosis, access to costly seeded fault data sets and example implementations, a collaborative user forum for the exchange of information, and an automated tool for impartially evaluating the performance and effectiveness of PHM technologies. This program is specifically focused on development of the prototype software product to illustrate the feasibility of the techniques, methodologies, and approaches needed to verify and validate PHM capabilities. A team of JSF system suppliers, including Pratt and Whitney, Northrop Grumman, and Honeywell, is being assembled to contribute, provide feedback, and make recommendations to the product. The approach being pursued for assessing overall PHM system accuracy is to quantify the associated uncertainties at each of the individual levels, as illustrated in Fig. 2.8, and build up to the system level through the PHM architecture.

2.6 CBM/PHM IMPACT ON MAINTENANCE AND OPERATIONS: CASE STUDIES

2.6.1 Verification and Validation of CBM on Navy Shipboard Chiller

Georgia Tech researchers developed a modular integrated diagnostic and prognostic architecture for a shipboard chiller under the Navy's Prognostic En-

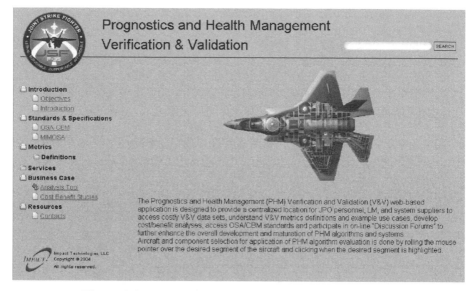

Figure 2.8   JSF PHM V&V by Impact and Georgia Tech.

hancements to Diagnostic Systems (PEDS) Program (Propes et al., 2002). A V&V study was carried out on major modules of this architecture. The scalable architecture is depicted in Fig. 2.9 and accounts for such functional blocks as interfacing and database management, system operating mode identification, fault-feature extraction, and diagnostic/prognostic algorithms. The modules form the constituent components of an open systems architecture accommodating a variety of development languages and allowing for such quality-of-service functions as flexibility, plug and play, and reconfigurability.

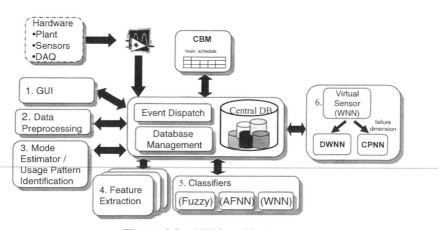

Figure 2.9   PEDS architecture.

The communication between modules is event-driven and does not require an overall schedule to manage the individual components. A user-friendly human-machine interface displays all sensor data, fault features, and diagnostic/prognostic results as real-time values and historical records.

A brief description of the PEDS main modules follows:

1. *Operating mode/usage pattern detection or mode identification (ModeID)*. Decides on the current operating mode and usage pattern and directs all the appropriate inputs to the corresponding classifier. Mode identification is helpful in diagnostic and prognostic procedures because fault or failure modes often are related to specific operating conditions or usage patterns.

2. *Feature extraction*. Involves preliminary processing of sensor measurements to obtain suitable parameters or indicators that reveal whether an interesting pattern is emerging. The extracted features represent the raw data enormously better in a compact form, under the assumption that the important structure in the data actually lies in much lower dimensional spaces. Features in various domains such as the time, frequency, and wavelet domains can be extracted and used in diagnostic and prognostic applications. The feature-extraction module also selects appropriate features for every specific subsystem because irrelevant or redundant features may result in serious performance degradation for fault detection and isolation, which can be viewed as a mapping from a set of given features into one of the predefined fault classes.

3. *Diagnosis*. Determines fault occurrence based on the fused decisions from three classifiers, namely, a fuzzy-logic classifier with a Mamdani inference engine, a static wavelet neural network (WNN), and an arrangement fuzzy neural network (AFNN). These algorithms are discussed in detail in Chap. 5.

4. *Prognosis*. Predicts the remaining useful life (RUL) of a failing component based on its current diagnostic results and operational data in the presence of uncertainty. The prognostic module consists of two components: a dynamic wavelet neural network (DWNN) (Vachtsevanos et al., 2000) and a confidence prediction neural networks (CPNN) (Khiripet and Vachtsevanos, 2001; Barlas et al., 2003). The DWNN functions as a predictor providing the time-to-failure information for CBM purposes. Long-time prediction results in higher uncertainty about the outcome. One way to describe the uncertainty is through confidence estimates. The CPNN determines the uncertainty distributions using multiple trends and confidence distributions. The prognostic algorithms are detailed in Chap. 6.

The PEDS architecture was applied to diagnose and predict the time evolution of faults seeded on a laboratory test-bed three-tank process called the *process demonstrator.* This facility offers a flexible platform for seeding a variety of fault modes safely and collecting time-series data under various operational conditions. Data collected from the process demonstrator are used to train and validate the PEDS algorithms at the system level, as well as at the component or subsystem level. The process demonstrator is a model of a continuous fluid process typically found in industry, as illustrated in Fig. 2.10. It consists of industrial-grade tanks, pipes, valves, pumps, mixers, and electric heaters. By activating these devices, water in the tanks can be stirred, heated, and circulate among the tanks. Honeywell smart transmitters are installed in the process, monitoring water level, water flow rate, and water temperature.

To simplify the problem, three fault modes are examined: tank 1 leakage, tank 1 level low, and valve FV108 in tank 1 stuck open. Tank 1 leakage is seeded by opening a faucet in tank 1 some time into the process of collecting one set of data. Tank 1 level low occurs when the water level in tank 1 is lower than a specified threshold. Valve FV108 in tank 1 stuck open is the valve of FV10 being stuck at 10 percent during the valve-opening process. The fault modes and the relevant features are shown in Table 2.9. Each fault is associated with one operating mode, for which one classifier and one prognosticator were implemented.

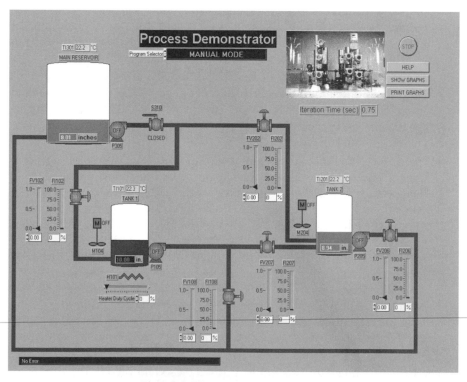

Figure 2.10   Process demonstrator.

TABLE 2.9 **Failure Modes and Features to Be Examined**

Failure Mode	Feature 1	Operating Mode
Tank 1 leakage	Tank 1 level slope	System idle
Tank 1 level low	Tank 1 level	System idle
FV108 stuck open	Flow rate between tanks 1 and 2	Transfer water from tank 1 to tank 2

Validation in this example employs an extensive set of data and Monte Carlo simulations to generate a sufficient basis for evaluating the performance and effectiveness of diagnostic and prognostic algorithms. The PEDS architecture is first decomposed into several major modules, and V&V begins by testing the individual modules. Eventually, moving up the hierarchy, the "system of systems" considers the component couplings or interactions. Verification is much less involved than the more complex statistical nature of the validation process, and software quality assurance tools are used to address programming errors and to verify the correctness of the software. A combination of white and black box testing strategies is necessary to carry out the verification as completely as possible. The statistical approach to system validation is depicted schematically in Fig. 2.11.

The approach requires a specially designed data set that replicates the pattern of inputs to be processed by the system. It establishes an operational profile, and the test data are constructed to reflect the full spectrum of this profile. The system is finally tested, and the numbers and properties of the failure modes are recorded. The reliability and other system performance metrics are computed after a sufficient number of tests are carried out. In the Monte Carlo simulation framework, the problem is cast as a cumulative probability distribution function, and a pseudo-random number generator is used to sample this distribution for a large number of trials, where each trial represents a solution to the original problem, based on the probability density function. According to the central-limit theorem, as the number of samples or trials approaches infinity, the average result approaches the true average value of the original problem. Although, in many cases, statistical methods are the only viable alternative to analytical models because of system complexity, they require a large volume of historical events, compounding the computational burden.

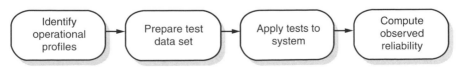

Figure 2.11 System validation process based on the statistical approach.

Diagnosis Validation. We present system validation results based on six sets of data collected from the process demonstrator when a tank 1 leakage fault is seeded. For each data set, the demonstrator runs at the system idle mode. At about 180 seconds, a small leakage is induced by opening the faucet to simulate the tank 1 leakage fault, and the data are acquired for 1364 to 3947 seconds for each time series. A typical curve of the water level of tank 1 is plotted in Fig. 2.12. Figure 2.13 shows one example of the classification results from application of the diagnostic modules, which is based on the tank 1 level data shown in Fig. 2.12.

Tables 2.10 through 2.13 show the diagnostic results of the tank 1 leakage fault from three classifiers implemented in the PEDS and the overall results combined using Dempster's rule of combination. Details of the Dempster-Shafer evidential theory are discussed in Sec. 5.4 of Chap. 5. The false-positive rate (FPR), false-negative rate (FNR), and the fault accuracy rate (FAR) are calculated for each data set. They are the percentages of the data points in which the diagnosed result was the same or differed from the actual fault or nonfault state. The thresholds of the classifiers were set at such values that the FPR is low with a reasonable FNR. The time to detection (TTD) is also given with a minimum value of 3 seconds, which is the time it takes to fetch the sensor data from the database, extract the features, and classify the results. We can see that the diagnostic modules perform very well for these data sets.

DWNN Performance Assessment Results. Figure 2.14 shows a typical prediction of the failure time of the faulted component (in this case tank 1

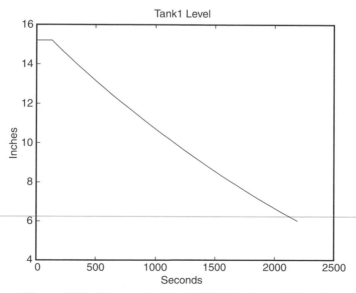

Figure 2.12   Time-series data D050604_16: tank 1 level.

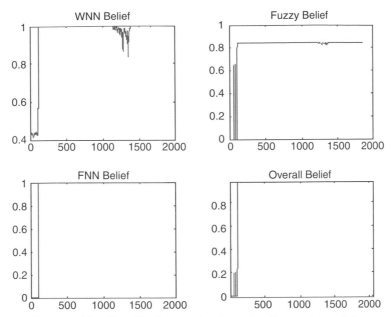

Figure 2.13   Classification results from the diagnostic modules.

TABLE 2.10   Diagnostic Results from the Fuzzy-Logic Classifier

Data Source	Threshold	FPR (%)	FNR (%)	FAR (%)	TTD (s)
D050604_16	0.79	0	0.3416857	99.67707	16
D050604_17	0.79	0	0.05743825	99.94256	9
D050604_18	0.79	0	0.2127659	99.8	3
D050604_19	0.79	0	0	100	3
D050704_10	0.79	0	0.9546539	99.90454	6
D050704_11	0.79	0	0.1536885	99.84631	9

TABLE 2.11   Diagnostic Results from the Wavelet Neural Network (WNN) Classifier

Data Source	Threshold	FPR (%)	FNR (%)	FAR (%)	TTD (s)
D050604_16	0.5	0.9708738	0.3418804	99.62325	18
D050604_17	0.5	0	0.1723148	99.82768	11
D050604_18	0.5	0	0.4252303	99.60053	7
D050604_19	0.5	0	0	100	3
D050704_10	0.5	0	0.1432665	99.85674	7
D050704_11	0.5	0	0.1793033	99.82069	11

TABLE 2.12   Diagnostic Results from the Arrangement Fuzzy Neural Network (AFNN) Classifier

Data Source	Threshold	FPR (%)	FNR (%)	FAR (%)	TTD (s)
D050604_16	0.9	0	0.3416857	99.67707	16
D050604_17	0.9	0	0.05743825	99.94256	9
D050604_18	0.9	0	0.2127659	99.8	4
D050604_19	0.9	0	0	100	3
D050704_10	0.9	0	0.1432665	99.85674	7
D050704_11	0.9	0	0.1281066	99.87189	8

due to the leakage) yielded by the DWNN. The horizontal axis is the time normalized by the actual failure time. The figure shows that the predicted failure time converges to the actual failure time eventually. In order to quantify the convergence rate, a *rising time* and a *settling time* are defined as the moments when the predicted value becomes within 20 and 10 percent of the actual value, respectively. Table 2.14 shows the rising and settling times of the prediction process based on different data sets. Normalized values of the rising and settling times are also given, which are the ratios of the rising and settling times of the prediction to the actual failure time. As shown in the table, the averaged values of the normalized rising and settling times are 10.46 and 27.15 percent, respectively. These values indicate that the DWNN makes accurate predictions about the failure time at a reasonably early stage of the prognostic process.

We also evaluate the DWNN prediction accuracy as defined in Chap. 7. Table 2.15 records the accuracy values of the prediction calculated at equally spaced time instances of 0.1, 0.2, ... , and 0.9 (time normalized by the actual failure time). The prediction accuracy is plotted in Fig. 2.15 as well. It can be seen that the accuracy improves as time progresses and that it converges to 1 eventually.

The V&V exercise presented in this case study provides a confidence level in terms of the capability of the diagnostic routines to detect a fault with a

TABLE 2.13   Overall Diagnostic Results of the Classifiers Combined Using Dempster-Shafer Theory

Data Source	Threshold	FPR (%)	FNR (%)	FAR (%)	TTD (s)
D050604_16	0.8	0	0.2849003	99.73089	16
D050604_17	0.8	0	0.05743825	99.94256	9
D050604_18	0.8	0	0.4252303	99.60053	7
D050604_19	0.8	0	0	100	3
D050704_10	0.8	0	0.09551098	99.90449	6
D050704_11	0.8	0	0.1280738	99.87193	8

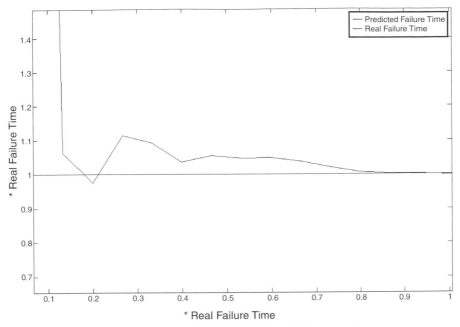

Figure 2.14   A typical DWNN prediction result.

reasonable time to detection. The prompt detection of a fault provides a sufficient time window that allows the prognostic algorithms to perform their intended tasks. The performance-assessment results of the prognostic component, based on the rising and settling times of the prediction curve, as well as the accuracy measure, suggest that the prognostic component was efficient and effective in predicting the time to failure of the faulty system. The V&V

TABLE 2.14   Prognostic Results of the Dynamic Wavelet Neural Network (DWNN)

Data Set	Rise Time (Seconds)	Normalized Rise Time	Settling Time (Seconds)	Normalized Settling Time	Failure Time (Seconds)
D05060416	100	6.67%	200	13.32%	1502
D05060417	500	34.82%	700	34.82%	1436
D05060418	100	7.33%	300	21.99%	1364
D05060419	100	6.61%	200	13.23%	1512
D05070410	100	4.81%	600	28.86%	2079
D05070412	100	2.53%	2000	50.67%	3947
Mean		10.46%		27.15%	

TABLE 2.15 Prediction Accuracy of the DWNN

Data	e^{D_t/D_0} at Different Time Instances During the Prediction Process								
	0.1	0.2	0.3	0.4	0.5	0.6	0.7	0.8	0.9
D05060416	0.9393	0.9763	0.8930	0.9116	0.9545	0.9526	0.9796	0.9934	0.9987
D05060417	0.5266	0.7542	0.8177	0.8853	0.9115	0.9192	0.9282	0.9458	0.9759
D05060418	0.3671	0.8674	0.9191	0.9479	0.9591	0.9647	0.9927	0.9985	0.9971
D05060419	0.9366	0.9810	0.9472	0.9700	0.9862	0.9980	0.9882	0.9882	0.9941
D05070410	0.0000	0.6951	0.8190	0.9061	0.9398	0.9430	0.9760	0.9758	0.9843
D05070412	0.0000	0.0000	0.0000	0.6414	0.8388	0.9273	0.9789	0.9980	0.9982
Mean values	0.4616	0.7123	0.7327	0.8771	0.9317	0.9508	0.9739	0.9833	0.9914

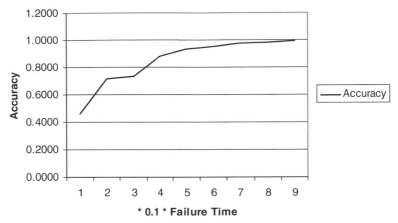

Figure 2.15   Accuracy of the DWNN prediction results.

results presented here are not intended to be comprehensive but rather represent an illustrative step in the ongoing process of determining the credibility of a deployed prognosis-enhanced diagnostic system.

2.6.2   Cost-Benefit Analysis of CBM on Nittany Trucking Company*

This study reported on possible maintenance cost savings through the application of CBM/PHM technologies to a fleet of 1000 eighteen-wheeled tractor-trailer trucks. The study's focus is on tire replacement, with the following data being made available from the company's data logs:

Total tires used	18,000
Mean life	85,000 miles
Standard deviation	15,000 miles
Cost	$350 per tire
Roadside repair cost	$500
Average tire usage	90,000 miles/year

We pose the following key question: When should the tires be changed? Figure 2.16 shows the effect of time-based maintenance assuming a normally distributed mean time between failures (MTBF); the figure also suggests the cumulative distribution, that is, component failures as a function of time. Figure 2.17 depicts a summary of the maintenance-cost analysis if time-based maintenance is practiced.

*Adapted from original concept developed by Ken Maynard at the Penn State Applied Research Lab and provided by Carl Byington of Impact Technologies.

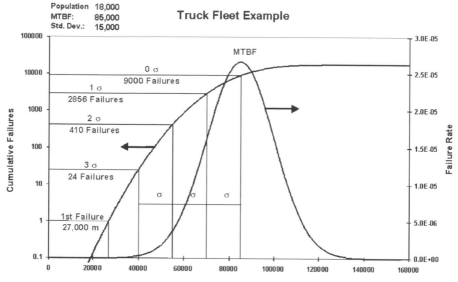

Figure 2.16   Effect of time-based maintenance.

If a sophisticated CBM/PHM approach is pursued, compared with the classical heuristic, "If you can see George's chin, replace the tire," by measuring the tread height over time and forecasting the time to minimum tread, suggested schematically in Fig. 2.18, then the tire costs and road failures may be reduced to the numbers shown in Fig. 2.19.

Thus, with CBM/PHM, we can achieve maintenance costs of one-half the best that statistical-based (traditional reliability) maintenance can offer. Of course, actual costs will vary depending on the details of the maintenance situation.

σ	Miles	Number of Tire Changes per Year	Cost of Tires	Number of Road Failures	Repair Cost	Total Cost
0	85000	1059	$ 370,588	9000	$ 4,500,000	$ 4,870,588
1	70000	1286	$ 450,000	2856	$ 1,427,905	$ 1,877,905
1.5	62500	1440	$ 504,000	1203	$ 601,271	$ 1,105,271
2	55000	1636	$ 572,727	410	$ 204,755	$ 777,482
2.5	47500	1895	$ 663,158	112	$ 55,888	$ 719,046
3	40000	2250	$ 787,500	24	$ 12,149	$ 799,649
3.8667	27000	3333	$ 1,166,667	1	$ 500	$ 1,167,167

Is this the answer? ➡

Figure 2.17   Maintenance-cost analysis.

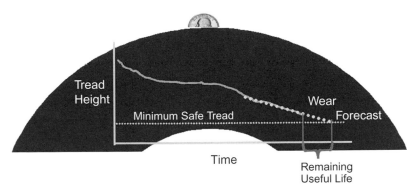

Figure 2.18 Applying a CBM/PHM approach resulting in substantial savings for the company.

2.7 CBM/PHM IN CONTROL AND CONTINGENCY MANAGEMENT

A substantial benefit may be realized from the application of CBM/PHM technologies to critical systems by taking advantage of the diagnostic and prognostic pronouncements and exploiting them in a feedback configuration to reconfigure or restructure system controls, replan missions, and sustain

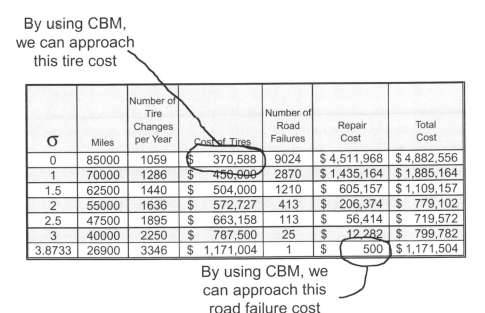

Figure 2.19 Revisiting with CBM/PHM maintenance cost analysis.

stable operating conditions in the presence of noncatastrophic failures. Failures and damage accommodation of critical system components (aircraft actuators, control surfaces, etc.) have been the subject of intense research activities in recent years. Automated contingency management (ACM) aims to optimize available assets and control authority so that the system (swarm of unmanned aerial vehicles, engines, etc.) continue to operate within an acceptable and stable regime (flight envelope, stable state, etc.).

Concepts of fault-tolerant control (FTC) and fault accommodation are promoted to assist not only the operator but also the system designer in optimizing system design, sensor requirements, component redundancy, active controls, etc. so that modern critical assets exhibit attributes of high confidence. Furthermore, notions of active/adaptive control are investigated in an attempt to design an added control authority (not part of the conventional controller design) that may extend the flight or stability envelope in critical situations. Consider, for example, the case of an unmanned aerial vehicle (UAV) that experiences a collective actuator failure in flight. Such an event invariably will drive the vehicle to an unstable state resulting in loss of the aircraft. If a controller is designed and implemented to regulate the rpms of the helicopter's main rotor and is activated immediately after the failure is detected and isolated, the UAV will remain stable and land safely during the emergency. The system designer stands to benefit from CBM/PHM by gaining a better understanding of the physics of failure mechanisms, the need for monitoring and sensing, issues of hardware and analytic redundancy, and means to design and build high confidence systems.

We will illustrate the ACM/FTC concepts proposed earlier with some examples. Key ACM/FTC assumptions include

- The failures or system parameter changes are unanticipated.
- After the appearance of a failure, the system operates in an emergency mode with another criterion until the failure is recovered.
- Nonstandard techniques, effectors, and configurations may be required.
- The available reaction time is small.
- The handling qualities of the restructured/reconfigured system may be degraded.
- The failure, in general, may influence the system's behavior and stability.

Sensor, component, and operational failures are considered. The last category refers to such failure modes for a typical rotorcraft as rotor stall, engine stall and surge, rapid battery discharge, loss of data link, etc. Figure 2.20 shows an engine perspective that illustrates the application of PHM at various levels of the control hierarchy. At the higher echelons, mission replanning and midlevel "intelligent" controls are exercised to safeguard the engine's operational integrity, whereas gain scheduling is practiced at the lowest level to ascertain system stability.

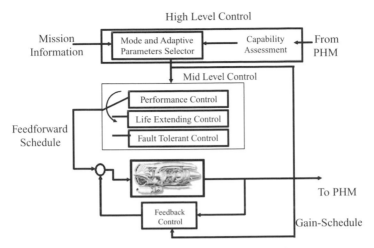

Figure 2.20 ACM, an engine perspective.

Next, consider the case of a vertical takeoff and landing (VTOL) UAV suffering a fault condition in flight. The vehicle is equipped with appropriate PHM sensors and active controllers designed to accommodate certain fault modes. A fault detection and identification (FDI) routine receives periodic sensor information, extracts "features" from the signals, and compares these features with known fault signatures. When a failure is identified, a fault-mode signal is generated that contains information about the nature of the fault. An FTC routine determines the reconfiguration actions necessary to maintain system stability and performance. These actions may include

- Starting new algorithms (controllers)
- Changing the source of certain signals (i.e., sensor signals and actuator commands)
- Changing set points
- Modifying local controller parameters

The FTC routine also sets new priority and quality-of-service (QoS) requirements for signals during the fault recovery process. Figure 2.21 is a schematic representation of the integration of PHM and control systems.

An overview of the FDI/FTC strategy that is applicable not only to rotorcraft but also to other dynamic systems is shown in Fig. 2.22 (Clements et al., 2000).

A wavelet neural network detection and classification algorithm serves as the FDI component. A three-level control reconfiguration hierarchy is suggested:

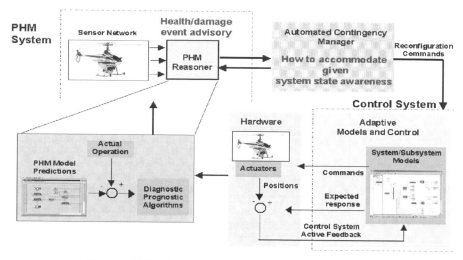

Figure 2.21   Integration of PHM and control systems.

- Redistribution of control authority between subsystems
- Updating of local controller set points (based on reconfigured overall system)
- Reconfiguration of local controllers

Example ACM/FTC strategies are listed in Table 2.16.

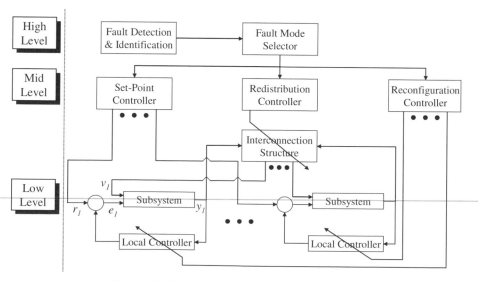

Figure 2.22   The FDI/FTC control hierarchy.

TABLE 2.16 ACM/FTC Strategies

Scenario	Description	Effect	Reconfiguration
1. Stuck collective	Main rotor collective is stuck in a known position.	Vertical acceleration can not be controlled by the collective.	Use rpms to control vertical acceleration.
2. Throttle stuck/ engine failure	Throttle is stuck in a known position or goes to zero (engine failure).	Rpms cannot be governed as is traditionally done.	Use altitude (or vertical acceleration) to compensate for loss of rpm control (i.e., autorotation).
3. Tail rotor stuck/ transmission failure	Tail rotor is stuck at a given pitch or stops rotating altogether.	Yaw acceleration cannot be controlled by the tail rotor.	Use thrust and collective control to control yaw acceleration.

A typical FDI/FTC example situation involves a stuck collective actuator while the helicopter is descending from 300 to 100 ft. Without reconfiguration, the vehicle would be unable to slow its descent, as shown in Fig. 2.23, and validated via computer simulations.

With FDI/FTC, the FDI routine detects the stuck collective actuator and outputs a fault-mode signal. The FTC routine receives the fault mode signal

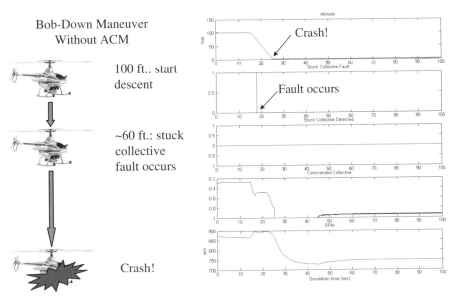

Figure 2.23 UAV bob-down maneuver without FDI/FTC.

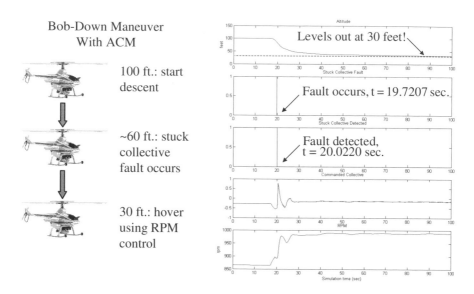

Figure 2.24   UAV bob-down maneuver with FDI/FTC.

and proceeds to reconfigure the system; that is, it starts the rpm controller routine, stops the nominal (collective) controller, and the rpm controller subscribes the sensor information and outputs low-level actuator commands. The vehicle hovers at 30 ft, now attaining a stable state, as shown in Fig. 2.24.

Researchers at the Georgia Institute of Technology have developed and validated/verified via simulation and flight testing this and other control reconfiguration algorithms under the Software Enabled Control (SEC) Program funded by DARPA. They used for this purpose the GTMax, an experimental VTOL UAV instrumented and owned by Georgia Tech. Flight-test video and other SEC contributions can be found at the Web site: *http:// controls.ae.gatech.edu/sec/.*

2.8   REFERENCES

I. Barlas, G. Zhang, N. Propes, and G. Vachtsevanos, "Confidence Metrics and Uncertainty Management in Prognosis," *MARCON,* Knoxville, TN, May 2003.

J. R. Burch, E. M. Clarke, D. E. Long, K. L. McMillian, and D. L. Dill, "Symbolic Model Checking for Sequential Circuit Verification," *IEEE Transactions on Computer-Aided Design of Integrated Circuits and Systems* 13(4):401–424, 1994.

C. Byington, R. Safa-Bakhsh, M. Watson, and P. Kalgren, "Metrics Evaluation and Tool Development for Health and Usage Monitoring System Technology," *AHS Forum 59,* Phoenix, Arizona, May 6–8, 2003.

E. Clarke and J. Wing, "Formal Methods: State of the Art and Future Directions," *CMU Computer Science Technical Report CMU-CS-96-178,* August 1996.

N. S. Clements, B. S. Heck, and G. Vachtsevanos, "Component-Based Modeling and Fault Tolerant Control of Complex Systems," *The 19th Digital Avionics Systems Conferences, 2000, Proceedings, DASC,* Vol. 2, October 2000.

Department of Defense (DoD), *Guide to Integrated Product and Process Development,* Version 1.0. Washington: Office of the Secretary of Defense, 1996.

W. P. Hughes, *Military Modeling for Decisions,* 3rd ed. Washington, D.C. Military Operations Research Society, 1997.

P. Kalgren, C. Byington, S. Amin, and P. Kallappa, "Bearing Incipient Fault Detection: Technical Approach, Experience, and Issues," *MFPT,* 2005.

S. Kay, *Fundamentals of Statistical Signal Processing,* Vol. II. Englewood Cliffs, NJ: Prentice-Hall, 1998.

N. Khiripet and G. Vachtsevanos, "Machine Failure Prognosis Using a Confidence Prediction Neural Network," *Proceedings of 8th International Conference on Advances in Communications and Control.* Crete, Greece, 2001.

L. J. Kusek and J. D. Keenan, "Trimming Ship Operating Support Costs: An Economic Analysis of the Integrated Condition Assessment System," *Center for Naval Research,* CRM 97-138, February 1998.

K. L. McMillian, *Symbolic Model Checking.* New York: Kluwer Academic Publication, 1993.

D. C. Montgomery, "Experimental Design for Product and Process Design and Development," *Journal of the Royal Statistical Society: Series D (The Statistican)* **48**(2):159–177, 1999.

R. Myers, A. Khuri, and W. Carter, Jr., "Response Surface Methodology: 1966–1988," *Technometrics* **31**(2):137–157, 1989.

R. F. Orsagh, M. J. Roemer, "Development of Metrics for Mechanical Diagnostic Technique Qualification and Validation," *COMADEM Conference,* Houston, TX, December 2000.

N. Propes, S. Lee, G. Zhang, I. Barlas, Y. Zhao, G. Vachtsevanos, A. Thakker, and T. Galie, "A Real-Time Architecture for Prognostic Enhancements to Diagnostic Systems," *MARCON,* Knoxville, TN, May, 2002.

D. Schrage, D. DeLaurentis, and K. Taggart, "FCS Study: IPPD Concept Development Process for Future Combat Systems," *AIAA MDO Specialists Meeting,* Atlanta, GA, September 2002.

L. P. Sullivan, "Quality Function Deployment," *Quality Progress* **19**(6):39–50, 1986.

G. Vachtsevanos, P. Wang, and N. Khiripet, "Prognostication: Algorithms and Performance Assessment Methodologies," *MARCON,* Knoxville, TN, 2000, pp. 801–812.

CHAPTER 3

SENSORS AND SENSING STRATEGIES

3.1 INTRODUCTION

Sensors and sensing strategies constitute the foundational basis for fault diagnosis and prognosis. Strategic issues arising with sensor suites employed to collect data that eventually will lead to online realization of diagnostic and prognostic algorithms are associated with the type, number, and location of sensors; their size, weight, cost, dynamic range, and other characteristic properties; whether they are of the wired or wireless variety; etc. Data collected by transducing devices rarely are useful in their raw form. Such data must be processed appropriately so that useful information may be extracted that is a reduced version of the original data but preserves as much as possible those characteristic features or fault indicators that are indicative of the fault events we are seeking to detect, isolate, and predict the time evolution of. Thus such data must be preprocessed, that is, filtered, compressed, correlated, etc., in order to remove artifacts and reduce noise levels and the volume of data to be processed subsequently. Furthermore, the sensor providing the data must be validated; that is, the sensors themselves are not subjected to fault conditions. Once the preprocessing module confirms that the sensor data are "clean" and formatted appropriately, features or signatures of normal or faulty conditions must be extracted. This is the most significant step in the condition-based maintenance/prognostics health management (CBM/PHM) architecture whose output will set the stage for accurate and timely diagnosis of fault modes. The extracted-feature vector will serve as one of the essential inputs to fault diagnostic algorithms.

Sensor suites are specific to the application domain, and they are intended to monitor such typical state awareness variables as temperature, pressure, speed, vibrations, etc. Some sensors are inserted specifically to measure quantities that are directly related to fault modes identified as candidates for diagnosis. Among them are strain gauges, ultrasonic sensors, proximity devices, acoustic emission sensors, electrochemical fatigue sensors, interferometers, etc., whereas others are of the multipurpose variety, such as temperature, speed, flow rate, etc., and are designed to monitor process variables for control and/or performance assessment in addition to diagnosis. More recently we have witnessed the introduction of wireless devices in the CBM/PHM arena.

Major questions still remain in the sensor and sensing domain. A few analytic techniques are available to assist in optimizing the placement and determining the type and number of sensors for a given system configuration. Sensor fault detection and validation techniques are lagging behind those used for other system components, that is, actuators, structures, etc. Data processing for feature extraction is primarily an art relying on the designer's intuition and experience as to where the fault information is and how to go about extracting it. Feature selection and extraction may pursue several paths. Features can be extracted in the time, frequency, wavelet, or even chaotic domains depending on where the dominant fault information resides. Features should be uncorrelated, ranked, and normalized according to their contribution in determining the fault event. The final feature vector should be optimal in terms of criteria relating to its discriminating ability to detect and isolate a particular fault while meeting computational requirements and other constraints. This chapter addresses some fundamental concerns in the important area of sensors and sensing strategies for CBM/PHM systems. We briefly review methodologic notions and introduce recent advances in sensor technologies that have an impact on the design of CBM/PHM systems. Sensor technologies for process control and performance monitoring are amply covered in the published literature and do not constitute the major subject of this chapter. We proceed in the next chapter to detail sensor data-processing methodologies while emphasizing aspects of feature extraction. We conclude with a brief description of database management tools and protocols that constitute the basis for an integrated CBM/PHM architecture.

3.2 SENSORS

In this chapter and the next one we address fundamental concepts in sensing and data processing. We will elaborate on transducing principles and focus on sensor systems that find extensive utility in the CBM/PHM arena. Chapter 4 will be devoted to analytic technologies required to process sensor data and extract useful information from them that can be used for fault diagnosis and prognosis and to a discussion of databases and database management strategies that constitute the data warehouse and the efficient management of databases to facilitate data and analysis results storage and transport from one CBM/PHM module to another.

A class of PHM sensors that finds significant inroads into the application domain includes accelerometers (for vibration measurements), eddy-current ultrasonic and acoustic emission sensors (crack and other anomaly detection), and transducing devices that take advantage of recent advances in microelectromechanical systems (MEMS) and fiberoptics. We will focus in this section on only a few typical and innovative sensors that are finding widespread use or promise to expand their utility in PHM-related applications. The concept of smart sensors has appealed not only to the PHM community but also to many other performance-monitoring and process-control domains. We will introduce certain notions of smart sensing devices in conjunction with a brief discussion of sensor networking issues. Protocols and standards for sensors, interfacing, and networking have been established and are updated continuously because the sensing and networking field is affecting more and more significantly every aspect of monitoring and control of complex industrial processes, aerospace systems, etc.

Sensor hardware issues for CBM/PHM applications relate to the utility of smart sensors and transducers, MEMS, networking and interfacing, DSP-based multitasking, and multiprocessing platforms for the efficient, reliable, and cost-effective implementation of software routines. The hardware configuration must be capable of measuring multiple quantities and, eventually, merge information together by making use of statistical relationships, principal-component analysis (PCA) tools, filtering, etc. to reduce the effects of noise and disturbance. Dynamic metrology systems must be nonintrusive, operating online with a high bandwidth, reliable, and easy to operate. An integrated hardware/software dynamic metrology architecture aims at establishing the fundamentals and operation principles of sensors and signal-processing methodologies for online wear diagnosis for a variety of mechanical, electromechanical, and other engineered systems.

Sensors and sensing strategies constitute a major technology with a broad application domain from engineering and the sciences to medicine and other disciplines. They have found extensive treatment in textbooks and technical publications over the past decades. Our specific emphasis in this book is to review sensor types and sensing technologies that have a direct impact on CBM/PHN practice. It generally is agreed that two classes of sensors are making significant inroads into system monitoring for fault diagnosis and prognosis. The first one refers to classic or traditional transducers aimed at monitoring mechanical, structural, performance and operational, and electrical/electronic properties that relate to failure mechanisms of mechanical, structural, and electrical systems. In this category, we include devices that measure fluid and thermodynamic, thermal, and mechanical properties of a variety of systems or processes—gas turbines, ground vehicles, pumps, aerospace systems, etc. The second important category refers to sensor systems that are placed almost exclusively to interrogate and track system properties that are related directly to their failure mechanisms. We refer to such devices as CBM/PHM sensors. We will summarize briefly in this chapter a few sensors belonging to the first category and expand on new and innovative sensor suites that are specifically designed for CBM/PHM applications.

3.2.1 Transducing Principles

A *transducer* is defined as a device that receives energy from one system and retransmits it, often in a different form, to another system. On the other hand, a *sensor* is defined as a device that is sensitive to light, temperature, electrical impedance, or radiation level and transmits a signal to a measuring or control device. A *measuring device* passes through two stages while measuring a signal. First, the *measurand* (a physical quantity such as acceleration, pressure, strain, temperature etc.) is sensed. Then, the *measured signal* is transduced into a form that is particularly suitable for transmitting, signal conditioning, and processing. For this reason, output of the transducer stage is often an electrical signal that is then digitized. The sensor and transducer stages of a typical measuring device are represented schematically in Fig. 3.1.

Usually a transduction element relies on a physical principle that a change in the measurand value alters the characteristics of the transducer output (the electrical signal). There are only a handful of physical principles behind many thousands of available transducers on the market. Allocca and Stuart (1984)

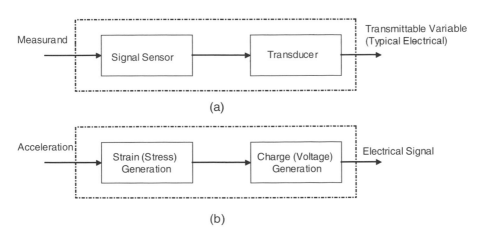

(a)

(b)

Figure 3.1 (*a*) Schematic representation of a measuring device. (*b*) Operation of a piezoelectric accelerometer.

provide a complete list of transducers classified according to different physical principles. In this book we will give a few examples that are closely relevant in mechanical systems diagnosis and prognosis.

A desirable feature for practical transducers is to have a static (nondynamic) input/output relationship so that the output instantly reaches the input value (or the measured variable). In this case, the transducer transfer function is a pure gain. This happens when the transducer time constants are small (i.e., the transducer bandwidth is high). However, in practice, the dynamic effects of the transducer are sometimes nonneglegible owing to limitations on the physical size of the components and the types of materials used in the construction. Usually the electrical time constant of the transducer is an order of magnitude smaller than the mechanical time constant (de Silva, 1989). When choosing a transducer system, it is important to note the performance specifications such as rise time, instrument bandwidth, linearity, saturation, etc.

3.2.2 Mechanical/Structural Sensor Systems

Mechanical and structural sensor systems have been studied extensively, and a large number of such devices are currently in use to monitor system performance for operational state assessment and tracking of fault indicators. A number of mechanical quantities—position, speed, acceleration, torque, strain, etc.—are commonly employed in dynamic systems. Conventional devices for measuring these quantities are available commercially (*www. sensorland.com/TorqPage001.html, www.oceanasensor.com/*), and their operation has been amply described in textbooks and technical publications (Allocca and Stuart, 1984; de Silva, 1989). We will use only two mechanical sensor systems—accelerometers for vibration monitoring and strain gauges— to illustrate their operational characteristics and highlight their utility in CBM/ PHM system applications. The reader is referred to the voluminous literature for other mechanical sensors of interest. Structural component monitoring and interrogation have been attracting interest in recent years, and ultrasonic sensing is a typical means to measure relevant quantities of interest. We will introduce recent advances in this area.

Accelerometers: Vibration Measurements. Recent years have seen an increased requirement for a greater understanding of the causes of vibration and the dynamic response of failing structures and machines to vibratory forces. An accurate, reliable, and robust vibration transducer therefore is required to monitor online such critical components and structures. Piezoelectric accelerometers offer a wide dynamic range and rank among the optimal choices for vibration-monitoring apparatus. They exhibit such desirable properties as

- Usability over very wide frequency ranges
- Excellent linearity over a very wide dynamic range
- Electronically integrated acceleration signals to provide velocity and displacement data.
- Vibration measurements in a wide range of environmental conditions while still maintaining excellent accuracy
- Self-generating power supply
- No moving parts and hence extreme durability
- Extremely compact plus a high sensitivity-to-mass ratio

Piezoelectric accelerometers are used to measure all types of vibrations regardless of their nature or source in the time or frequency domain as long as the accelerometer has the correct frequency and dynamic ranges. A basic simplified model of a piezoelectric accelerometer is shown in Fig. 3.2 (Serridge and Licht, 1987).

The piezoelectric element shown has an equivalent stiffness κ and constitutes the active element of the transducer. It acts as a spring connecting the base mass m_o to the seismic mass m_s. A force equal to the product of the seismic mass and the acceleration it experiences when subjected to this vibratory force acts on the piezoelectric element, producing a charge proportional to the applied force. As the seismic mass accelerates with the same magnitude and phase as the accelerometer base over a wide frequency range, the output of the accelerometer is proportional to the acceleration of the base and hence to the acceleration of the structure into which the device is mounted. From the equation of motion of the simple model shown in Fig. 3.2, we easily may arrive at the following relationship for the $\mathcal{R}$ ratio $\mathcal{R}/\mathcal{R}_o$

$$\frac{\mathcal{R}}{\mathcal{R}_o} = \frac{1}{1 - \left(\dfrac{\omega}{\omega_n}\right)^2}$$

where $\mathcal{R}_o$ is the displacement of the masses at frequencies below the natural resonance frequency of the accelerometer ($\omega \ll \omega_n$), assuming that the displacement varies sinusoidally, and $\mathcal{R}$ is the displacement at higher frequencies. Thus the displacement between the base and the seismic mass increases as the forcing frequency becomes comparable with the natural resonance frequency of the accelerometer. Consequently, the force on the piezoelectric element and the electrical output from the accelerometer also increase. The frequency range over which the accelerometer can be used is determined by the mounted resonance frequency of the device. There is a tradeoff between high-frequency performance and transducer sensitivity. Piezoelectric materials typically in use include lead titanate, lead zirconate titanate, quartz crystal, etc. Figure 3.3 shows a typical Bruel and Kjaer piezoelectric accelerometer,

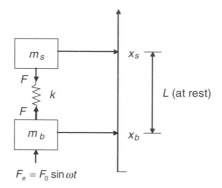

Figure 3.2   Simplified model of an accelerometer.

whereas Fig. 3.4 depicts a triaxial accelerometer manufactured by Bruel and Kjaer.

Strain Gauges. A strain-gauge sensor is based on a simple principle from basic electronics that the resistance of a conductor is directly proportional to its length and resistivity and inversely proportional to its cross-sectional area. Applied stress or strain causes the metal transduction element to vary in length and cross-sectional area, thus causing a change in resistance that can be measured as an electrical signal. Certain substances, such as semiconductors, exhibit the piezoresistive effect, in which application of strain greatly affects their resistivity. Strain gauges of this type have a sensitivity approximately two orders greater than the former type.

The transducer usually is used within a Wheatstone bridge arrangement, with either one, two, or all four of the bridge arms being individual strain gauges, so that the output voltage change is an indication of measurand (the strain) change.

Figure 3.3   A Bruel and Kjaer piezoelectric accelerometer (type 7704A-50).

Figure 3.4   Miniature triaxial DeltaTron accelerometer (types 4524).

Strain-Gauge Positioning. Four strain gauges were positioned and wired to sense bending modes and blade vibration (Lackner, 2004). They are all positioned so that they are parallel with the leading and trailing edges of the blade. Two strain gauges are on one side of the blade, and the other two strain gauges are mirror images on the other side. The strain gauges are all positioned at approximately one-third the span of the blade. Two of the gauges are at one-quarter of the chord, and two are at three-quarters of the chord. The positions of each of the strain gauges are shown in Fig. 3.5. The strain gauges are highlighted in black, and the dashed line indicates flipping from one side of the blade to the other.

Strain-Gauge Wiring Scheme. The strain gauges were wired as a full Wheatstone bridge. A diagram of a Wheatstone bridge is shown in Fig. 3.6. Each resistor represents one of the strain gauges.

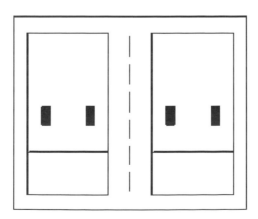

Figure 3.5   Strain-gauge positions for bending blade.

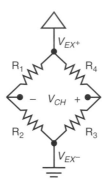

Figure 3.6 Wheatstone bridge.

For a full bridge, all four of the resistors correspond to each of strain gauges. When a strain gauge undergoes tension, the resistance of the strain gauge increases. Likewise, when it undergoes compression, the resistance decreases. The four strain gauges can be connected to each other in such a way that they make a full bridge. Consider the case of bending. When the blade bends, the two gauges on one side of the blade undergo a tension force, whereas the two gauges on the opposite side undergo a compressive force. V_{EX} is the excitation voltage, which is provided by the amplifier to power the bridge. If the gauges are wired such that the two gauges that experience tension are R_1 and R_3 in the Wheatstone bridge and the two gauges that experience compression are R_2 and R_4 in the Wheatstone bridge, then a voltage will appear across V_{CH}. This voltage, V_{CH}, is the output voltage signal.

A full bridge is an advantageous setup for the strain gauges for a variety of reasons. First, a full bridge eliminates centrifugal stress effects. When the blade is spun, centrifugal forces produce a tension force throughout the blade. However, all four strain gauges are at the same spanwise location, so they experience the same tension force. Since each gauge experiences the same stress, the resistance of each gauge increases the same amount. The result is that the bridge is unaffected by the centrifugal stress, and V_{CH} is zero. Thermal effects are canceled out in the same way. As a spinning blade heats up owing to rotor and aerodynamic friction, the blade also tends to lengthen somewhat. Again, all four of the gauges experience the same force, so the output of the bridge is zero.

The output of the strain-gauge amplifier is a simple voltage signal that can be connected to a BNC cable. The BNC cable then can be connected to an oscilloscope to view the strain output or to the data-acquisition system to take strain-gauge data.

Ultrasonic Sensor Systems.* Ultrasonic sensor systems are being considered for monitoring the health of critical structures such as airplanes, bridges,

*Contributed by Professors Jennifer Michaels and Thomas Michaels, School of ECE, Georgia Tech.

and buildings. Ultrasonic methods are particularly suitable for structural health monitoring both because ultrasonic waves travel long distances and thus have the potential to monitor a large volume of material and because ultrasonic methods have proven useful for nondestructive inspection (NDI) of such structures during maintenance. There are three main types of ultrasonic waves that are suitable for structural health monitoring: guided waves, bulk waves, and diffuse waves. Regardless of the type of wave, the strategy is to monitor changes and then detect, localize, and characterize damage based on the nature of the change. This strategy of looking for changes can enable detection sensitivity to be similar to that of NDI despite the limitation of fixed sensors.

Figure 3.7 shows examples of ultrasonic transducers mounted to metallic components. In Fig. 3.7a, a simple piezoelectric disk is encapsulated in epoxy and glued to the surface of an aluminum component. The nominal frequency is 2.25 MHz, but the disks are broadband transmitters and receivers, effectively operating from lower than 50 kHz to above 3 MHz. This same type of disk also can be packaged in a housing with a connector. Figure 3.7b shows commercially available angle-beam transducers attached to a specimen; the U-shaped aluminum blocks are for mechanical protection. These transducers have a center frequency of 10 MHz and can be used to monitor very small cracks.

Guided Waves. Guided ultrasonic waves require one or more interfaces to guide the wave propagation and can propagate long distances while maintaining sufficient energy for useful reception. Rayleigh waves, or surface waves, are suitable for monitoring surface or near-surface defects; their penetration depth is approximately a wavelength. Lamb waves, or plate waves, are suitable for monitoring platelike structures such as metallic and composite aircraft skins; however, Lamb waves are dispersive and can be difficult to analyze. Guided waves typically are generated in situ by mounting low-profile

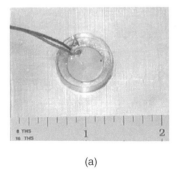
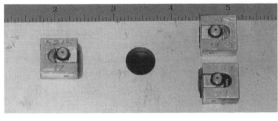

(a)   (b)

Figure 3.7 Ultrasonic transducers attached to metallic parts. (*a*) Piezoelectric disks encapsulated in epoxy and bonded to the part. (*b*) Commercially available angle-beam transducers protected with an aluminum U-block and bonded to the part.

transducers on an accessible surface. Each transducer can be as simple as a piezoelectric disk or patch or also can be a shaped device designed to propagate specific plate modes.

Defects can be detected and localized from in situ guided waves using a variety of methods. Cook and Bertholet (2001) used Rayleigh waves to monitor crack growth and applied simple time-domain differencing to estimate crack size. Leonard et al. (2000) and Gao et al. (2005) have applied tomographic reconstruction methods to Lamb waves and have obtained images of damage. Giurgiutiu and Bao (2004) and Wilcox (2003) have applied phased-array techniques to compact guided-wave arrays. Recently, Michaels and Michaels (2006) of Georgia Tech have generated images of damage in plates using very sparse arrays of piezoelectric disks. Figure 3.8a shows differenced waveforms owing to the scattered sound field from a drilled hole in an aluminum plate, and Fig. 3.8b shows the resulting image of the hole using a sparse array of only four transducers.

Bulk Waves. Bulk waves refer to longitudinal and shear elastic waves that propagate in bulk materials. These are the conventional ultrasonic waves typically used for NDI methods, whereby the transducer is moved either manually or automatically to sweep the ultrasonic beam path through the volume

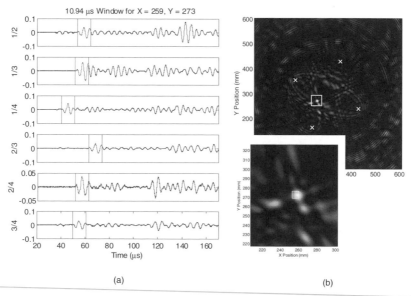

(a)   (b)

Figure 3.8 (*a*) Ultrasonic Lamb wave signals scattered from a drilled hole in an aluminum plate with the marked regions indicating scattering from the hole. (*b*) The resulting phased-array image constructed from these six signals. The solid circles are plotted at the locations of the attached transducers, the open circle is centered at the actual hole location, and the inset is a zoomed view of the hole.

of material being inspected. Bulk-wave methods are suitable for monitoring of "hot spots" in structures where the location of damage already has been identified through either analysis or experience. For example, Michaels et al. (2005) and Mi et al. (2005) have developed an angle-beam shear-wave method for monitoring crack growth originating from fastener holes. As shown in Fig. 3.9, the incident shear wave converts to a creeping wave that travels around the fastener hole, and any cracks block this wave and keep it from propagating to the receiver. As the crack opens under load, this effect is accentuated and permits the crack area to be estimated. Figure 3.10 shows a typical plot of estimated crack area versus fatigue cycles.

Diffuse Waves. The concept of a diffuse-wave field refers to a wave field in which all modes are equally excited. Practically speaking, diffuse or diffuse-like waves are those in which specific echoes cannot be identified, and energy essentially has "filled" the structure in which they are propagating. The acoustoultrasonic (AU) NDI method refers to generation and analysis of diffuse-like waves that are recorded for a long period of time and eventually decay to negligible levels. The appeal of diffuse waves is that they are very easy to generate, and they are very sensitive to any structural changes. The downside of diffuse waves is that it is very difficult to discriminate changes owing to damage from those owing to benign effects such as environmental changes and transducer coupling. However, the situation is somewhat more tractable for structural health monitoring as compared with NDI because the fixed nature of the transducers minimizes variability owing to transducer coupling and positioning. Recent work at Georgia Tech ((Lu and Michaels, 2005; Michaels and Michaels, 2005; Michaels et al., 2005) has been successful at discriminating artificial damage in aluminum specimens from environmental changes. One of the most effective discriminators is the local temporal coherence, which is a measure of the similarity in shape of two signals as a

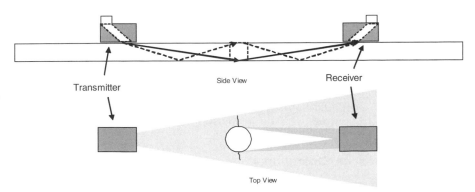

Figure 3.9 Angle-beam through-transmission method for in situ monitoring of fatigue cracks originating from fastener holes.

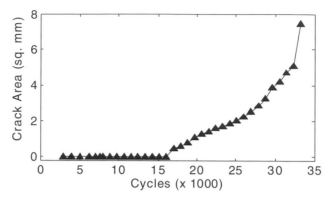

Figure 3.10 Plot of crack area versus fatigue cycles for monitoring of fastener holes.

function of time. Figure 3.11 shows local temporal coherence plots illustrating progression of damage in an aluminum specimen. Damage causes a distinctive loss of coherence compared with a baseline waveform, whereas coherence is maintained throughout the decay of the reverberating diffuse-like signals for modest environmental changes.

Other Ultrasonic Methods. Research is ongoing in applying other ultrasonic NDI methods to structural health monitoring. Ultrasonic velocity and attenuation have been used for NDI to characterize material properties; a corresponding health monitoring implementation is very possible and is expected to be effective in detecting gradual deterioration of material properties over the volume being interrogated. Nonlinear ultrasonic methods have shown promise for detection of microdamage; however, the technique is very difficult to implement, even in the laboratory, and an in situ implementation would be even more difficult.

System Architecture. Various ultrasonic sensing methods can be implemented in a synergistic manner to detect a wide variety of damage both locally and globally. Figure 3.12 illustrates the architecture of a complete ultrasonic health monitoring system. Current and historical data are analyzed and fused using both data-based and physics-based methods to obtain the best possible diagnosis of the current structural state. This state information then is used for prognosis of the structure's future performance.

3.2.3 Performance/Operational Sensors

System performance and operational data are monitored routinely in all industrial establishments, utility operations, transportation systems, etc. for

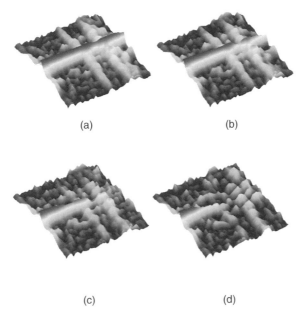

(a)   (b)

(c)   (d)

Figure 3.11 Local temporal coherence of diffuse waves from a damaged component compared with a baseline signal from an undamaged component. Plots (*a*) through (*d*) show the progressive loss of coherence as an artificial defect is enlarged.

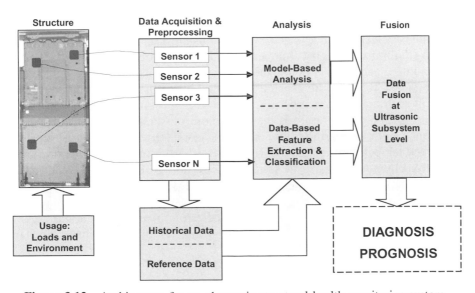

Figure 3.12 Architecture for an ultrasonic structural health monitoring system.

process control, performance evaluation, quality assurance, and fault diagnosis purposes. A large number of sensor systems have been developed and employed over the years. The list includes devices that are intended to measure such critical properties as temperature; pressure; fluid, thermodynamic, and optical properties; and biochemical elements, among many others. Sensors based on classic measuring elements—inductive, capacitive, ultrasound—have found extensive applications. Semiconductor materials usually constitute the building blocks for a large number of sensors. Magnetic-field (Hall sensors) devices are coupled with electrical sensors to monitor parameters of magnetic materials. Fiberoptic sensors are treated in a separate section of this chapter. More recently, biochemical sensors have begun taking central stage, and their detection principles and requirements are described in the technical literature (Guardia, 1995). Characteristic chemical-sensor properties with potential application to structural fault diagnosis include liquid and solid electrolytic sensors, photochemical sensors, humidity sensors, and field-effect and mass-sensitive devices. As an example of their principles of operation, consider conductivity sensors. In these devices, the interaction of the gas with the solid (semiconducting metal oxide or organic semiconductor) causes a change in conductivity. A change in resistance also can be caused by a change in the temperature of the sensor material. An ample number of reference textbooks and technical publications thoroughly cover the operating principles, design methods, and applications of these sensor systems. We will restrict our attention in this book only to one class of sensors—temperature sensors—for illustrative purposes and as primary candidates for fault diagnosis.

Temperature Sensors/Thermography. Temperature variations in many mechanical, electrical, and electronic systems are excellent indicators of impending failure conditions. Temperatures in excess of control limits should be monitored and used in conjunction with other measurements to detect and isolate faults. Temperature sensing has found numerous applications over the years in such areas as engineering, medicine, environmental monitoring, etc. Except for time, temperature is the most closely watched physical parameter in science and engineering. Temperature sensors are used widely to measure and track temperature levels and variations. The thermal properties of materials from living organisms to solids, liquids, and gases are significant indicators of the state of such objects. For CBM/PHM system design, temperature variations—hot spots, large temperature gradients, and temperatures in excess of specified levels—are considered key indicators of incipient failure conditions. A number of devices based on heat-transfer principles have been in use.

Among the most common temperature-sensing devices are resistance temperature detectors (RTDs), whose principle of operation is variation of the resistance of a platinum wire or film as a function of temperature. Platinum usually is employed because of its stability with temperature and the fact that its resistance tends to be almost linear with temperature. Industrial platinum resistance thermometer elements typically have a temperature range of −200 to 650°C. Specialty RTDs are available up to 850°C (*www.watlow.com/products/sensors/se_rtd.cfm*). SensorTec (*www.sensortecinc.com/*) offers a full line of RTDs with temperature ranges from −200 up to 650°C. Resistance thermometer elements are available in a number of forms (*www.peaksensors.co.uk/*). The most common are (1) wire wound in a ceramic insulator (high temperatures to 850°C) and (2) wire encapsulated in glass (resists the highest vibration and offers most protection to the platinum thin film, with platinum film on a ceramic substrate; inexpensive mass production). Figure 3.13 shows an industrial platinum RTD series type 21A straight-probe assembly) with an operating range of 195° to 482°C. (*www.rdfcorp.com/products/industrial/r-21_01.shtml*).

In use, an RTD normally is connected to form one arm of a Wheatstone bridge measuring arrangement, allowing high measurement accuracy, particularly in a four-wire connection. The transducer's quite low resistance (∼100 Ω) may necessitate in certain circumstances such a four-wire element configuration to compensate for the resistance of the connecting leads. Resistance thermometers require a small current to be passed through in order to determine the resistance. This can cause self-heating, and manufacturer's limits always should be followed along with heat-path considerations in design. Care also should be taken to avoid any strains on the resistance thermometer in its application.

The major advantages of RTDs are stability, repeatability, and accuracy. They also have a few disadvantages, such as cost (platinum is expensive), temperature range (usually −200 to 650°C), and slow response to temperature changes. In addition, the device can be damaged by vibration and shock.

A *thermistor* is a thermally sensitive resistor whose electrical resistance varies directly with temperature. Thermistors are ceramic semiconductors exhibiting a negative (NTC) or positive (PTC) temperature coefficient, and typically they are used within the −200 to 1000°C temperature range for NTCs and 60 to 180°C for PTCs. NTCs should be chosen when a continuous change

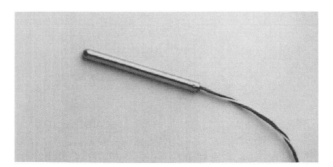

Figure 3.13   Industrial platinum RTD series 21A.

of resistance is required over a wide temperature range, and PTCs should be chosen when a drastic change in resistance is required at a specific temperature or current level. Hermetically sealed thermistors have excellent long-term stability with a wide range of relatively high resistance values and a fast response time, and they are available in small sizes (0.005-in. diameter). Figure 3.14 shows Honeywell ISO-curve glass bead and glass probe thermistors that are designed for use as precision sensing elements where curve-matched interchangeability is required for precise temperature control and precision temperature indication. These highly reliable products consist of a matched pair of hermetically sealed thermistors and may be selected to tolerances only limited to the system's capability to test them (*http://content.honeywell.com/ sensing/prodinfo/thermistors/isocurves.asp*). Positive temperature coefficients of 0.7 percent per degree centigrade are typical, with dissipation constants of the order of 2.5 mW/°C. Figure 3.15 shows an array of scaled thermistors from U.S. Sensor (*www.ussensor.com/pgthermistors.html*). Military-grade surface-mount thermistors are manufactured by Quality Thermistor, Inc. (*www.thermistor.com/ptctherms.cfm*).

Thermocouples are by far the most widely used sensors in industry. They are very rugged and can be used from subzero temperatures to temperatures well over 4000°F (2000°C). The sensing of temperature by thermocouples is based on the Seebeck, or thermoelectric, effect in which two dissimilar materials are joined, and if the two junctions so formed are at different temperatures, a current flows around the circuit. When the circuit is opened, an electromotive force (emf) appears that is proportional to the observed Seebeck current. Thermocouples respond to temperature differences rather than absolute temperature; their output is a function of the temperatures at two points, the so-called hot and cold junctions. Common thermocouple wire pairs are listed in standard tables specifying millivolt output versus temperature with a reference junction temperature of 0°C. Their sensitivities range from 5 to 80 μV/°C. Watlow supplies thermocouples with a wide temperature range from −200 to 2200°C (−328 to 4000°F) (*www.watlow.com/products/sensors/se_ therm.cfm*).

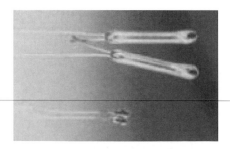

Figure 3.14 ISO-curve glass bead and glass probe thermistors.

Figure 3.15 Standard leaded epoxy-coated thermistors.

Iron-constantan (ANSI symbol J), copper-constantan (ANSI symbol T), chromel-alumel (ANSI symbol K), and chromel-constantan (ANSI symbol E) are typical thermocouple materials. Figure 3.16 shows a thermocouple construction (or assembly) one can expect to see in the real world. The thermocouple is a heavy-duty thermocouple assembly with metal-alloy protection tubes, series A33 from NANMAC (*www.nanmac.com/handbook/a33.htm*).

It is well known that such defects as cracks in metallic structures generate heat when subjected to mechanical vibrations. The increased hysteresis of the defect generates more thermal activity than the surrounding undamaged region. The thermal waves launched from the defect propagate to the surface, and the resulting thermal pattern may reveal the presence of the defect when synchronized with an appropriate excitation signal. This nondestructive evaluation (NDE) technique is employed for inspection purposes and uses a method called *thermoelastic stress analysis* (TSA) and *lock-in thermography* (LT). The technique has been demonstrated successfully on subsurface cracks in a single-layer structure up to 20 mm thick by El-Hajjar and Haj-Ali (2004) at the Georgia Institute of Technology. Figure 3.17 provides infrared images of the progression of fatigue damage around a rivet hole. It is evident from the images that the rivet impedes the heat transmitted directly through it.

L   12" is standard

↑ Optional Adjustable Mounting Fitting

1/2 NPT   Optional Mounting Flange

Figure 3.16 Heavy-duty thermocouple assembly with metal-alloy protection tubes, series A33.

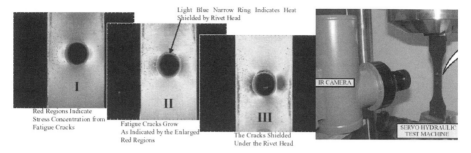

Figure 3.17   Infrared images of the progression of fatigue damage around a rivet hole in a 7050-T7451 test specimen.

However, the heat generated by the damaged region (stress concentration) is captured vividly by the TSA method.

3.2.4   Electrical Measurements

Electromechanical, electrical, and electronic systems constitute a major component of our industrial engine. They are the dominant element in such areas as transportation systems, utilities, biomedical instrumentation, communications, computing, etc. Electric drives and electrical/electronic components/ systems are currently attracting the interest of the CBM/PHM community, and sensors monitoring their performance are becoming the designer's focused activity. College textbooks and courses in measurement present an excellent treatment of classic sensor systems aimed at tracking such quantities as voltage, current, power, torque, and their derivatives—harmonic content, for example—as key indicators of degrading performance. For this reason, in this section we will elaborate on electrical measurements. Feature or condition indicators from electrical and other similar measurements for fault diagnosis are highlighted throughout this book.

A number of sensor systems have been developed and applied in the recent past in an attempt to interrogate in-situ critical components and systems for fault diagnosis and prognosis. Transducing principles based on eddy-current response characteristics, optical and infrared signal mentoring, microwaves, and others have been investigated. We will outline the basic principles and possible application domains of a few of those devices.

Eddy-Current Proximity Probes.   Response characteristics of induced eddy currents in conducting media are monitored for changes in their behavior owing to material anomalies, cracks, shaft or mating-part displacements, etc. Eddy-current proximity probes are a mature technology that has been used for protection and management of rotating machinery. They are employed commonly in high-speed turbomachinery to observe relative shaft motion directly, that is, inside bearing clearances of fluid-film interfaces. One can search for eddy-current proximity sensors at *http://proximity-sensors. globalspec.com/SpecSearch/Suppliers/Sensors_Transducers_Detectors/ Proximity_Presence_Sensing/Eddy_Current_Proximity_Sensors.*   Manufacturers of eddy-current proximity sensors include Panasonic Electric Works (formerly Aromat; *http://pewa.panasonic.com/acsd/*), Lion Precision (*www. lionprecision.com/*), and Metrix Instrument Co. (www.metrix1.com/). Figure 3.18 shows a typical proximity sensor.

Proximity sensors are used to detect lateral shaft motion for absolute or relative misalignment, bearing defects, cracks, and other material anomalies by generating *xy* orbit plots from the response of three such devices strategically located in the structure under test. Figure 3.19*a* presents the eddy-current amplitude, power spectral density, and orbit plot of three proximity sensors interrogating a normal or unfaulted specimen. Figure 3.19*b*, though, shows a similar situation when a fault is present. Zou et al. (2000) describe the application of eddy-current proximity sensing to the detection of a crack in a seal/rotor drive-shaft arrangement.

Microelectromechanical System (MEMS) Sensors. Of interest also are sensor systems that can be produced inexpensively, singly or in an array, while maintaining a high level of operational reliability. Microelectromechanical systems (MEMS) and sensors based on fiberoptic technologies are finding popularity because of their size, cost, and ability to integrate multiple transducers in a single device. Micromachined MEMS devices in silicon or other materials are fabricated in a batch process with the potential for integration with electronics, thus facilitating on-board signal processing and other "smart" functions. A number of MEMS transducer and sensor systems have been manufactured in the laboratory or are available commercially, monitoring such critical parameters as temperature, pressure, acceleration, etc.

Analog Devices, Inc. (ADI), produces a three-axis integrated MEMS (iMEMS) Motion Signal Processing technology that will enable placement of very small, low-power three-axis (*xyz*) accelerometers that combine

Figure 3.18   An eddy-current proximity sensor (GL-18H/18HL series by SUNX).

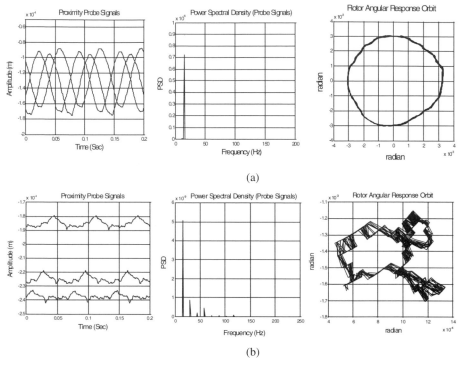

Figure 3.19   Eddy-current signals, power spectral densities, and orbit plots for (*a*) unfaulted and (*b*) faulted conditions.

a tiny, robust motion sensor and advanced signal-conditioning circuitry on a single low-power chip (*www.analog.com/en/subCat/0,2879, 764%255F800%255F0%255F%255F0%255F,00.html*). Honeywell manufactures MEMS gyro-based inertial sensors such as HG1900, which uses Honeywell MEMS gyros and RBA500 accelerometers. These sensors require only a low-voltage input power source and give delta angle and delta velocity as their output (*http://content.honeywell.com/dses/products/tactical/hg1900.htm*).

3.2.5 Fiberoptic Sensors

Fiber optics have penetrated the telecommunications and other high-technology sectors in recent years. They find utility in the sensor field because of their compact and flexible geometry, potential for fabrication into arrays of devices, batch fabrication, etc. Fiberoptic sensors have been designed to measure strain, temperature, displacement, chemical concentration, and acceleration, among other material and environmental properties. Their main advantages include small size, light weight, immunity to electromagnetic and radio frequency interference (EMI/RFI), high- and low-temperature endurance, fast response, high sensitivity, and low cost. Fiberoptic technologies are based on extrinsic Fabry-Perot interferometry (EFPI), chemical change in the fiber cladding, optical signal changes owing to fiber stress and deformation, etc. Manufacturers of fiberoptic sensors include Luna Innovations (*www. lunainnovations.com/r&d/fiber_sensing.asp*), KEYENCE (*http://world. keyence.com/products/fiber_optic_sensors/fs.html*), and Techno-Sciences (*www.technosci.com/capabilities/fiberoptic.php*).

Luna Innovations has developed fiberoptic temperature sensors based on EFPI. The basic physics governing the operation of these sensors makes them relatively tolerant of or immune to the effects of high-temperature, high-EMI, and highly corrosive environments. These sensors are able to operate in harsh environments. The basic physics also leads to a very stable, accurate, and linear temperature sensor over a large temperature range. These sensors are also quite small and therefore ideal for applications where restricted space or minimal measurement interference is a consideration. The size also leads to a very small time response as compared with other temperature measurement techniques (Fig. 3.20).

3.3 SENSOR PLACEMENT

Placement of sensors to monitor fault indicators or signatures accurately and robustly in a critical military or industrial system constitutes an essential function of the PHM/CBM design process. Sensors typically are placed on the basis of experience taking into consideration the estimated number of sensors needed, types of measurements to be collected, physical access for sensor placement, etc. Guidelines are available for sensor locations for some military and aerospace systems (Kammer, 1991). A few analytic techniques have been suggested in recent years as to how to determine the number, type, and location of sensors. Most of this work stems from the chemical industry, where the need to minimize the number of sensors employed for a particular

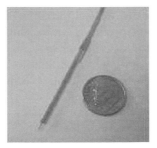

Figure 3.20   A typical fiberoptic temperature sensor.

process is obvious owing to the size, complexity, and number of parameters to be tracked in such complex chemical processes (Alonso et al., 2004; Chmielewski et al., 2002).

The sensor suite installed to monitor fault signatures must, of necessity, possess attributes of durability, robustness, accuracy, and sensitivity while responding to the dynamic range of the variables/parameters being monitored. The sensor's useful life must be much longer than the estimated lifespan of the components/subsystems it is intended to measure. Optimizing the task of selecting an appropriate sensor suite and identifying the best location for the physical devices is contributing to the technoeconomic success and viability of PHM/CBM systems. Traditionally, sensors are placed to meet control and performance monitoring objectives (Anthony and Belinda, 2000; Al-Shehabi and Newman, 2001; Giraud and Jouvencel, 1995; Chen and Li, 2002). It is instructive to take advantage of such sensors in a fault diagnosis monitoring scheme because they can provide useful information relating to fault behaviors of critical system variables. Fault signatures, though, may require additional PHM/CBM-specific sensors; typical devices in this class are accelerometers for vibration monitoring, proximity sensors, acoustic emission sensors, ultrasonic transducers, etc. More recently, research on sensor placement has focused on two different levels: the component level and the system level. At the component level, attempts have been reported regarding placement at the component's range, for example, a bearing or an object in three-dimensional view (Anthony and Belinda, 2000; Naimimohasses et al., 1995). For complex, large-scale systems consisting of multiple components/subsystems, a fault may propagate through several components. With a large number of possible sensor locations, selection of an optimal location, as well as the number and types of sensors, poses a challenging problem that must be addressed at the system level.

A possible systems-theoretic approach views the optimal sensor placement problem from various perspectives, but the underlying theoretical framework relies on estimation techniques, sensitivity analysis, observability concepts, and optimization tools. We are basically seeking to estimate through measurements the value of a critical variable, maximize the sensor sensitivity to fault signals while minimizing false alarms, ascertain that indeed such variables are "observable," and finally, optimize the number, types, and locations of the sensors selected. As an illustration of how such methods can address the optimal sensor placement problem, we summarize below a systems approach developed at the Georgia Institute of Technology (Zhang, 2005). It is intended only to motivate further research into an important PHM/CMB topic. Other approaches may exploit different techniques, seeking answers to the same questions raised in this section.

The foundational concept of this approach is a fault detectability metric that expresses the sensor's capability to detect a fault. A graph-based technique, called the *quantified directed model* (QDM), is employed to model fault propagation from one component or subsystem to the next. An appropriate figure of merit is finally defined that is aimed at maximizing fault detectability while simultaneously minimizing the required number of sensors and achieving optimal sensor placement. The sensor selection/localization architecture is shown in Fig. 3.21. It entails several functional modules: requirements analysis, system modeling, a figure of merit (FOM), optimization, and performance assessment.

Four main requirements must be met to place sensors optimally for fault diagnostic purposes: detectability, identifiability, fault-detection reliability, and a requirement associated with limited resources. Meanwhile, sensor uncertainty needs to be addressed because it is always related to the sensor measurement capability. Sensor fault detectability is the extent to which a sensor can detect the presence of a particular fault. It depends on such factors as signal-to-noise ratio, time-to-detection-to-time-to-failure ratio, fault detection sensitivity, and symptom-duration-to-time-to-failure ratio. Some of these terms are not common in the engineering lexicon and must be explained further. *Time to detection* is the time span between the initiation of a fault and its detection by a sensor, whereas *time to failure* is the duration between initiation of the fault and the time when the failure occurs. *Fault detection sensitivity* of a sensor is defined as the ratio of the change in the fault measurement to the change in the sensor measurement. The influence of these terms on overall fault detectability is intuitively obvious. Each term can be defined empirically, and their collection (in additive or multiplicative form) comprises the sensor fault detectability metric.

Uncertainty plays an important role in the definition of fault detection sensitivity and signal-to-noise ratio. Signal-to-noise ratio may be estimated by composing probabilistic distributions from all possible noise sources and then employing the law of propagation of uncertainty. An FMECA study is used widely to identify fault modes, their severity, and their frequency of

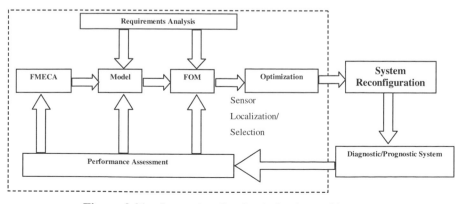

Figure 3.21 Sensor localization/selection architecture.

occurrence, as detailed in Sec. 2.3 in Chap. 2. It also forms the basis of identification of fault symptoms and the sensors that are potentially capable of monitoring the initiation and evolution of a fault.

A candidate system model to be employed in sensor placement that may represent fault signal propagation events faithfully is based on graph theoretic notions such as the directed graph (DG) (Raghuraj et al., 1999), signed directed graph (SDG) (Bhushan and Rengaswamy, 2000; Kramer and Palowitch, 1987), and the quantified directed graph (QDG) (Zhang, 2005). A directed branch in a DG model connects one node to another and shows the fault signal propagation direction between the nodes. In the SDG model, each branch has a sign (+ or −) associated with it that indicates whether the fault signals at the two nodes change in the same (+) or opposite (−) direction. Neither models include any quantitative information about fault signal propagation. In the QDG model, each node stands for a possible sensor location and is assigned a singal-to-noise ratio in decibels with it. Each directed branch has a plus or minus sign associated with it, as well as the fault propagation time and the fault propagation gain between the two nodes connected by the branch. Based on the QDG model, the possible types and locations of sensors in the fault propagation path can be determined by scanning all the sensors in the path. The Figure of Merit (FOM) is in the form of the weighted sum of the fault detectability and the number of sensors. The weights are adjustable and are determined mainly by such considerations as failure severity and the probability of failure occurrence.

Finally, the optimization problem is set as an integer nonlinear programming problem. Several commercially available solvers can be employed to arrive at the optimal or suboptimal sensor number and location. Unfortunately, to detect and identify a large number of faults by placing sensors at various locations, the computational burden becomes prohibitive. A reduction of this effort can be achieved by calling on a particle-swarm optimization method (Eberhart and Kennedy, 1995), where sensors are selected for each fault on the basis of a heuristic search algorithm. The performance-assessment module's objective is to estimate the fault-detecting error rate. The specific performance metrics for each fault include a single false-positive rate and a single false-negative rate from which the average false-positive and false-negative rates are computed. The mathematical details of how to realize these metrics and an illustrative example application of the approach to a continuous-batch demonstration process can be found in Zhang (2005).

3.4 WIRELESS SENSOR NETWORKS

Wireless sensor networks (WSNs) are important in applications where wires cannot be run owing to cost, weight, or accessibility. Properly designed WSNs can be installed and calibrated quickly and can be up and running in a very short time frame. Systems designed at the University of Texas at Arlington (UTA) Automation & Robotics Research Institute (ARRI) can be installed in machinery spaces in a few hours. More details about WSNs can be found in Lewis (2004).

Figure 3.22 shows the complexity of WSNs, which generally consist of a data-acquisition network and a data-distribution network monitored and controlled by a management center. The plethora of available technologies makes even the selection of components difficult, let alone the design of a consistent, reliable, robust overall system. The study of WSNs is challenging in that it requires an enormous breadth of knowledge from a large variety of disciplines.

3.4.1 Network Topology

The basic issue in communication networks is the transmission of messages to achieve a prescribed message throughput (quantity of service) and quality of service (QoS). QoS can be specified in terms of message delay, message due dates, bit error rates, packet loss, economic cost of transmission, transmission power, etc. Depending on QoS, the installation environment, economic considerations, and the application, one of several basic network topologies may be used.

A communication network is composed of nodes, each of which has computing power and can transmit and receive messages over communication links, wireless or cabled. The basic network topologies are shown in Fig. 3.23

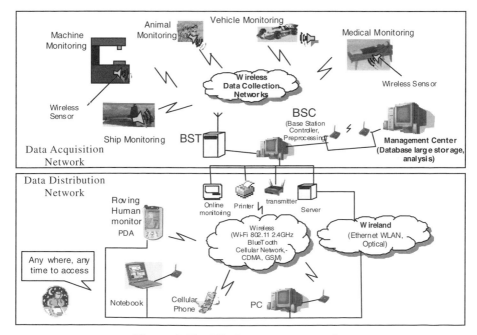

Figure 3.22 Wireless sensor networks.

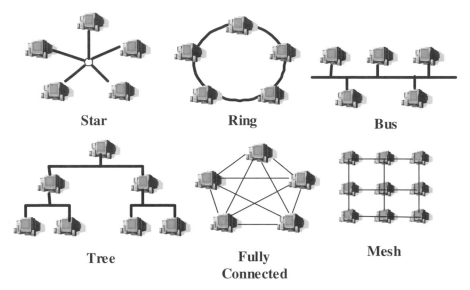

Figure 3.23   Basic network topologies.

and include fully connected, mesh, star, ring, tree, and bus. A single network may consist of several interconnected subnets of different topologies. Networks are further classified as local-area networks (LANs), for example, inside one building, or wide-area networks (WANs), for example, between buildings.

3.4.2   Power Management

With the advent of ad hoc networks of geographically distributed sensors in remote environments (e.g., sensors dropped from aircraft for personnel/vehicle surveillance), there is a focus on increasing the lifetimes of sensor nodes through power generation, power conservation, and power management. Current research centers on designing small MEMS radio-frequency (rf) components for transceivers, including capacitors, inductors, etc. The limiting factor now is in fabricating micro-sized inductors. Another thrust is in designing MEMS power generators using technologies including solar, vibration (electromagnetic and electrostatic), thermal, etc. (Fig. 3.24 and Fig. 3.25).

Radio-frequency identification (RF-ID) devices are transponder microcircuits having an inductor-capacitor tank circuit that stores power from received interrogation signals and then uses that power to transmit a response. Passive tags have no on-board power source and limited on-board data storage, whereas active tags have a battery and up to 1 Mb of data storage. RF-ID devices operate in a low-frequency range of 100 kHz to 1.5 MHz or a high-frequency range of 900 MHz to 2.4 GHz, which has an operating range of up to 30 m. RF-ID tags are very inexpensive and are used in manufacturing

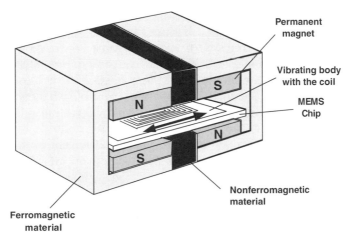

Figure 3.24 MEMS power generator using vibration and electromagnetic method.

and sales inventory control, container shipping control, etc. RF-ID tags are installed on water meters in some cities, allowing a metering vehicle simply to drive by and read the current readings remotely. RF-ID tags are also used in automobiles for automatic toll collection.

Meanwhile, software power management techniques can decrease the power consumed by rf sensor nodes significantly. Time Division Multiple Access (TDMA) is especially useful for power conservation because a node can power down or "sleep" between its assigned time slots, waking up in time to receive and transmit messages.

The required transmission power increases as the square of the distance between source and destination. Therefore, multiple short message transmission hops require less power than one long hop. In fact, if the distance be-

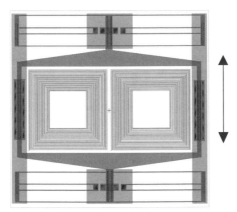

Figure 3.25 MEMS fabrication layout of power-generator dual vibrating coil showing folded-beam suspension.

tween source and destination is R, the power required for single-hop transmission is proportional to R^2. If nodes between source and destination are taken advantage of to transmit n short hops instead, the power required by each node is proportional to R^2/n^2. This is a strong argument in favor of distributed networks with multiple nodes, that is, nets of the mesh variety.

A current topic of research is *active power control,* whereby each node cooperates with all other nodes in selecting its individual transmission power level (Kumar, 2001). This is a decentralized feedback control problem. Congestion is increased if any node uses too much power, but each node must select a large enough transmission range that the network remains connected. For n nodes randomly distributed in a disk, the network is asymptotically connected with probability 1 if the transmission range r of all nodes is selected using

$$r \geq \sqrt{\frac{\log n + \gamma(n)}{\pi n}}$$

where $\gamma(n)$ is a function that goes to infinity as n becomes large.

3.4.3 Wireless Communication Protocols

IEEE ratified the IEEE 802.11 specification in 1997 as a standard for wireless LANs (WLANs). Current versions of 802.11 (i.e., 802.11b) support transmission up to 11 Mb/s. WiFi, as it is known, is useful for fast and easy networking of PCs, printers, and other devices in a local environment, for example, the home. Current PCs and laptops as purchased have the hardware to support WiFi. Purchasing and installing a WiFi router and receivers is within the budget and capability of home PC enthusiasts.

Bluetooth was initiated in 1998 and standardized by the IEEE as a wireless personal–area network (WPAN) in specification IEEE 802.15. Bluetooth is a short-range rf technology aimed at facilitating communication of electronic devices between each other and with the Internet, allowing for data synchronization that is transparent to the user. Supported devices include PCs, laptops, printers, joysticks, keyboards, mice, cell phones, personal digital assistants (PDAs), and other consumer products. Mobile devices are also supported. Discovery protocols allow new devices to be hooked up easily to the network. Bluetooth uses the unlicensed 2.4-GHz band and can transmit data up to 1 Mb/s, can penetrate solid nonmetal barriers, and has a nominal range of 10 m that can be extended to 100 m. A master station can service up to seven simultaneous slave links. Forming a network of these networks, for example, a piconet, can allow one master to service up to 200 slaves.

Currently, Bluetooth development kits can be purchased from a variety of suppliers, but the systems generally require a great deal of time, effort, and knowledge for programming and debugging. Forming piconets has not yet been streamlined and is unduly difficult.

Home rf was initiated in 1998 and has similar goals to Bluetooth for WPANs. Its goal is shared data/voice transmission. It interfaces with the Internet as well as the public switched telephone network. It uses the 2.4-GHz band and has a range of 50 m, suitable for home and yard. A maximum of 127 nodes can be accommodated in a single network. *IrDA* is a WPAN technology that has a short-range, narrow-transmission-angle beam suitable for aiming and selective reception of signals.

3.4.4 Commercially Available Wireless Sensor Systems

Many vendors now produce sensors of many types commercially that are suitable for wireless network applications. See, for instance, the Web sites of SUNX Sensors, Schaevitz, Keyence, Turck, Pepperl & Fuchs, National Instruments, UE Systems (ultrasonic), Leake (IR), and CSI (vibration). Table

TABLE 3.1 Measurements for Wireless Sensor Networks

	Measurand	Transduction Principle
Physical properties	Pressure	Piezoresistive, capacitive
	Temperature	Thermistor, thermomechanical, thermocouple
	Humidity	Resistive, capacitive
	Flow	Pressure change, thermistor
Motion properties	Position	Electromagnetic, GPS, contact sensor
	Velocity	Doppler, Hall effect, optoelectronic
	Angular velocity	Optical encoder
	Acceleration	Piezoresistive, piezoelectric, optical fiber
Contact properties	Strain	Piezoresistive
	Force	Piezoelectric, piezoresistive
	Torque	Piezoresistive, optoelectronic
	Slip	Dual torque
	Vibration	Piezoresistive, piezoelectric, optical fiber, sound, ultrasound
Presence	Tactile/contact	Contact switch, capacitive
	Proximity	Hall effect, capacitive, magnetic, seismic, acoustic, rf
	Distance/range	Electromagnetic (sonar, radar, lidar), magnetic, tunneling
	Motion	Electromagnetic, IR, acoustic, seismic (vibration)
Biochemical	Biochemical agents	Biochemical transduction
Identification	Personal features	Vision
	Personal ID	Fingerprints, retinal scan, voice, heat plume, vision motion analysis

3.1 shows which physical principles may be used to measure various quantities. MEMS sensors are now available for most of these measurands.

Many wireless communications nodes are available commercially, including Lynx Technologies and various Bluetooth kits, as well as the Casira devices from Cambridge Silicon Radio (CSR). We discus two particular products here.

Crossbow Berkeley Motes may be the most versatile wireless sensor network devices on the market for prototyping purposes (Fig. 3.26). Crossbow (*www.xbow.com/*) makes three Mote processor radio module families—MICA (MPR300) (first generation), MICA2 (MPR400), and MICA2-DOT (MPR500) (second generation). Nodes come with five sensors installed—temperature, light, acoustic (microphone), acceleration/seismic, and magnetic. These are especially suitable for surveillance networks for personnel and vehicles. Different sensors can be installed if desired. Low power and small physical size enable placement virtually anywhere. Since all sensor nodes in a network can act as base stations, the network can self-configure and has multihop routing capabilities. The operating frequency is the industrial, scientific, and medical (ISM) band, either 916 or 433 MHz, with a data rate of 40 kb/s and a range of 30 to 100 ft. Each node has a low-power microcontroller processor with speed of 4 MHz, a flash memory with 128 kB, and SRAM and EEPROM of 4 kB each. The operating system is Tiny-OS, a tiny microthreading distributed operating system developed by U.C. Berkeley with a NES-C (Nested C) source-code language (similar to C). Installation of these devices requires a great deal of programming. A workshop is offered for training.

Microstrain's X-Link Measurement System (*www.microstrain.com/*) may be the easiest system to get up and running and to program (Fig. 3.27). The

Figure 3.26   Berkeley Crossbow Motes.

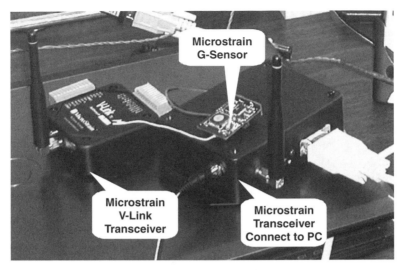

Figure 3.27 Microstrain wireless sensors.

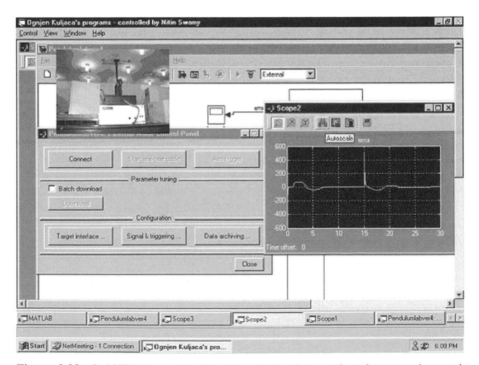

Figure 3.28 LabVIEW user interface for wireless Internet-based remote site monitoring.

frequency used is 916 MHz, which lies in the U.S. license-free ISM band. The sensor nodes are multichannel, with a maximum of eight sensors supported by a single wireless node. There are three types of sensor nodes—S-link (strain gauge), G-link (accelerometer), and V-link (supports any sensor generating voltage differences). The sensor nodes have a preprogrammed EPROM, so a great deal of programming by the user is not needed. Onboard data storage is 2 MB. Sensor nodes use a 3.6-V lithium-ion internal battery (9-V rechargeable external battery is supported). A single receiver (base station) addresses multiple nodes. Each node has a unique 16-bit address, so a maximum of 2^{16} nodes can be addressed. The rf link between base station and nodes is bidirectional, and the sensor nodes have a programmable data-logging sample rate. The rf link has a 30-m range with a 19,200 baud rate. The baud rate on the serial RS-232 link between the base station and a terminal PC is 38,400. The LabVIEW interface is supported.

3.4.5 Decision-Making and User Interface

Many software products are available to provide advanced digital signal processing (DSP), intelligent user interfaces, decision assistance, and alarm functions. Among the most popular, powerful, and easy to use is National Instruments LabVIEW software. Available are tool kits for camera image processing, machinery signal processing and diagnosis, sensor equipment calibration and testing, feedback control, and more. Figure 3.28 shows a LabVIEW user interface for monitoring machinery conditions over the Internet for automated maintenance scheduling functions. Included are displays of sensor signals that can be selected and tailored by the user. The user can prescribe bands of normal operation, excursions outside of which generate alarms of various sorts and severity.

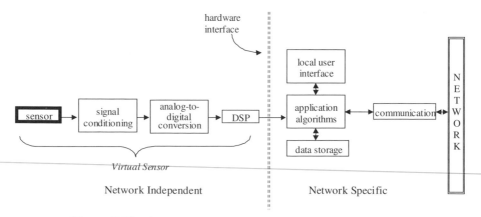

Figure 3.29 A general model of a smart sensor (IEEE, 2001).

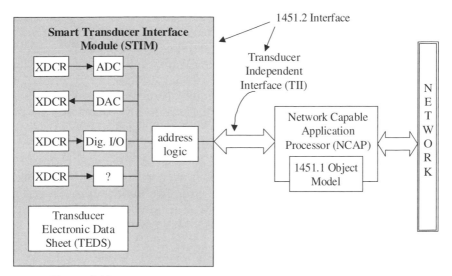

Figure 3.30 IEEE 1451, "Standard for Smart Sensor Networks."

3.5 SMART SENSORS

A *smart sensor* is a sensor that provides extra functions beyond those necessary for generating a correct representation of the sensed quantity (Frank, 2000). Desirable functions for sensor nodes include ease of installation, self-identification, self-diagnosis, reliability, time awareness for coordination with other nodes, some software functions and DSP, and standard control protocols and network interfaces (IEEE, 2001). A general model of a smart sensor is shown in Fig. 3.29. Objectives for smart sensors include moving the intelligence closer to the point of measurement; making it cost-effective to integrate and maintain distributed sensor systems; creating a confluence of transducers, control, computation, and communications toward a common goal; and seamlessly interfacing numerous sensors of different types.

3.5.1 IEEE 1451 and Smart Sensors

There are many sensor manufacturers and many networks on the market today. It is too costly for manufacturers to make special transducers for every network on the market. Different components made by different manufacturers should be compatible. Therefore, in 1993, the Institute of Electrical and Electronics Engineers (IEEE) and the National Institute of Standards and Technology (NIST) began work on a standard for smart sensor networks. IEEE 1451, "Standard for Smart Sensor Networks," was the result. The objective of this standard is to make it easier for different manufacturers to develop smart sensors and to interface those devices with networks. Figure 3.30 shows the basic architecture of IEEE 1451 (Conway and Hefferman, 2003). Major components include STIM, TEDS, TII, and NCAP, as detailed in the figure. A major outcome of IEEE 1451 studies is the formalized concept of a *smart sensor*.

Recent advances in micromachining, communications, and computing technologies are providing the impetus for the design and fabrication of new sensor systems with increased quality and reliability and lower cost. A significant breakthrough has resulted from the ability to perform a large number of signal-processing functions on board the sensor structure. Figure 3.31 shows a single smart sensor with processing elements that facilitate conversion of signals from analog to digital form and perform such major processing functions as amplification, filtering, Fast Fourier Transform (FFT) analysis, scaling, linearization, etc. (Kanoun and Tränkler, 2004). Self-calibration and

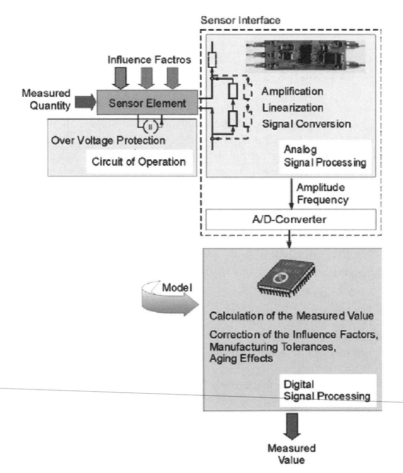

Figure 3.31 Signal processing by individual sensors. (*Courtesy of Kanoun, 2004.*)

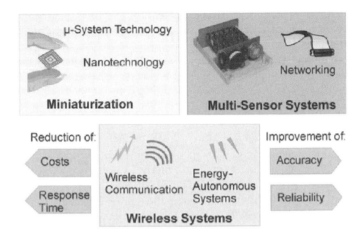

Figure 3.32 Trends in smart sensor technologies. (*Courtesy of Kanoun 2004*).

temperature compensation of sensors are increasing their reliability and usability.

The notion of a single smart sensor is migrating toward multisensor systems aided by miniaturization, dramatic cost reductions, wireless technologies, networking, and communications. Networked embedded sensor systems are promising to improve our monitoring capabilities significantly in large distributed systems and increase the accuracy and reliability of measurements through multisensor fusion, advanced data mining, and other intelligent techniques. Future trends in smart sensor systems include large-scale silicon-based devise fabrication, micromachining, advanced on-board signal processing, power harvesting at the sensor site, wireless systems, and the widespread introduction of networked/wireless embedded sensors. The fabrication of microelectromechanical systems (MEMS) in silicon and other materials, potentially in addition to electronic devices, offers significant advantages because of batch fabrication, potential for integration with electronics, fabrication of arrays of sensors, and small size (mm–cm) of individual devices. Micromachining has been developing rapidly owing to "piggybacking" on IC technology infrastructure. Recent advances in this area emphasize the use of high-temperature materials (ceramics) as the packaging medium, all required electronics and power external to the actual transducer (Fonseca et al., 2001). These new technologies are opening up possibilities for online monitoring of hot spots in high-temperature environments (engine, etc.) Figure 3.32 depicts a view of future smart sensor trends suggested by Kanoun (2004).

3.6 REFERENCES

J. A. Allocca and A. Stuart, *Transducers: Theory and Applications.* Englewood Cliff, NJ: Prentice-Hall, 1984.

A. A. Alonso, C. E. Frouzakis, and I. G. Kevrekidis, "Optimal Sensor Placement for State Reconstruction of Distributed Process Systems," *AIChE Journal* **50**(7):1438–1452, 2004.

A. G. Al-Shehabi and B. Newman, "Aeroelastic Vehicle Optimal Sensor Placement for Feedback Control Applications Using Mixed Gain-Phase Stability," *Proceedings of the American Control Conference,* Vol. 3, 2001, pp. 1848–1852, Arlington, VA.

L. F. Anthony and B. K. Belinda, "Sensor Location in Feedback Control of Partial Differential Equation Systems," *Proceedings of the 2000 IEEE International Conference on Control Applications,* Vol. 1, 2005, pp. 536–541, Anchorage, AK.

M. Bhushan and R. Rengaswamy, "Design of Sensor Location Based on Various Fault Diagnostic Observability and reliability Criteria," *Computers and Chemical Engineering* **24**:735–741, 2000.

S. Y. Chen and Y. F. Li, "A Method of Automatic Sensor Placement for Robot Vision in Inspection Tasks," *Proceedings of the IEEE International Conference on Robotics and Automation,* Vol. 3, 2002, pp. 2545–2550, Washington, DC.

D. J. Chmielewski, T. Palmer, and V. Manousiouthakis, "On the Theory of Optimal Sensor Placement," *AIChE Journal* **48**(5):1001–012, 2002.

D. A. Cook and Y. H. Berthelot, "Detection of Small Surface-Breaking Fatigue Cracks in Steel Using Scattering of Rayleigh Waves," *NDT&E International* **34**:483–492, 2001.

P. Conway and D. Heffernan, "CAN and the New IEEE 1451 Smart Transducer Interface Standard" 6th International CAN conference, pp. 197–203, Torino, Italy, 1999.

R. C. Eberhart and J. Kennedy, "A New Optimizer Using Particle Swarm Theory," *Proceedings of the Sixth International Symposium on Micromachine and Human Science,* Nagoya, Japan, 1995, pp. 39–43.

R. F. El-Hajjar and R. M. Haj-Ali, "Infrared (IR) Thermography for Strain Analysis in Fiber Reinforced Plastics," *Experimental Techniques, Society for Experimental Mechanics (SEM)* **28**(2):19–22, 2004.

M. A. Fonseca, J. M. English, M. von Arx. and M. G. Allen, "High Temperature Characterization of Ceramic Pressue Sensors," *11th International Conference on Solid-State Sensors and Actuators.* Munich, Germany, June 10–14, 2001, pp. 486–489.

R. Frank, *Understanding Smart Sensors,* 2d ed., Norwood, MA: Artech House, 2000; *www.artechhouse.com.*

H. Gao, Y. Shi, and J. L. Rose, "Guided Wave Tomography on an Aircraft Wing with Leave in Place Sensors," in D. O. Thompson and D. E. Chimenti (eds.), *Review of Progress in Quantitative Nondestructive Evaluation,* Vol. 760, Golden, Colorado, American Institute of Physics, 2005, pp. 1788–1794.

C. Giraud and B. Jouvencel, "Sensor Selection: A Geometrical Approach," *Proceedings of the IEEE/RSJ International Conference,* vol. 2, 1995, pp. 555–560, Pittsburg, PA.

V. Giurgiutiu and J. Bao, "Embedded-Ultrasonics Structural Radar for in Situ Structural Health Monitoring of Thin-Wall Structures," *Structural Health Monitoring* **3**(2):121–140, 2004.

M. de la Guardia, "Biochemical Sensors: The State of the Art," *Microchim Acta* **120**(1–4):243–255, 1995.

IEEE 1451, "Standard Smart Transducer Interface," Sensors Expo, Philadelphia, October 2001; *http://ieee1451.nist.gov/Workshop_04Oct01/1451_overview.pdf.*

D. C. Kammer, "Sensor Placement for On-Orbit Modal Identification and Correlation of Large Space Structures," *Journal of Guidance, Control, and Dynamics* **14**:251–259, 1991.

O. Kanoun and H. Tränkler, "Sensor Technology Advances and Future Trends," *IEEE Transactions on Instrumentation and Measurement* **53**(6):1497–1501, 2004.

M. A. Kramer, and B. L. Palowitch, "A Rule-Based Approach to Fault Diagnosis Using the Signed Directed Graph," *AIChE Journal* **33**(7):1067, 1987.

P. R. Kumar, "New Technological Vistas for Systems and Control: The Example of Wireless Networks," *IEEE Control Systems Magazine,* February 2001, pp. 24–37.

M. Lackner, "Vibration and Crack Detection in Gas Turbine Engine Compressor Blades Using Eddy Current Sensors," master thesis, Massachusetts Institute of Technology, August 2004.

K. R. Leonard, E. V. Malyarenko, and M. K. Hinders, "Ultrasonic Lamb Wave Tomography," *Inverse Problems* **18**(6):1795–1808, 2000.

F. L. Lewis, *Applied Optimal Control and Estimation: Digital Design and Implementation.* Englewood Cliffs, NJ: Prentice-Hall, TI Series, 1992.

F. L. Lewis, "Wireless Sensor Networks," in D. J. Cook and S. K. Das (eds.), *Smart Environments: Technologies, Protocols, Applications.* New York: Wiley, 2004.

Y. Lu and J. E. Michaels, "A Methodology for Structural Health Monitoring with Diffuse Ultrasonic Waves in the Presence of Temperature Variations," *Ultrasonics* **43**:717–731, 2005.

B. Mi, J. E. Michaels, and T. E. Michaels, "An Ultrasonic Method for Dynamic Monitoring of Fatigue Crack Initiation and Growth," *Journal of the Acoustical Society of America,* accepted August 2005.

J. E. Michaels and T. E. Michaels, "Enhanced Differential Methods for Guided Wave Phased Array Imaging Using Spatially Distributed Piezoelectric Transducers," in D. O. Thompson and D. E. Chimenti (eds.), *Review of Progress in Quantitative Nondestructive Evaluation,* Vol. 25, Brunswick, Maine, American Institute of Physics, 2006.

J. E. Michaels, T. F. Michaels, B. Mi, A. C. Cobb, and D. M. Stobbe, "Self-Calibrating Ultrasonic Methods for In-Situ Monitoring of Fatigue Crack Progression," in D. O. Thompson and D. E. Chimenti (eds.), *Review of Progress in Quantitative Nondestructive Evaluation,* Vol. 24B. Golden, Colorado, American Institute of Physics, 2005, pp. 1765–1772.

J. E. Michaels and T. E. Michaels, "Detection of Structural Damage from the Local Temporal Coherence of Diffuse Ultrasonic Signals," *IEEE Transactions on Ultrasonics, Ferroelectrics, and Frequency Control,* accepted October 2005.

J. E. Michaels, Y. Lu, and T. E. Michaels, "Methods for Quantifying Changes in Diffuse Ultrasonic Signals with Applications to Structural Health Monitoring," in T. Kundu (ed.), *Proceedings of SPIE, Health Monitoring and Smart Nondestructive Evaluation of Structural and Biological Systems III,* Vol. 5768. San Diego, CA, International Society for Optical Engineering, 2005, pp. 97–105.

R. Naimimohasses, D. M. Barnett, D. A. Green, and P. R. Smith, "Sensor Optimization Using Neural Network Sensitivity Measures," *Measurement Science & Technology* **6**:1291–1300, 1995.

R. Raghuraj, M. Bhushan, and R. Rengaswamy, "Locating Sensors in Complex Chemical Plants Based on Fault Diagnostic Observability Criteria," *AIChE Journal* **45**(2): 310–322, 1999.

M. Serridge, and T. R. Licht, *Piezoelectri Accelerometer and Vibration Preamplifier Handbook.* Bruel and Kjaer, Naerum, Denmark, 1987.

C. W. de Silva, *Control Sensors and Actuators.* Englewood Cliffs, NJ: Prentice-Hall, 1989.

P. Wilcox, "Omni-Directional Guided Wave Transducer Arrays for the Rapid Inspection of Large Areas of Plate Structures," *IEEE Transactions on Ultrasonics, Ferroelectrics, and Frequency Control,* **50**(6):699–709, 2003.

G. Zhang, "Optimum Sensor Localization/Selection in a Diagnostic/Prognostic Architecture," Ph.D. thesis, Electrical and Computer Engineering, Georgia Institute of Technology, 2005.

M. Zou, J. Dayan, and I. Green, "Dynamic Simulation and Monitoring of a Noncontacting Flexibly Mounted Rotor Mechanical Face Seal," *IMechE, Proc. Instn. Mech. Engrs* **214**(C9):1195–1206, 2000.

CHAPTER 4

SIGNAL PROCESSING AND DATABASE MANAGEMENT SYSTEMS

4.1 INTRODUCTION

A major challenge has confronted condition-based maintenance/prognostics and health management (CBM/PHM) researchers and practitioners over the past years: How are system functional changes (symptoms) monitored or measured in terms of measurable system states (outputs)? We addressed this issue in Chap. 3 and explored means to monitor such measurable quantities as vibration, temperature, etc. How do we relate fault (failure) mechanisms to the fundamental "physics" of complex dynamic systems? It is well known that fault (failure) modes induce changes in the energy (power) of the system, its entropy, power spectrum, signal magnitude, chaotic behavior, etc. Where is the *information?* It is possibly in the time domain, the frequency domain, or the wavelet domain. How do we *extract* useful information from massive raw databases? We will address these fundamental questions in this chapter.

Data are rarely useful or usable in their raw form. Consider, for example, vibration data sampled at 100 kHz. Such large amounts of data are unmanageable unless they are processed and reduced to a form that can be manipulated easily by fault diagnostic and prognostic algorithms. An economist suggested that "if you beat on data long enough, you can make them confess to anything." Our objective in processing raw sensor data parallels the economist's assertion with one major difference: We want data to confess only to *true* or *correct* information.

We will focus on technologies that will assist us in selecting and extracting the best "features" or condition indicators that contain as much information as possible. The eventual goal is for these features to be employed optimally so that we can maximize fault detection and isolation accuracy while minimizing false alarms. Thus data collected by sensing devices must be processed appropriately so that useful information may be extracted from them that constitutes a reduced version of the original data but preserves as much as possible the characteristic features or fault indicators that are indicative of the fault events we are seeking to detect, isolate, and predict. Such data must be preprocessed, that is, filtered, compressed, correlated, etc., in order to remove artifacts and reduce noise levels and the volume of data to be processed subsequently. Furthermore, the sensor providing the data must be validated; that is, the sensors themselves are subject to fault conditions. Once the preprocessing module confirms that the sensor data are "clean" and formatted appropriately, features or signatures of normal or faulty conditions must be extracted. This is the most significant step in the CBM/PHM architecture, whose output will set the stage for accurate and timely diagnosis of fault modes. The extracted-feature vector will serve as one of the essential inputs to fault diagnostic algorithms.

4.2   SIGNAL PROCESSING IN CBM/PHM

CBM/PHM technologies are developed and applied to a large variety of machines, systems, and processes in the transportation, industrial, and manufacturing arenas. Rotating equipment has received special attention because of its critical operating regimes and frequent failure modes and the availability of measurements (vibration, temperature, etc.) intended to allow detection and isolation of incipient failures. Other important categories include structural components subjected to stress, corroding surfaces, hot spots in machinery under thermal stress, etc.

Sensor measurements that affect fault diagnosis for such critical systems may be viewed from two different perspectives: static or process-related measurements such as temperature, speed, position, pressure, and flow rate and those that are characterized by their high bandwidth (i.e., they contain high-frequency components). In the latter category we list ultrasonic measurements, vibration measurements, acoustic recordings, and alternating current or voltage.

Processing of sensor data with the intention of extracting fault features or condition indicators in each of these two categories may require different processing tools. For example, temperature measurements may be processed in the time domain, and features such as peak amplitude, energy, and rate of change may be appropriate indicators of the condition or health of the equipment under test. Analysis in the frequency or wavelet domain may be more suitable for vibration or ultrasonic signals.

Further, we might distinguish between one-dimensional or single-point measurements, for example, temperature for a single-CT probe or vibration for a single-axis accelerometer, and measurements taken over a surface or volume, such as those monitoring structural surfaces for corrosion or hot spots or oil debris in a lubricant. Again, these situations require different monitoring devices and corresponding analysis tools.

In the following sections we will review some classic signal preprocessing and processing techniques for signal conditioning and feature extraction, and we will focus on novel approaches for the extraction of useful information from high-bandwidth signals that originate primarily from rotating equipment.

Signal processing generally involves two steps. The first step usually is called *signal preprocessing* and is intended to enhance the signal characteristics that eventually may facilitate the efficient extraction of useful information, that is, the indicators of the condition of a failing component or subsystem. The tools in this category include filtering, amplification, data compression, data validation, and denoising and generally are aimed at improving the signal-to-noise ratio. The second step aims to extract the features or condition indicators from the preprocessed data that are characteristic of an incipient failure or fault.

We will illustrate the preprocessing step with a few examples of hardware and software techniques used commonly for the types of signals derived from CBM/PHM sensors. There is a rich technical literature in the signal processing domain that expands greatly on other preprocessing schemes. Our major objective in this chapter is to focus on algorithms and methods that will process sensor data appropriately and intelligently to arrive at an optimal feature vector for fault diagnosis and prognosis purposes.

4.3 SIGNAL PREPROCESSING

4.3.1 Signal Conditioning

Signal conditioning to improve the signal-to-noise ratio and extract useful information from raw data may be performed with hardware and/or software. We will consider simple examples of both methods.

Signals coming from sensors can be very noisy, of low amplitude, biased, and dependent on secondary parameters such as temperature. Moreover, one may not always be able to measure the quantity of interest but only a related quantity. Therefore, signal conditioning usually is required. Signal conditioning is performed using electronic circuitry, which may be built using standard very large scale integration (VLSI) fabrication techniques in situ with microelectromechanical systems (MEMS) sensors. A reference for signal conditioning, analog-to-digital conversion, and filtering is Lewis (1992).

A real problem with MEMS sensors is undesired sensitivity to secondary quantities such as temperature. Temperature compensation often can be built directly into a MEMS sensor circuit or added during the signal conditioning stage.

A basic technique for improving the signal-to-noise ratio (SNR) is *low-pass filtering,* since noise generally dominates the desirable signals at high frequencies. Shown in Fig. 4.1 is an analog low-pass filter (LPF) that also amplifies, constructed from an operational amplifier. Such devices are fabricated easily using VLSI semiconductor techniques. The time constant of this circuit is $\tau = R_2 C$. The transfer function of this filter is $H(s) = ka/(s + a)$, with 3-dB cutoff frequency given by $a = 1/\tau$ rad and gain given by $k = R_2/R_1$. Here, s is the Laplace transform variable. The cutoff frequency should be chosen larger than the highest useful signal frequency of the sensor.

Alternatively, one may use a digital LPF implemented on a computer after sampling. A digital LPF transfer function and the associated difference equation for implementation is given by

$$\hat{s}_k = K \frac{z + 1}{z - \alpha} s_k$$

and

$$\hat{s}_{k+1} = \alpha s_k + K(s_{k+1} + s_k)$$

respectively. Here, z is the z-transform variable treated as a unit delay in the time domain, s_k is the measured signal, and $\hat{s}_k$ is the *filtered* or *smoothed variable* with reduced noise content. The filter parameters are selected in terms of the desired cutoff frequency and the sampling period (Lewis, 1992).

It is often the case that one can measure a variable s_k (e.g., position) but needs to know its rate of change v_k (e.g., velocity). Owing to the presence of

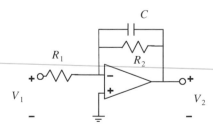

Figure 4.1   Analog low-pass filter.

noise, one cannot simply take the difference between successive values of s_k as the velocity. A filtered velocity estimate given by $v_{k+1} = \alpha v_k + K(s_{k+1} - s_k)$ both filters out noise and gives a smooth velocity estimate.

Often, changes in resistance must be converted to voltages for further processing. This may be accomplished by using a Wheatstone bridge (de Silva, 1989). Suppose that $R_1 = R$ in Fig. 4.2 is the resistance that changes depending on the measurand (e.g., strain gauge), and the other three resistances are constant (quarter-bridge configuration). Then the output voltage changes according to $\Delta V_0 = V_{ref} \Delta R/4R$. We assume a balanced bridge so that $R_2 = R_1 = R$ and $R_3 = R_4$. Sensitivity can be improved by having two sensors in situ such that the changes in each are opposite (e.g., two strain gauges on opposite sides of a flexing bar). This is known as a *half bridge*. If R_1 and R_2 are two such sensors and $\Delta R_1 = -\Delta R_2$, then the output voltage doubles. The Wheatstone bridge also may be used for differential measurements (e.g., for insensitivity to common changes of two sensors), to improve sensitivity, to remove zero offsets, for temperature compensation, and to perform other signal conditioning.

Specially designed operational amplifier circuits are useful for general signal conditioning (Frank, 2000). Instrumentation amplifiers provide differential input and common-mode rejection, impedance matching between sensors and processing devices, calibration, etc. SLEEPMODE amplifiers (Semiconductor Components Ind., LLC) consume minimum power while asleep and activate automatically when the sensor signal exceeds a prescribed threshold.

4.3.2 Denoising and Extracting Weak Signals

Signal preprocessing is an important step toward enhancing data reliability and thereby improving the accuracy of signal analysis methods by enhancing the signal-to-noise ratio (SNR). Traditional denoising methods require some information and pose several assumptions regarding the signals that must be separated from noise. Wavelet theory has been applied successfully as a denoising tool, and both hardware and software implementations of wavelet-based denoising methods are available. We will discuss wavelets further in

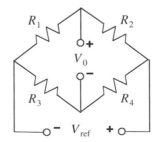

Figure 4.2 Wheatstone bridge.

Sec. 4.4.4 as fault feature extractors. When applying wavelets to improve the SNR, it is sufficient to know only what family of classes the target signal belongs to rather than the exact class. An orthogonal wavelet transform can compress the "energy" of a signal in a very few large coefficients, leaving the energy of the white noise dispersed throughout the transform as small coefficients. Soft-thresholding (Donoho, 1995) and wavelet-shrinkage denoising (Zheng and Li, 2000) are two popular denoising methods. Other signal denoising applications for fault detection include adaptive threshold denoising in power systems (Yang and Liao, 2001), denoising of acoustic emission signals to detect fatigue cracks in rotor heads (Menon et al., 2000), improving SNR of knee-joint vibration signals (Krishnan and Rangayyan, 1999), structure fault detection in fighter aircraft by denoising the residual signal (Hu et al., 2000), and short transient detection for localized defects in roller bearings by removing the background noise (Pineyro et al., 2000).

In traditional filter-based denoising methods, the frequency components outside a certain range often are set to zero, resulting in loss of information owing to impulses in the signal spanning a wide frequency range. Wavelet-based denoising methods only remove small coefficients, and thus the impulse information is retained because it is usually expressed as large coefficients in the transformed representation.

4.3.3 Vibration Signal Compression

With an increasing number of diagnostic sensors and higher sampling rates, the amount of data acquired from industrial systems tends to be voluminous and, in most cases, difficult to manage. Thus, for fault diagnostic systems, especially those implemented online or other Internet-based remote techniques, data compression is one of the most successful applications of wavelets, including both one-dimensional signal and two-dimensional image compression.

The data compression and decompression tasks generally involve five major steps: transform, thresholding, quantization/encoding, decoding, and reconstruction. The compression performance is measured via two metrics: the *compression ratio,* defined as the ratio of the number of bits of the original data to the number of bits in the compressed data, and the *normalized mean squared error* (Staszewski, 1998b), stated as

$$\text{MSE}(x) = \frac{100}{N\sigma_x^2} \sum_{i=1}^{N} (x_i - \hat{x}_i)^2$$

where σ_x is the standard deviation of $x(t)$, and N is the number of sample points in the signal.

Compression of mechanical vibration signals to obtain compression ratios of up to 10 to 20 (Shen and Gao, 2000), compression of vibration signals from motor bearing rings (Tanaka et al.,1997), and compression of gearbox vibration spectra for storage, transmission, and fault feature selection for condition monitoring (Staszewski, 1997) are some of the examples from wavelet-based compression techniques.

Wavelet compression has been shown to be more effective on nonstationary signals. A more compactly supported wavelet, and therefore a wavelet function that is less smooth, is more suitable for nonstationary and irregular signals. Similar to wavelets used for denoising, they achieve better compression performance because they do not discard singularities in the signal, unlike other frequency transform FT-based techniques.

4.3.4 Time Synchronous Averaging (TSA)

Conditioning of a vibration signal coming from rotating equipment such as a gearbox may range from signal correction, based on the data-acquisition unit and amplifiers used, and mean value removal to time synchronous averaging and filtering. We will focus here on time averaging of the signal because most of the available data-analysis techniques for transmission systems are based on the *time synchronous averaging* (TSA) of vibration signals (McClintic et al., 2000; McFadden, 1987, 1989, 1991). It is assumed that a pulse signal synchronized to the rotation of a gear indicating the start of an individual revolution is available. Numerous revolutions are ensemble averaged, resulting in an averaged data segment with a length corresponding to a single revolution. The TSA technique is intended to enhance the vibration frequencies, which are multiples of the shaft frequency, which in many cases is mainly vibration related to the meshing of the gear teeth after averaging out other components such as random vibrations and external disturbances. The resulting averaged data show vibration characteristics of a gear in the time domain over one complete revolution and differences in the vibration produced by individual gear teeth. Advanced local damage to gear teeth often can be detected by direct inspection of the time averaged data (McFadden, 1991).

Since the rotational speed of a transmission typically varies slightly during normal operation, the numbers of the data samples per revolution are different for a given sampling frequency. Interpolation is required to make the sample numbers per revolution the same before ensemble averaging can be carried out. Interpolation transforms the vibration signal from the time domain to the angle domain and redefines the sampling frequency to be a function of angular position rather than time.

In this section the TSA technique is compared with other vibration data preprocessing tools in the frequency domain. Since interpolation, as required by the TSA, is computationally demanding and time-consuming, it is undesirable, especially for online real-time vibration monitoring. A simple technique is introduced that achieves the same result as the TSA method with much lower computational complexity because interpolation is not required (Wu et al., 2005). An analysis and comparison between these different preprocessing techniques are presented here.

The data segment $x_m(n)$, for $n = 0, 1, \ldots, N - 1$, is defined as the vibration data of revolution m of a total number of M revolutions after interpolation, and $\bar{x}(n)$, for $n = 0, 1, \ldots, N - 1$, as the time-averaged data. The discrete Fourier transform (DFT) of the time-averaged data is given by

$$
X(k) = \sum_{n=0}^{N-1} \bar{x}(n) e^{-j2\pi(kn/N)} \qquad k = 0, 1, \ldots, N - 1
$$

$$
= \frac{1}{M} \sum_{n=0}^{N-1} \sum_{m=1}^{M} x_m(n) e^{-j2\pi(kn/N)}
$$

A new data preprocessing technique involves taking the DFT of each revolution of the vibration data (no interpolation required) and then ensemble averaging the complex numbers (including the amplitude and phase information) of the corresponding indices. The rationale behind this technique is as follows: The frequency resolution of the DFT is given by $\Delta f = 1/T_{\text{shaft}} = f_{\text{shaft}}$, where T_{shaft} is the time length of each individual revolution and may vary slightly from one revolution to another. If the period of the gear shaft rotation T_{shaft} is longer, the data volume sampled in a revolution is greater. The frequency resolution then is finer, whereas the sampling frequency is the same, resulting in a higher number of frequency indices. However, the index of the meshing frequencies and their sidebands does not change. For example, the meshing frequencies of a planetary gear with 228 gear teeth are always $228 \times f_{\text{shaft}}$ and its integer multiples, and the indices in DFT are always 228 and its integer multiples. We can average the frequencies with common and corresponding indices and discard the extra frequencies of higher indices.

The result obtained using this preprocessing technique is equivalent to the DFT of the time synchronous averaged (TSA) data (which will be explained in the following paragraphs), whereas the interpolation and resampling as required in time averaging are not needed. In some sense, we obtain the TSA data in the frequency domain. This processing step reduces the computational complexity because, in most cases, the DFT (FFT) is simpler and faster than resampling the vibration data, especially when the sampling frequency is high.

Since the DFT operation is linear, the DFT of the ensemble averaging of the interpolated data is equivalent to the ensemble averaging of the DFT of the interpolated data, as shown below:

$$
X(k) = \sum_{n=0}^{N-1} \bar{x}(n) e^{-j2\pi(kn/N)} \qquad k = 0, 1, \ldots, N - 1
$$

$$
= \frac{1}{M} \sum_{n=0}^{N-1} \sum_{m=1}^{M} x_m(n) e^{-j2\pi(kn/N)}
$$

$$
= \frac{1}{M} \sum_{m=1}^{M} \sum_{n=0}^{N-1} x_m(n) e^{-j2\pi(kn/N)}
$$

Figure 4.3 illustrates the DFT of a TSA data set, which can be calculated by averaging the DFT, as shown in the right part of the figure. In the figure, the data points connected by the dashed lines are to be averaged. Since the time length of each individual revolution varies, the frequency resolution of the DFT of each individual revolution varies as well.

Figure 4.4 shows the averaging of the DFT for multiple revolutions without interpolation. From the illustrations in the frequency domain, it can be seen that the results are equivalent at the common indices to the DFT of the TSA data shown in Fig. 4.3. Since the variation in the rotational speed of a rotor shaft is very small (usually the difference in the sample numbers between two revolutions of raw data is about 0.05 percent), the difference between the DFT of the TSA data and the averaging of the DFT of the raw data (without interpolation) is at the end of the spectrum axis and is negligible when the sampling rate is very high (e.g., 100 kHz).

Another perspective is provided by viewing the sampling of individual revolutions in the angle domain. The frequency in this case is defined in the angle domain instead of the time domain. It has a unit of rotor shaft frequency instead of hertz; that is, the frequencies are multiples of the shaft frequency. It is a very useful representation because the tooth frequency is always a specific multiple of the shaft frequency no matter how the rotational speed of the shaft changes. True frequency values may change with shaft rotational speed; however, the frequency values as multiples of the shaft frequency may not change. This allows invariant characteristics of the vibration data across multiple revolutions to be extracted and enhanced. Figure 4.5 shows the data samples in the angle domain and the corresponding frequency domain of the interpolated data, as in the TSA. Figure 4.6 shows the data samples in the angle domain and the corresponding frequency domain of the data without interpolation. Again, we can see that in the frequency domain the two tech-

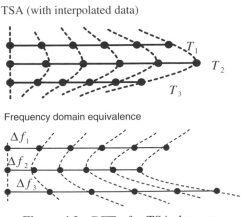

Figure 4.3 DFT of a TSA data set.

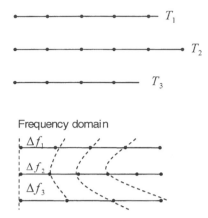

Figure 4.4 Multiple-revolution data without interpolation in the time and frequency domains.

niques are equivalent at the nontrivial frequencies. The code of the TSA in the frequency domain is given in the Appendix.

4.4 SIGNAL PROCESSING

4.4.1 Feature Selection and Extraction

Feature or condition indicator selection and extraction constitute the cornerstone of accurate and reliable fault diagnosis. The classic image-recognition and signal-processing paradigm of *data* → *information* → *knowledge* becomes most relevant and takes central stage in the fault diagnosis case particularly because such operations must be performed online in a real-time environment. Figure 4.7 presents a conceptual schematic for feature extraction and fault-mode classification.

Fault diagnosis depends mainly on extracting a set of features from sensor data that can distinguish between fault classes of interest and detect and isolate a particular fault at its early initiation stages. These features should be

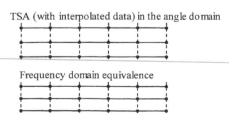

Figure 4.5 TSA in the angle domain and the corresponding frequency domain.

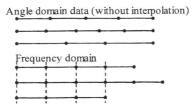

Figure 4.6   Data without interpolation in the angle and frequency domains.

fairly insensitive to noise and within fault class variations. The latter could arise because of fault location, size, etc. in the frame of a sensor. "Good" features must have the following attributes:

- Computationally inexpensive to measure
- Mathematically definable
- Explainable in physical terms
- Characterized by large interclass mean distance and small interclass variance
- Insensitive to extraneous variables
- Uncorrelated with other features

Much attention has focused in recent years on the feature-extraction problem, whereas feature selection has relied primarily on expertise, observations, past historical evidence, and our understanding of fault signature characteristics. In selecting an optimal feature set, we are seeking to address such questions as, Where is the information? and How do fault (failure) mechanisms relate to the fundamental "physics" of complex dynamic systems? Fault modes may induce changes in

- The energy (power) of the system
- Its entropy

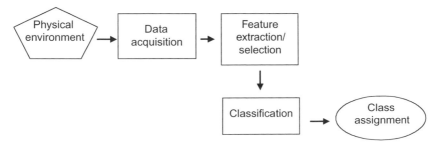

Figure 4.7   A general scheme for feature extraction and fault-mode classification.

- Power spectrum
- Signal magnitude
- Chaotic behavior
- Other

Feature selection is application-dependent. We are seeking those features, for a particular class of fault modes, from a large candidate set that possess properties of fault distinguishability and detectability while achieving a reliable fault classification in the minimum amount of time. We will elaborate on these concepts in this section. Feature extraction, on the other hand, is viewed as an algorithmic process where, given sensor data, features are extracted in a computationally efficient manner while preserving the maximum information content. Thus the feature-extraction process converts the fault data into an N-dimensional feature space such that one class of faults is clustered together and can be distinguished from other classes. However, in general, not all faults of a class need N features to form a compact cluster. It is only the faults that are in the overlapping region of two or more classes that govern the number of features required to perform classification.

Although prevailing feature-selection practice relies on expertise and/or heuristic evidence, it is possible to approach this challenging problem from an analytic perspective by setting it as an optimization problem: Maximize detection/isolation accuracy while minimizing false alarms. A multiobjective function may be defined with such target criteria as fault distinguishability and detectability/isolability that is constrained by minimum processing requirements from real-time applications. Additionally, features must be ordered in a standard form, whereas, for fault classification purposes, a *degree of certainty* metric may be called on to limit the number of false positives and false negatives. We elaborate on these notions below. The *distinguishability measure* quantifies a feature's ability to differentiate between various classes by finding the area of the region in which the two classes overlap. The smaller the area, the higher is the ability of the feature to distinguish between the classes. Suppose that there are two classes w_1 and w_2 that need to be classified based on a feature A. Let there be a training data set that includes N_1 samples belonging to w_1 and N_2 samples belonging to w_2. Assuming a Gaussian distribution for the training data belonging to each class (a valid assumption for a large training data set), the distinguishability measure of feature A for classes w_1 and w_2 is

$$
d_A^{w_1 w_2} = \int_{-\infty}^{T} \frac{1}{\sqrt{2\pi v_2}} \exp\left[-\frac{(z - \mu_2)^2}{2\sigma_2^2} \right] dz
$$

$$
+ \int_{T}^{\infty} \frac{1}{\sqrt{2\pi v_1}} \exp\left[-\frac{(z - \mu_1)^2}{2\sigma_1^2} \right] dz
$$

where μ_1 is the mean of the w_1 training data set, μ_2 is the mean of the w_2 training data set, σ_1 is the standard deviation about μ_1, and σ_2 is the standard deviation about μ_2. Figure 4.8 illustrates the feature distinguishability measure graphically. The shaded region in the figure is the region for feature A in which the two classes w_1 and w_2 overlap. In order to calculate the distinguishability measure, T needs to be determined. This is done by minimizing this expression, that is, differentiating it with respect to T and setting the result to zero.

Detectability is the extent to which the diagnostic scheme can detect the presence of a particular fault; it relates to the smallest failure signature that can be detected and the percentage of false alarms. An expression that captures these notions is

$$ d_k(a_i) = \max_{a_1,\ldots,a_n} \left[\frac{F_k(a_i)}{N(a_i)s} \left(1 - \lambda |L_{\psi_{a_i}} - L_k|\right) \right] \qquad i = 1, \ldots, n $$

and $\quad D_k(a_i) = 1 - e^{-d_k(a_i)}$

where $F_k(a_i)$ = magnitude of the feature in w_i for the scale a_i for the kth fault
$\qquad N(a_i)$ = mean square value of the random noise in w_i for the scale a_i
$\qquad L_{\psi_{a_i}}$ = length of the wavelet function for the scale a_i
$\qquad s$ = factor for decimation e.g. scale factor in fractal scanning
$\qquad L_k$ = length of feature for the kth fault
$\qquad \lambda$ = constant less than 1

The *identifiability* measure tracks the similarity of features as they identify a fault mode k but also distinguishing it from other fault classes, $p \neq k$. It can be expressed as

$$ I_k(a_i) = \frac{1}{n} \left[\sum_{i=1}^{n} \mathrm{sim}(w_{i,k}, \beta_{ai,k}) - \sum_{i=1}^{n} \sum_{p=1,p\neq k}^{M} \mathrm{sim}(w_{i,p}, \beta_{ai,k}) \right] \qquad i = 1, \ldots, n $$

where sim is a *similarity* metric defined as follows: Let $X, Y \in [0, 1]$ be two sets with elements x_i and y_i; then a measure of similarity is defined by

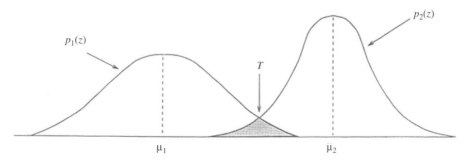

Figure 4.8 The feature distinguishability measure.

$$\text{sim}(X, Y) = \frac{1}{K} \sum_{i=1}^{K} \frac{1 - |x_i - y_i|}{1 + \alpha|x_i - y_i|}$$

And $w_{i,k}$ are fault features or signatures, and $\beta_{ai,k}$ are corresponding fault modes. Figure 4.9 illustrates the similarity metric.

Identifiability goes one step further in distinguishing between various failure modes once the presence of a failure has been established. It targets questions such as the source, location, type, and consequence of a failure and the distinguishability between sensor, actuator, and system component failures.

For supervised fault classification approaches, the prior information about the fault classes is available in the form of a training data set. Using the training data set, the distinguishability power of each feature can be determined. Thus it is reasonable assumption that the knowledge base contains information about the distinguishability of features of each fault, as well as the time necessary to measure the features. It is generally true that the time required to derive a set of mass functions differs widely from one feature class to another as much as from one sensor to another similar device. In order to achieve a reliable classification in the minimum amount of time, the features should be aligned in an optimal order such that features that require less processing time and provide greater distinguishability between classes should be first in the feature-ordering scheme. Assuming that an ordered feature set has been obtained, fuzzy classification is performed using one feature at a time from the ordered feature set. Other classification schemes are also possible. Each classification result, based on a feature, is converted

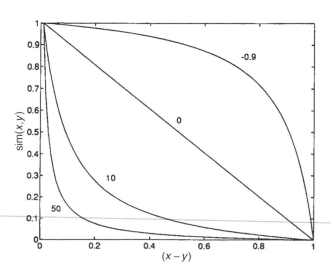

Figure 4.9 The similarity metric.

into a body of evidence. Thus, for each feature, the classifier provides a piece of evidence. To determine the degree of confidence (DOC) achieved based on the latest evidence and all the previous evidence, the DOC associated with the combined body of evidence is calculated. The total evidence is aggregated until the required DOC is achieved. The process of fuzzy classification is terminated as soon as enough evidence is obtained to achieve a prespecified DOC. Thus the suggested scheme controls the information-gathering process and uses an amount of information that is sufficient for the task at hand. Figure 4.10 shows the approach for a multisensor environment. The knowledge base, as shown in Fig. 4.10, represents the database including the optimal feature ordering and the training data set for the fuzzy classifier. The procedure is generic in the sense that it can be applied in a multisensor environment to perform sensor fusion at the feature level or only the features extracted from a single sensor.

In order to obtain the optimal order of the available features, the processing time for classification and the distinguishability for each feature need to be evaluated. These evaluations can be performed as an offline learning process using the training data set. Feature ordering considers features from all available sensors, thus fusing information at the feature level. An optimal feature ordering is performed for each pair of classes. Let there be c fault classes that need to be classified. The total number of ordered feature sets required is $c(c - 1)/2$. An ordered feature set F_{ij} represents the feature order that maximizes the distinguishability between the ith and the jth classes while

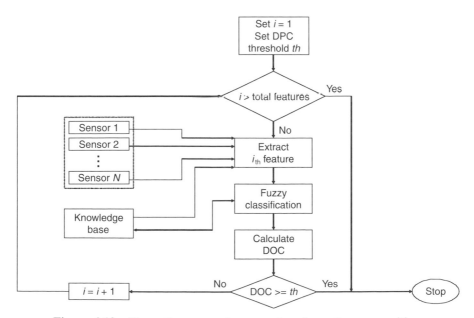

Figure 4.10   The active perception procedure for pattern recognition.

keeping the processing time low. The DOCs for all possible classes are calculated based on the evidence provided by the current features and all previous ones. If the highest DOC value is less than the prespecified threshold, the next feature is selected so that it maximizes the distinguishability between the class with the highest DOC and the one with the second-highest DOC. Thus feature ordering for an unknown fault is modified dynamically online during the classification process.

Ranking features as to how close each one is in determining a specific fault typically is based on Fisher's discriminant ratio (FDR):

$$\text{FDR} = \frac{(\mu_1 - \mu_2)^2}{\sigma_1^2 + \sigma_2^2}$$

where μ_1, σ_1 = mean standard deviation of fault feature 1
μ_2, σ_2 = mean and standard deviation of fault

4.4.2   Signal Processing in the Time Domain

To use information from CBM/PHM sensors for diagnosis, one must know how to compute statistical moments, which we describe in this section. The fourth moment, kurtosis, has been found to be especially useful in fault diagnosis (Engel et al., 2000) because it captures the bang-bang types of motion owing to loose or broken components.

Given a random variable, its pth moment is computed using the ensemble average

$$E(x^p) = \int x^p f(x) \, dx$$

with $f(x)$ the probability density function (PDF). If a random variable is ergodic, then its ensemble averages can be approximated by its time averages, which we now describe.

Moments about Zero and Energy. Given a discrete time series x_k, the pth moment over the time interval $[1, N]$ is given by

$$\frac{1}{N} \sum_{k=1}^{N} x_k^p$$

The first moment is the (sample) mean value

$$\bar{x} = \frac{1}{N} \sum_{k=1}^{N} x_k$$

and the second moment is the moment of inertia

$$\frac{1}{N} \sum_{k=1}^{N} x_k^2$$

As N becomes larger, these better approximate the corresponding ensemble averages.

The energy of the time series over the time interval $[1, N]$ is given by

$$\sum_{k=1}^{N} x_k^2$$

and the root-mean-square (RMS) value is

$$\sqrt{\frac{1}{N} \sum_{k=1}^{N} x_k^2}$$

Moments about the Mean. Given a time series x_k, the pth moment about the mean over the time interval $[1, N]$ is given by

$$\frac{1}{N} \sum_{k=1}^{N} (x_k - \bar{x})^p$$

The second moment about the mean is the (sample) variance

$$\sigma^2 = \frac{1}{2} \sum_{k=1}^{N} (x_k - \bar{x})^2$$

where σ is the standard deviation. Unfortunately, the sample variance is a biased estimate of the true ensemble variance, so in statistics one often uses instead the unbiased variance:

$$s^2 = \frac{1}{N-1} \sum_{k=1}^{N} (x_k - \bar{x})^2$$

The third moment about the mean contains information about the symmetry of a distribution. A commonly used measure of symmetry is the *skewness:*

$$\frac{1}{N\sigma^3} \sum_{k=1}^{N} (x_k - \bar{x})^3$$

Skewness vanishes for symmetric distributions and is positive (negative) if the distribution develops a longer tail to the right (left) of the mean $E(x)$. It measures the amount of spread of the distribution in either direction from the mean. Both probability density functions (PDFs) in Fig. 4.11 have the same

Figure 4.11   Skewness, a measure of the symmetry of a distribution.

mean and standard deviation. The one on the left is positively skewed. The one on the right is negatively skewed.

Also useful is the *kurtosis,* defined as

$$\frac{1}{N\sigma^4} \sum_{k=1}^{N} (x_k - \bar{x})^4 - 3$$

The kurtosis measures the contribution of the tails of the distribution, as illustrated in Fig. 4.12. The PDF on the right has higher kurtosis than the PDF on the left. It is more peaked at the center, and it has fatter tails. It is possible for a distribution to have the same mean, variance, and skew and not have the same kurtosis measurement.

MATLAB (MATLAB, 2004) provides these related functions:

- m = moment(X,order) returns the central moment of X specified by the positive integer order.
- k = kurtosis(X) returns the sample kurtosis of X.
- y = skewness(X) returns the sample skewness of X.
- m = mean(X) calculates the sample average.
- y = std(X) computes the sample standard deviation of the data in X.
- y = var(X) computes the variance of the data in X.

Correlation, Covariance, Convolution.   The (auto)correlation is given by

$$R_x(n) = \frac{1}{N} \sum_{k=1}^{N} x_k x_{k+n}$$

Figure 4.12   The kurtosis, a measure of the size of the side lobes of a distribution.

Note that $NR_x(0)$ is the energy. The (auto)covariance is given as

$$P_x(n) = \frac{1}{N} \sum_{k=1}^{N} (x_k - \bar{x})(x_{k+n} - \bar{x})$$

Note that $P_x(0) = \sigma^2$.

Given two time series x_k, y_k the cross-correlation is

$$R_{xy}(n) = \frac{1}{N} \sum_{k=1}^{N} x_k y_{k+n}$$

and the cross-covariance is

$$P_{xy}(n) = \frac{1}{N} \sum_{k=1}^{N} (x_k - \bar{x})(y_{k+n} - \bar{y})$$

Often used in statistics is Pearson's correlation, which is defined as

$$r = \frac{\Sigma x_k y_k - \dfrac{\Sigma x_k \, \Sigma y_k}{N}}{\sqrt{\left[\Sigma x_k^2 - \dfrac{(\Sigma x_k)^2}{N} \right]\left[\Sigma y_k^2 - \dfrac{(\Sigma y_k)^2}{N} \right]}}$$

The discrete-time convolution for N point sequences is given as

$$x*y\,(n) = \sum_{k=0}^{N-1} x_k y_{n-k}$$

which is nothing more than polynomial multiplication. It is interesting to note that the correlation is expressed in terms of the convolution operator by

$$R_{xy}(n) = \frac{1}{N} \sum_{k=1}^{N} x_k y_{k+n} = \frac{1}{N} x(k)*y(-k)$$

Again, MATLAB (MATLAB, 2004) provides these related functions:

- C = cov(X) computes the covariance matrix. For a single vector, cov(x) returns a scalar containing the variance. For matrices, where each row is an observation and each column a variable, cov(X) is the covariance matrix.
- c = xcorr(x,y) (in Signal Processing Toolbox) estimates the cross-correlation sequence of a random process. Autocorrelation is handled as a special case.
- w = conv(u,v) convolves vectors u and v.

Moving Averages and Time Windows. The properties of most time sequences of interest change with time. One captures this by windowing the data. Figure 4.13 shows a time series x_k windowed by a window of size N that moves with the current time n. Thus it takes the most current N samples of x_k and discards previous (old) samples.

To capture the fact that the mean of a windowed sample may change with time, one defines the moving average (MA) as

$$\bar{x}(n) = \frac{1}{N} \sum_{k=n-(N-1)}^{n} x_k$$

The MA is considered a function of the current time n and implicitly of the window size N. As N decreases, the MA changes faster.

The moving variance is given by

$$\sigma(n)^2 = \frac{1}{N} \sum_{k=n-(N-1)}^{n} [x_k - \bar{x}(n)]^2$$

Windowed versions of other moments are defined similarly.

4.4.3 Frequency-Spectrum Information for Fault Diagnosis

In many situations, especially with rotating machinery (Eisenmann and Eisenmann, 1998), the frequency spectrum of measured time signals, such as vibration, carries a great deal of information useful in diagnosis. A rotating machine might have a base rotational frequency of 120 Hz, a component owing to a three-bladed fan at 360 Hz and components owing to a 12-toothed gear at 1440 Hz. A broken gear tooth, for instance, might cause a large increase in the component at 1440 Hz. The frequency spectrum of time signals is computed easily using the discrete Fourier transform (DFT).

Discrete Fourier Transform. Given an N-point time series $x(n)$, the DFT is given as

$$X(k) = \sum_{n=1}^{N} x(n)e^{-j2\pi(k-1)(n-1)/N} \qquad k = 1, 2, \ldots, N$$

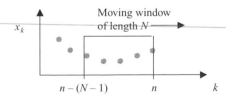

Figure 4.13 Windowing of time series.

The inverse DFT is given by

$$x(n) = \frac{1}{N} \sum_{k=1}^{N} X(k) e^{j2\pi(k-1)(n-1)/N} \qquad n = 1, 2, \ldots, N$$

The DFT is periodic with period N. The inverse DFT yields a time function that is periodic with period N. There is an efficient algorithm for computing the DFT known as the *fast fourier transform* (FFT) that is available in software packages such as MATLAB. More information about the DFT is given in Oppenheim and Schafer (1975).

Scaling the Frequency Axis. To plot frequency in radians, the frequency axis should be scaled to run from 0 to 2π. To accomplish this, one uses the frequency variable

$$\omega = \frac{2\pi}{N}(k - 1)$$

The FFT first harmonic frequency is given by

$$\Omega = \frac{2\pi}{NT}$$

with T the sampling period. We will illustrate the frequency-scaling technique in the examples concerned with sampling and relating the continuous and discrete time and frequency scales.

Example 4.1 FFT of Time Window. Let $x(n)=$ 1, 1, 1, 1, 1, 1, 1, 1, 0, 0, 0, . . . , as shown in Fig. 4.14, with eight ones followed by zeros. Take first the DFT of this $x(n)$ with length 32 using the MATLAB code:

```
>> x=[1 1 1 1 1 1 1 1]'
>> y=fft(x,32)
>> k=1:32
>> plot(k,abs(y))
```

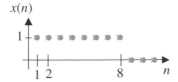

Figure 4.14   Window time series.

The result is shown in Fig. 4.15. Note that the resolution of the FFT may not be acceptable. To correct this, pad the time signal with extra zeros at the end to arrive at a DFT of length 512 using the MATLAB code:

```
>> y=fft(x,512)
>> k=1:512
>> w=2*pi*(k-1)/512
>> plot(w,abs(y))
```

The results are shown in Fig. 4.16. The resolution obviously has improved substantially. Note that we have scaled the frequency to run from 0 to 2π rads. The DFT phase is shown in Fig. 4.17.

Windowed DFT

Moving Windowed DFT. Most signals of interest in fault diagnosis have time-varying characteristics, including time-varying spectra. The best-known examples outside the CBM/PHM arena are speech signals and financial market stock prices. Therefore, one usually computes a moving windowed DFT at time N using only the past m values of the signal:

$$X(k, N) = \sum_{n=N-m+1}^{N} x(n)\, e^{-j2\pi(k-1)(n-1)/N}$$

Now, the moving windowed DFT is a function of two variables, the time index N and the frequency k. It must be plotted as a three-dimensional plot. A sample time-varying DFT is shown in Fig. 4.18.

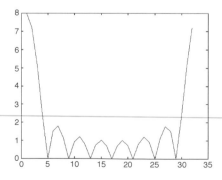

Figure 4.15   Magnitude of DFT of eight-ones sequence of length 32.

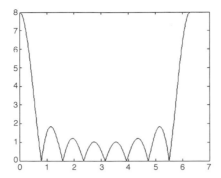

Figure 4.16 Magnitude of DFT of eight-ones sequence of length 512.

In some applications, one does not consider the variation in time for every sample. Instead, the time index is sectioned into bins of length m, and the DFT is performed only for values of $N = im$, with i an integer (Fig. 4.19). This is common, for example, in speech processing (Oppenheim and Schafer, 1975). Although the bin windowed DFT requires less computation than the moving windowed DFT, it is often difficult to select a suitable bin size.

Properties of DFT. The DFT offers several useful properties, some of which are given below. Some of them are given here.

Parseval's theorem states that

$$\sum_{n=1}^{N} x^2(n) = \frac{1}{N} \sum_{k=1}^{N} |X(k)|^2$$

The *convolution theorem* states that the transform of the convolution

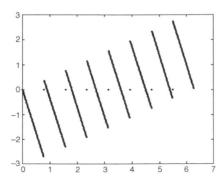

Figure 4.17 Phase of DFT of eight-ones sequence of length 512.

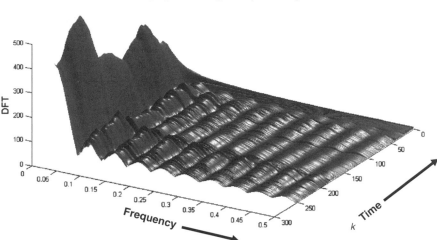

Figure 4.18 Short time Fourier transform, bin windowed DFT.

$$x^*y(n) = \sum_{i=1}^{N} x_i y_{n-i}$$

is given by the product of the transforms $X(k)Y(k)$.

The *correlation theorem* says that the transform of the correlation

$$R_{xy}(n) = \frac{1}{N} \sum_{i=1}^{N} x_i y'_{i+n}$$

is given by the product $X(k)Y'(k)/N$, with the prime denoting complex conjugate transpose. Note that normally we shall deal with real sequences in the time domain.

A special case of this theorem says that the transform of the autocorrelation

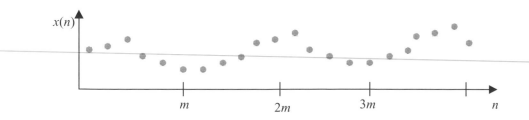

Figure 4.19 Taking the DFT of each signal block (bin) of length m samples.

$$R_x(n) = \frac{1}{N} \sum_{i=1}^{N} x_i x'_{i+n}$$

is given by

$$\Phi_x(k) = \frac{1}{N} X(k) X'(k) = \frac{1}{N} |X(k)|^2$$

which is known as the *power spectral density* (PSD). It is often easier to see details in the frequency response using the PSD than using $X(k)$.

Example 4.2   Using FFT to Extract Frequency Component Information.
This example is based on the example in the *MATLAB Reference Book* (MATLAB, 1994) under the listing "FFT." A time signal as measured from a noisy accelerometer sensor placed on rotating machinery might appear as shown in Fig. 4.20. This type of signal is common in rotating machinery. The plot of $y(t)$ shows that the noise makes it impossible to determine the frequency content of the signal.

The FFT is a very efficient and fast method to compute the DFT for time series whose lengths are an exact power of 2. Therefore, to analyze the spectrum, take the FFT of the first 512 samples using the MATLAB command

```
>> Y=fft(y,512);
```

The DFT magnitude spectrum is plotted in Fig. 4.21 using the MATLAB command plot(abs(Y)).

The frequency must be scaled to be meaningful (Oppenheim and Schafer 1975). Suppose that the sample time is $T = 0.001$ s. The index k varies from 1 to 512, whereas the frequency f of the DFT varies from 0 to the sampling frequency

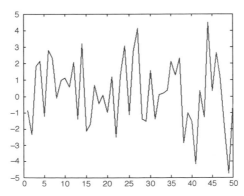

Figure 4.20   Plot of measured signal $y(t)$ with noise.

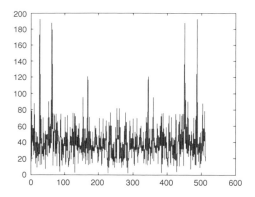

Figure 4.21   Magnitude spectrum of noise-corrupted signal.

$$f_s = \frac{1}{T}$$

Therefore, a meaningful frequency axis is given by

$$f = \frac{f_s}{N}(k-1) = \frac{1}{NT}(k-1)$$

Now redo the plot using

```
>> f=1000*(0:255)/512
>> plot(f,abs(Y(1:256)))
```

The result is shown in Fig. 4.22. Only the first half of the spectrum is needed because it is symmetric.

Note the presence of spikes at 50 and 120 Hz. The peaks are better detected using the PSD, which gives the energy at each frequency. The PSD is computed using

```
>> Py=Y.*conj(Y)/512;
>> plot(f,Py(1:256))
```

and shown in Fig. 4.23. Note how much more prominent the peaks are in the PSD. Clearly, the measured data contain frequency components at 50 and 120 Hz.

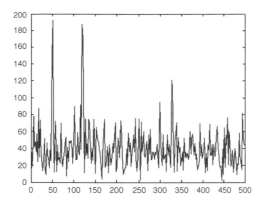

Figure 4.22 Magnitude spectrum of measured signal with scaled frequency.

4.4.4 Signal Processing in the Wavelet Domain

Although FFT-based methods are powerful tools for fault diagnosis, they are not suitable for nonstationary signals. For analysis in the time-frequency domain, the Wigner-Ville distribution (WVD) and the short time Fourier transform (STFT) are the most popular methods for nonstationary signal analysis. However, WVD suffers from interference terms appearing in the decomposition, and STFT cannot provide good time and frequency resolution simultaneously because it uses constant resolution at all frequencies. Moreover, no orthogonal bases exist for STFT that can be used to implement a fast and effective STFT algorithm.

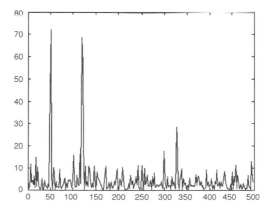

Figure 4.23 PSD of measured signal with scaled frequency.

In 1984, Morlet put forward a novel concept of wavelets for multiscale analysis of a signal through dilation and translation processes aimed at extracting time-frequency features. Table 4.1 shows the comparative performance of different time-frequency analysis methods.

The wavelet transform is a linear transform that uses a series of oscillating functions with different frequencies as window functions $\psi_{\alpha,\beta}(t)$ to scan and translate the signal $x(t)$, where α is the dilation parameter for changing the oscillating frequency, and β is the translation parameter. At high frequencies, the wavelet reaches a high time but a low frequency resolution, whereas at low frequencies, it reaches a low time but a high frequency resolution, making it more suitable for nonstationary signal analysis. The basis function for the wavelet transform is given in terms of a translation parameter β and a dilation parameter α with the mother wavelet represented as

$$\psi_{\alpha,\beta}(t) = \frac{1}{\sqrt{\alpha}} \, \psi\!\left(\frac{t - \beta}{\alpha}\right)$$

The wavelet transform $W_{\psi}(\alpha, \beta)$, of a time signal $x(t)$ is given by

$$W_{\psi}(\alpha, \beta) = \frac{1}{\sqrt{\alpha}} \int_{-\infty}^{+\infty} x(t)\psi'\left(\frac{t - \beta}{\alpha}\right) dt$$

TABLE 4.1   Comparing Different Time-Frequency Analysis Methods

Methods	Resolution	Interference	Speed
Continuous wavelet transform (CWT)	Good frequency and low time resolution for low-frequency components; low frequency and high time resolution for high-frequency components	None	Fast
Short time fourier transform (STFT)	Depending on window function used, either good time or good frequency resolution	None	Slower than CWT
Wigner-Ville distribution (WVD)	Good time and frequency resolution	Severe	Slower than STFT
Choi-Williams distribution (CWD)	Good time and frequency resolution	Less than WVD	Very slow
Cone-shaped distribution (CSD)	Good time and frequency resolution	Less than CWD	Very slow

Source: Peng and Chu (2004).

where $\psi'(t)$ is the complex conjugate of $\psi(t)$. A number of different real and complex valued functions can be used as analyzing wavelets. The modulus of the wavelet transform shows how the energy of the signal varies with time and frequency—a physical interpretation of the transform definition that we take advantage of in analyzing certain fault signals.

There are several ways in which wavelets can be applied for signal analysis. Some of the most common techniques are summarized below.

Time-Frequency Analysis. As suggested earlier, wavelets are used most commonly for time-frequency analysis of nonstationary signals. In engineering applications, the square of the modulus of the CWT, termed *scalogram* and shown in Fig. 4.24, is used widely for fault diagnosis:

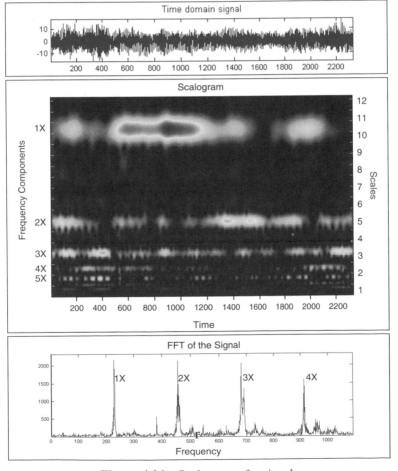

Figure 4.24   Scalogram of a signal.

$$S_x(\alpha, \beta; \psi) = |W_\psi(\alpha, \beta)|^2$$

As can be seen from the figure, the FFT of the signal shows the presence of different harmonics, but it is not clear how its energy is distributed in the time domain. The scalogram shows not only the different harmonics but also their time evolution. Furthermore, it can be seen that wavelets have a higher resolution at low frequencies than at high frequencies. The complex wavelets also yield a phase plot that is much more difficult to interpret than scalograms but can provide important information about signal discontinuities and spikes.

Scalograms have been used for detecting edge cracks on cantilever beams (Zhang et al., 2001), cracks on rotors (Zou et al., 2002; Adewusi and Al-Bedoor, 2001), cracks in beam structures (Quek et al., 2001), and cracks in metallic and composite structures (Bieman et al., 1999; Deng et al., 1999; Staszewski et al., 1999; Staszweski, 1998a). They also have been used to detect rub impact, oil whirl, and coupling misalignment in roller bearings (Peng et al., 2002a, 2003).

Fault Feature Extraction. Several types of features can be extracted from wavelet-based methods, which can be categorized roughly into wavelet coefficient-based, wavelet energy-based, singularity-based, and wavelet function-based methods.

Coefficient-Based Features. Owing to compact support of the basis functions used in wavelet transforms, wavelets have a good energy concentration property. Most wavelet coefficients c_{mn} are very small and can be discarded without significant error in the signal, allowing for a smaller set to be used as fault features. Thresholding is a simple and typical method employed to choose these coefficients:

$$A(c_{mn}) = \begin{cases} c_{mn} & c_{mn} > \theta \\ 0 & c_{mn} \leq \theta \end{cases}$$

where θ is the threshold.

These coefficients then can be used as fault-feature vectors to train classification algorithms. Other methods for selecting these coefficients include statistics-based criteria (Yen and Lin, 1999, 2000), wavelet packets to obtain a reduced set of coefficients, and principal component analysis (PCA) techniques to reduce the size of the feature space (Akbaryan and Boshnoi, 2001).

Energy Based Features. The most important advantage of wavelets versus other methods is their ability to provide an image visualization of the energy of a signal, making it easier to compare two signals and identify abnormalities or anomalies in the faulty signal. Based on these observations, suitable features can be designed either using image-processing techniques or simply exploiting the energy values directly. For example, in a study of a helicopter's planetary gear system (Saxena et al., 2005), it was observed that the energy distribution among the five planets became asymmetric as a fault (crack) appeared on the gear plate (Fig. 4.25). This observation led to a feature based on the increasing variance among energy values associated with the five planets as time evolves. Similarly, other features can be designed that characterize a visible property in a numerical form.

Singularity-Based Features. Singularities can be discerned from wavelet phase maps and can be used as features for detecting discontinuities and impulses in a signal. Singularity exponents, extracted from the envelope of vibration signals, have been used to diagnose breakers' faults (Xu et al., 2001). Other applications reported in the technical literature relate to detection of shaft center orbits in rotating mechanical systems (Peng et al., 2002b).

Some of the most interesting applications of wavelet-based features include fault detection in gearboxes (Chen and Wang, 2002; Essawy et al., 1997; Hambaba and Huff, 2000; Zheng et al., 2002, Yen and Lin, 2000), fault detection in rolling element bearings (Altmann and Mathew, 2001; Shiabata et al., 2000), and fault diagnosis in analog circuits (Aminian and Aminian, 2001), among others.

Wavelet energy-based features often cannot detect early faults because slight changes in the signal result in small energy changes. Coefficient-based features are more suitable for early fault detection. Similarly, singularity-based methods are not very robust to noise in the signal, and denoising must be carried out before calculating any such features.

Singularity Detection. Fault information is often represented as peaks or discontinuities in the signals that can be detected easily as singularity points. The most standard method for singularity detection is the modulus maxima method to compute Hölder exponents that can be used as fault features (Mallat

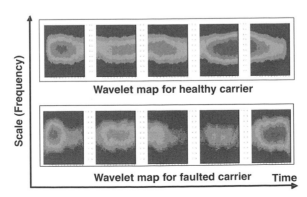

Figure 4.25 Energy-based method for feature calculation using a scalogram.

and Hwang, 1992). These methods reveal not only the anomalies in the signal but also the location of the faults in some cases such as power systems (Dong et al., 1997). Singularity detection also has been used in fault diagnosis for electrohydraulic servo systems (Chen and Lu, 1997), in vibration analysis in a reciprocating compressor valve (Lin et al., 1997), and in detecting the position of rub faults occurring in rotating machines (Zhang et al., 2001).

The wavelet should be sufficiently regular (long filter impulse response) in order to detect all the singularities. Also, as suggested previously, signal preprocessing and denoising must be carried out before using singularity detection methods.

System and Parameter Identification. An alternative way of detecting system faults is by observing the changes in the system parameters, or model parameters, such as natural frequency, damping, stiffness, etc. Both time- and frequency-based methods have been developed for fault diagnosis by observing the impulse-response and frequency-response functions of the systems under study. For instance, wavelet transforms were used to extract the impulse response of structural dynamic systems (Robertson et al., 1998a, 1995a; Freudinger et al. 1998) and to identify structural modes, mode shapes, and damping parameters for these systems (Robertson et al., 1998b, 1995b). Similarly, He and Zi (2001) and Ruzzene et al. (1997) identified viscous damping ratios and natural frequencies of a hydrogenerator axle, and Ma et al. (2000) used a wavelet-based method to identify dynamic characteristics of bearings.

Impulse-response functions are computationally expensive and very sensitive to noise, whereas frequency-response functions suffer from energy leakage and spectrum overlap problems. Wavelet-based methods take advantage of their time-frequency representation, compact support base, and effective signal filtering capabilities to overcome these problems. Several methods have been developed for estimating these response functions for both linear and nonlinear systems.

Example 4.3   Complex Morlet Wavelet. One of the most responsive wavelets for vibration signal analysis is the Morlet wavelet. The complex Morlet wavelet is defined below in the time domain and then in the frequency domain:

$$\psi_{\text{Morlet}}(t) = \frac{1}{\sqrt{\pi f_b}} \cdot e^{j2\pi f_c t - (t^2/f_b)}$$

$$\Psi_{\text{Morlet}}(f) = e^{\pi^2 f_b (f - f_c)^2}$$

where f_c is the center frequency and f_b is the bandwidth (variance). Choosing a suitable combination of bandwidth and center frequency parameters is a design question that depends on the signal under analysis. Figure 4.26 shows the effect of the bandwidth parameter on Morlet wavelets in both the time

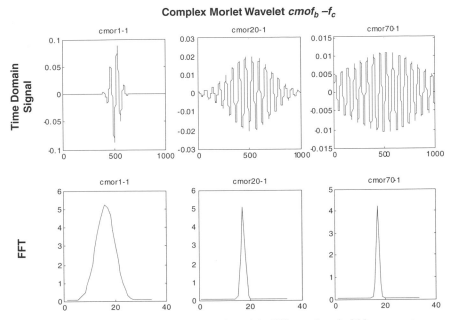

Figure 4.26 Complex Morlet wavelet with different bandwidth parameters.

and frequency domains. More mathematical details about Morlet wavelets and their properties with regard to their implementation can be found in Teolis (1998) and Walnut (2002).

Computer Implementation. For the diagnosis of faults in rotating mechanical systems, where the vibration signal is a periodic function of the shaft frequency, the computer implementation can be made simpler and faster using frequency-domain signal properties. The selected wavelet $\psi(t)$ should be such that $\psi(-t) = \psi'(t)$ so that, by using the convolution theorem, the previous equation on page 122 can be written as

$$W_\psi(\alpha, \beta) = \frac{1}{\sqrt{\alpha}} F^{-1} \left[X(f)\overline{\Psi}(f/\alpha) \right]$$

where F represents the Fourier transform, and $\overline{\Psi}(f)$ represents the $F[\psi'(t)]$, and thus calculation of the wavelet transform can be implemented by an inverse Fourier transform, taking advantage of the efficient fast Fourier transform (FFT) algorithm. Based on this equation, it can be shown that the center frequency of the wavelet is proportional to the scale parameter α (Wang and McFadden, 1996). The narrower the wavelet, the wider is its representation in the time domain. This increases the frequency overlap between the wavelets of adjacent high scales, thus creating a redundancy. Wang and McFadden

(1996) further show that choosing the scale parameters in an effective way not only helps in reducing such redundancy but also makes the wavelet map computations faster using notions of circular convolution and FFT techniques. The latter coincides with the concept of discrete wavelet transform (DWT), as described in the preceding section.

4.5   VIBRATION MONITORING AND DATA ANALYSIS

Rotating equipment is used frequently in engineering systems, and it is important to monitor and maintain its healthy status. Vibration monitoring and data analysis are an effective and efficient approach to providing, in real time, condition indicators for machine components such as bearings and gears for diagnostic and prognostic purposes. We will focus in this section on common CBM/PHM data-analysis methodologies developed to address rotating equipment.

4.5.1   A General Approach to Vibration Analysis

This subsection lists a few "typical" methods of vibration analysis (SKF, 2000). They were documented by experienced SKF field application engineers. The intent is to lay the foundation for a better understanding of machinery analysis concepts and to show the reader what is needed to perform an actual analysis on specific machinery.

FFT Spectrum Analysis. An FFT spectrum is an incredibly useful tool. Since certain rotating machinery problems occur at certain frequencies, we analyze the FFT spectrum by looking for amplitude changes in certain frequency bands.

Envelope Detection. The object of enveloping is to filter out the high-frequency rotational vibration signals and to enhance the repetitive components of a bearing's defect signals occurring in the bearing defect frequency range. Envelope detection is most common in rolling-element bearing and gear mesh analysis, where a low-amplitude and repetitive vibration signal may be hidden by the machine's rotational and structural vibration noise. It can be used to detect a defect on the outer race of a rolling-element bearing. Each roller rolls over the defect as it goes by, which causes a small vibration signal at the bearing's defect frequencies.

Spectral Emitted Energy (SEE). SEE provides very early bearing and gear mesh fault detection by measuring acoustic emissions generated by metal as it fails or generated by other specific conditions. Circumstances that can cause acoustic emissions include

- Bearing defects
- Contaminated lubrication
- Lack of lubrication
- Dynamic overloading
- Microsliding/fretting
- Bearing friction
- Cavitation/flow
- Electrically generated signals
- Metal cutting
- Compressor rotor contact
- Electrical arcing
- Cracking or tin cry

SEE technology uses a special acoustic emission sensor that "listens" for ultrasonic emissions that occur when bearing elements degrade (these acoustic emissions occur in the 150- to 500-kHz range). This type of signal is not considered vibration as much as it is considered high-frequency sound; however, *vibration* is the commonly used industrial term.

Since SEE technology measures the ultrasonic noise (acoustic emissions) created when metal degrades, it is the best tool for detecting bearing problems in their earliest stages, when the defect is subsurface or microscopic and not causing any measurable vibration signal. For that matter, SEE measurements are also very effective for detecting any machine condition that produces acoustic emissions, such as corrosion and friction due to fretting, cavitation, sliding, or friction events.

If SEE values rise gradually above normal, there is usually no need to replace the bearing immediately. SEE detection provides enough prewarning for the maintenance person to make operational or lubrication corrections and potentially save the bearing or effectively extend its life.

If SEE values rise, monitor the bearing more closely (shorten measurement intervals and perform multiparameter monitoring). Trend the bearing's condition with SEE—enveloping, temperature, and vibration measurements—to best analyze the problem and predict the best time for corrective action. A good understanding of the machine and a logical approach to problem solving is needed to help direct repair action.

Phase Measurement. Phase is a measurement, not a processing method. Phase measures the angular difference between a known mark on a rotating shaft and the shaft's vibration signal. This relationship provides valuable information on vibration amplitude levels, shaft orbit, and shaft position and is very useful for balancing and analysis purposes.

High-Frequency Detection (HFD). HFD provides early warning of bearing problems. The HFD processing method displays a numerical overall value for high-frequency vibration generated by small flaws occurring within a high-frequency band pass (5 to 60 kHz). The detecting sensor's resonant frequency is within the band pass and is used to amplify the low-level signal generated by the impact of small flaws. Because of its high-frequency range, the HFD measurement is made with an accelerometer and displays its value in G's. The HFD measurement may be performed as either a peak or RMS overall value.

4.5.2   Engine Blade Crack Time-of-Arrival Data Analysis: A Case Study

High vibrations with complicated resonances can cause turbine engine blades to have large deflections at a broad range of frequencies, resulting in blade cracking and sometimes in-flight failures. To prevent these potentially life-threatening situations from occurring and to help lower the operating and maintenance costs of fleets, it is important to look into key enabling technologies that not only provide early warning of potential disasters but also are substantive indicators of the real-time health of engine components and the overall system.

Pratt and Whitney and Hood Technology (Tappert 2003) conducted research in noncontacting blade vibration and disk deformation monitoring to detect abnormal behavior, particularly looking for cracks in turbine engine blades and disks. The objective of the health monitoring system is to automatically predict future machine conditions as well as to detect the faults. Owing to the harsh working environment in an engine, intrusive sensors such as strain gauges and slip rings are not reliable enough to measure the actual stress on a blade in an operating engine so as to continuously monitor component health. Nonintrusive and high-temperature sensors are preferred that can measure blade vibration or deflection from outside the engine casing. These sensors measure blade deflection by detecting the time of arrival of every blade on every revolution. These nonintrusive measurements require no interruption at all in the gas path and remove the sensor itself from the major environmental effects of the engine: temperature, foreign objects, blade rub, moisture, corrosion, etc.

In this case study we will show the application of nonintrusive sensors and real-time data analysis in measuring disk deformation and unusual blade stretch that are indications of blade crack (Tappert, 2001). Under normal conditions, each blade arrives with the same circumferential arrival interval. If a blade has a crack or is plastically deformed, the interblade spacing (IBS) will be changed, or blades will bend "out of position" and be a good indicator of the fault. Optical sensors, eddy-current probes, and capacitive sensors were used in the testing conducted by Hood Technology. Figure 4.27 shows Hood Technology's Blade Vibration Monitoring System, including sensors, sensor conditioning, and signal conditioning and electronics needed for reliable detection of blade passage (Tappert, 2003).

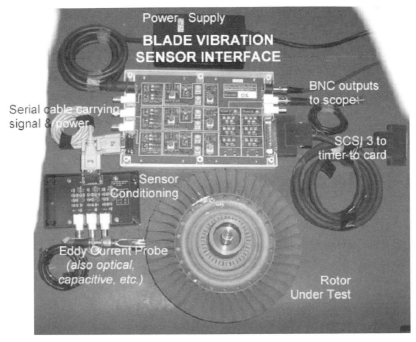

Figure 4.27 Hood Technology's Blade Vibration Monitoring System includes sensors, sensor conditioning, and signal conditioning electronics for reliable detection of blade passage. (Tappert, 2003.)

The article under test is a postservice F100 twelfth-stage HPC disk with 90 blades and three flaws (cracks) seeded on the antirotation tangs inboard of the blades. The flaws are located 18 slots apart, approximately near blades 13, 67, and 85. For this test, two radial capacitive probes, two simple seven-fiber light probes, and two eddy-current sensors were used. The capacitive probes were dead from the beginning of the test and were not used. The eddy-current probes gave the cleanest and most repeatable signal during cold cycles. However, electromagnetic interference (EMI) from the heater strongly compromised the data. The simple seven-fiber light probes, one on the leading edge and one on the trailing edge, were used throughout the test. The measurements were monitored in real time in order to determine when to terminate the test prior to failure.

The test consisted of 1213 stress cycles. For each stress cycle, the speed of the blade disk was swept from about 6500 to 13,000 rpm, as shown in Fig. 4.28. The data-acquisition software acquires and saves raw binary information. The low-cycle-fatigue data-processing software first performs tasks such as identification of revolutions with missing or extra pulses so that they are eliminated and making sure that the data are sorted by blade. Figure 4.29 shows the circumferential interblade spacing (IBS) data (between blades 67

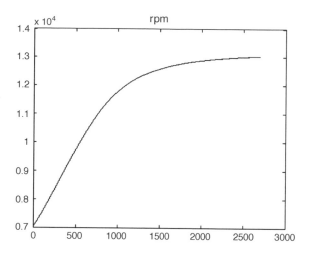

Figure 4.28   Rpm sweep.

and 68 at cycle 1130) provided by one of the optical sensors measuring the time of arrival (TOA) of each blade. The blade lag increases as the speed of the blade disk increases. In other words, blade deformation scales with the stress.

The lighter shaded line in Fig. 4.29 shows the $A + B \times (rpm)^2$ least-mean-squares fit to the data. Note here that from this point on, the rpm value is

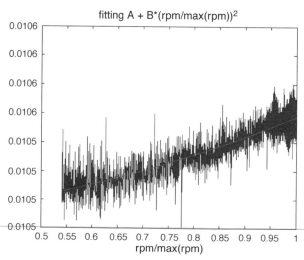

Figure 4.29   The lighter shaded line shows the rpm2 fit of the circumferential position (IBS) data (blade 67 at cycle 1130) from optical sensor s8 (trailing-edge sensor).

normalized to the maximum rpm. The parameters A and B are related to the elastic and plastic deformation of the blades, respectively. The least-mean squares-regression yields two numbers (A and B) per blade per cycle per sensor. Figure 4.30 shows parameters A and B of the 90 blades throughout the 1213 cycles during the test. At cycle 574, the testing operation was shut down, and the blade disk was disassembled for inspection purpose. The parameters show offsets at cycle 574 that need to be removed in data processing.

As an indicator of the crack growth from cycle to cycle, the combination of the two parameters is used, which is an evaluation of the two-parameter curve fit at the top rpm ($=1$) used in the fit. It is the most consistent condition indicator for this test. The IBS evaluated at the high speed then is translated into individual blade position relative to the mean of all blades. A transformation matrix is used for this purpose, and it provides the relationship between interblade spacing and single-blade time of arrival, assuming that

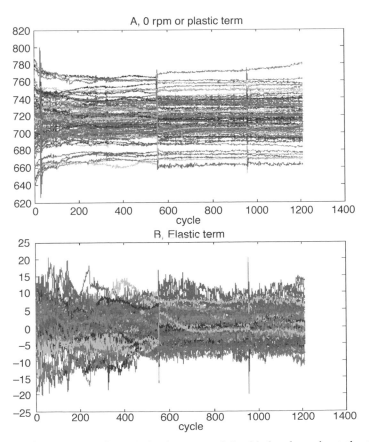

Figure 4.30 The plastic and elastic terms of the blades throughout the test.

there is no collective motion of all blades. The matrix is given by T = [-eye(Nblades) + [zeros(Nblades-1,1),eye(Nblades-1);1,zeros(1,Nblades-1)];ones(1,Nblades)], where Nblades=90. Figure 4.31 shows the positions of the blades referenced to the positions at cycle 200 and with the offsets at cycle 574 removed. Figure 4.32 shows the blade deflection as a function of cycle and blade numbers plotted in three-dimensional view starting from the reference cycle. Evidence of each of the three seeded flaws was observed.

A finite-element analysis (FEA) model of the cracked disk provides knowledge of expected disk deformation patterns that may be used to predict crack growth in a seeded-fault engine test. With knowledge of the relationship between maximum blade tip deflection and crack size, the crack size can be estimated and tracked directly.

In looking for disk cracks in a rotating engine or in a spin pit, frequently radial, axial, and circumferential vibration all are measured. The system searches for asymmetric disk deformations caused by growing cracks. Comparison of the measured deformations with precalculated predictions enables the system to discriminate among several hypothesized crack types.

4.5.3 Gearbox Vibration Monitoring and Analysis: A Case Study

Planetary Gear Plate Crack and the Seeded-Fault Testing. Epicyclic gears are important components of many rotorcraft transmission systems. An *epicyclic* gear system is defined by a sun/planet configuration in which an inner "sun" gear is surrounded by two or more rotating "planets." Planetary

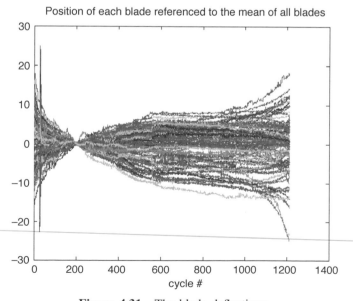

Figure 4.31 The blade deflections.

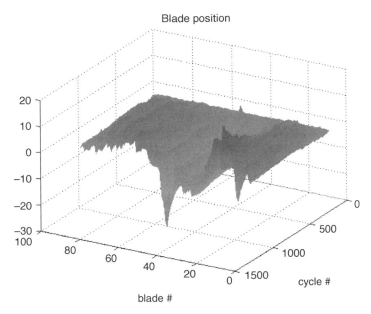

Figure 4.32 Blade deflections as a function of cycle and blade numbers.

systems are a subset of epicyclic gears defined by a stationary outer ring gear. Torque is transmitted through the sun gear to the planets, which ride on a planetary carrier. The planetary carrier, in turn, transmits torque to the main rotor shaft and blades. Figure 4.33 shows the H-60 main rotor transmission with a planetary gear system. Figure 4.34 is a picture of a carrier plate with five posts for five planets. The Web site *www.ticam.utexas.edu/~jedrek/ Research/gear_proj.html* has an animation of the meshing motion of a planetary gear train.

A crack was found recently in the planetary carrier of the main transmission gear of the U.S. Army's UH-60A Blackhawk helicopter (Keller and Grabill, 2003) and the U.S. Navy's Seahawk helicopter. Figure 4.35 shows the carrier plate with a 3.25-inch crack of the main transmission gear. Since the planetary carrier is a flight-critical component, a failure could cause an accident resulting in loss of life and/or aircraft. This resulted in flight restrictions on a significant number of the Army's UH-60A's. Manual inspection of all 1000 transmissions is not only costly in terms of labor but also time-prohibitive. There has been extensive work to develop a simple, cost-effective test capable of diagnosing this fault based on vibration data analysis techniques. A condition indicator (CI) or feature extracted from vibration data typically is used to describe gear health.

Compared with the measured vibrations produced by fixed-axis gear systems, those produced by epicyclic gear systems are fundamentally more complicated. For fixed-axis gear systems, the vibration signal consists mainly of

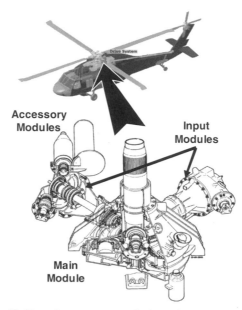

Figure 4.33 H-60 main rotor transmission with planetary gear system.

Figure 4.34 Main gearbox plate.

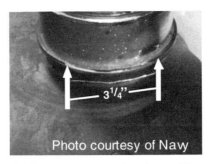

Figure 4.35   Planetary carrier plate with a crack.

the gear meshing frequency, which is $N_t \times f_{sh}$, with N_t being the gear tooth number, and f_{sh} being the shaft frequency and its harmonics. For an accelerometer mounted on the gearbox's housing, the vibration damage to individual gear teeth modulates the vibration signal, and in the frequency domain, the damage appears in the form of symmetric sidebands around the gear meshing frequency, which is still the dominant component. Many of the features for fixed-axis gears detect damage based on the signal amplitude at the gear meshing frequency and/or the modulating sidebands (Polyshchuk et al., 2002; McFadden, 1986).

For epicyclic gear systems, an accelerometer mounted on the housing will measure a periodic amplitude-modulated vibration signal as each planet rotates past the sensor. Thus even healthy gears will produce a vibration spectrum resulting in naturally occurring sidebands. Furthermore, because the vibrations produced by the planets generally are not in phase, the dominant-frequency component often does not occur at the fundamental gear meshing frequency. In fact, the gear meshing frequency component often is completely suppressed. This phenomenon was first recognized and explained by McFadden and Smith (1985) and later McNames (2002). In Keller and Graybill (2003), the features developed for fixed-axis gears were modified for application to an epicyclic gearbox. McFadden (1991), Blunt and Hardman (2000), Samuel and Pines (2001) also have conducted laboratory experiments and analytic studies on vibration monitoring of faults in individual planets of epicyclic gearboxes using windowing techniques.

In this case study we present the accelerometer data-analysis techniques and results for the seeded-fault test conducted under the DARPA Structural Integrity Prognosis System (SIPS) Program. The project was aimed mainly at developing, demonstrating, and validating a prognosis system for a cracked planetary carrier plate. A seeded-fault test was designed and conducted to provide a comprehensive database of accelerometer and crack measurements. The testing started with an initial 1.344-inch crack that propagated from a 0.25-inch starter electro-discharged-machining (EDM) notch in a carrier plate post. Figure 4.36 shows the EDM notched crack and the instrumentation, such

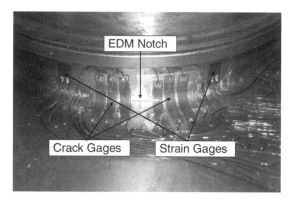

Figure 4.36 Seeded crack and the instrumentation.

as crack and strain gauges, to measure crack length and the stress of the plate during the test. The carrier plate was then stressed with a loading spectrum emulating a severe high-engine-torque mission and considered ground air ground (GAG), 1P geometric vibratory, 980-Hz gear vibratory, and transverse shaft bending. Snapshots of data were collected as the crack was growing. Vibration data were collected from accelerometers mounted on the external transmission casing, far removed from the seeded fault, as shown in Fig. 4.37. These sensors represent existing sensors from health and usage monitoring systems (HUMS) that are beginning to be installed in the Army's and Navy's rotorcraft fleet.

Vibration Data and Feature Extraction. Each snapshot of the test data is 5 seconds long and was acquired at a rate of 100 kHz. A raw tachometer signal synchronized to the revolution of the planetary gear carrier plate and the main rotor is also included. The signal indicates the start of an individual

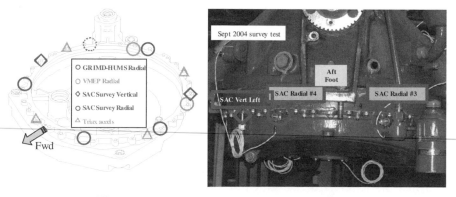

Figure 4.37 Ring-gear accelerometer locations.

revolution. The rotational frequency of the main rotor is slightly more than 4 Hz (it also varies a little over time), and therefore, the snapshots of the test data contain 20 complete revolutions. Primarily, the analysis results on the VMEP2 sensor signal are presented here. Figures 4.38 and 4.39 show one revolution of the VMEP2 raw data collected at GAG 260 of the test and the TSA data that resulted from ensemble averaging of the 20 revolutions of the raw data, respectively. Five peaks, caused by the five planets passing the sensor on the gearbox housing, are visible in the TSA data. The frequency spectrum of the TSA data is shown in Fig. 4.40, in which the scale along the x axis is the integer multiple of the shaft frequency. The figure shows that significant frequency components are around the meshing frequency ($228 \times f_{sh}$, with 228 being the tooth number of the ring gear, and f_{sh} being the rotational frequency of the planet carrier) and the harmonics up to the seventh.

Figure 4.41 shows the spectrum content around the fundamental meshing frequency. We can observe that every five indices apart there is an apparent component, and the dominant frequency is not the meshing frequency (228) but occurs at 230. The principal component of the spectrum is slightly removed from the gear meshing frequency; that is, the sidebands are nontrivial and asymmetric about the meshing frequency. MacFadden (1985) gave an explanation of this asymmetry, suggesting that it is caused by the superposition of vibrations with different phase angles produced by the planets at

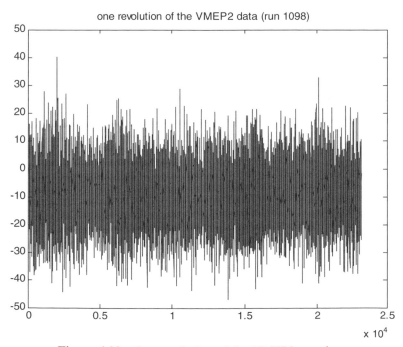

Figure 4.38   One revolution of the VMEP2 raw data.

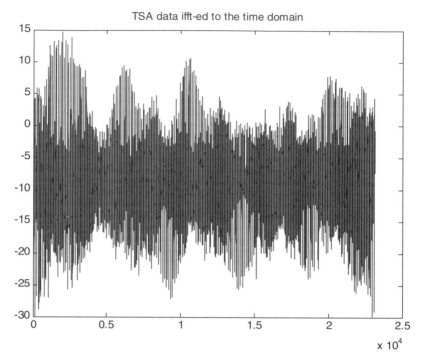

Figure 4.39 TSA data of the VMEP2 data.

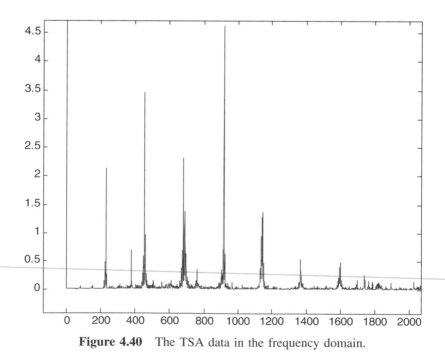

Figure 4.40 The TSA data in the frequency domain.

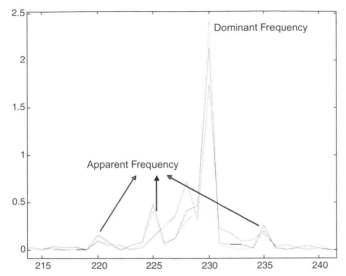

Figure 4.41 The spectrum around the fundamental meshing frequency.

different positions, although the motion of a single planetary gear past an accelerometer fixed on the housing produces symmetric sidebands about the tooth meshing frequency. Keller (2003) gave a good summary of the explanation and concluded that only terms with frequencies $(mN_t \pm n)$ that are multiples of the number of planets N_p appear for a healthy gear system and are measured by a sensor in the fixed frame, where m is the harmonic number, N_t is the tooth number (228 in this case), and n is an integer number. For the fundamental meshing frequency, $m = 1$, so the apparent frequencies occur at 225, 230, 235, etc.

In Fig. 4.41, three spectra for three sets of vibration data are plotted, with the black solid line for data at GAG 9, the dash dotted line for GAG 260, and the dashed line for GAG 639. As the test progresses, the crack size is supposed to grow from its initial value of 1.4 inch. The magnitude of the dominant frequency decreasing and the other apparent frequencies and magnitude increasing may be a good sign of crack growth and are quantified as features for fault diagnosis and prognosis.

Feature 1: Harmonic Ratio. For an epicyclic gearbox, the *regular meshing components* (RMCs) are defined as the dominant frequency and the other apparent frequencies around a specific gear meshing frequency. The *harmonic ratio* then is the ratio between the non-RMCs and the RMCs at a particular meshing harmonic. The overall harmonic ratio is the average of the feature for all harmonics. Figure 4.42 shows the feature values for the fundamental meshing frequency and harmonics 2, 3, 4, and 5 from top to bottom, respectively. Figure 4.43 shows the overall harmonic ratio as a function of the GAG

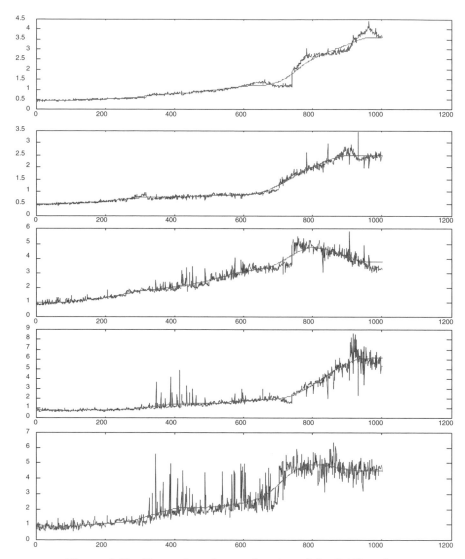

Figure 4.42   Harmonic ratio as a function of the GAG cycles.

cycles for the whole duration of the test. In the figures, the lighter shaded curves are the low pass filtered values of the features in order to describe the trend in feature change as the test progresses. The feature values are evaluated for vibrations acquired when the gear system is under the 100 percent load for GAGs 1 to 320 and 93 percent load for the rest.

Feature 2: Sideband Index. Here, *sideband index* (SI) is defined as the average of the first-order sidebands of the fundamental dominant frequency component (Keller, 2003) and is given by

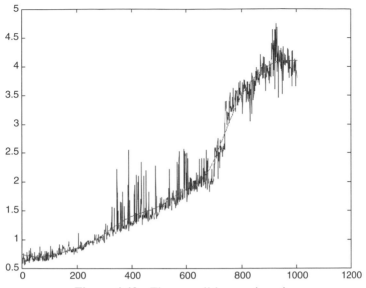

Figure 4.43 The overall harmonic ratio.

$$SI^e = \frac{RMC_{1,n_{\text{dominant}}-1}(x) + RMC_{1,n_{\text{dominant}}+1}(x)}{2}$$

where x is the TSA data of the vibration signal. Figure 4.44 shows the feature value.

Feature 3: Sideband Level Factor. The *sideband level factor* (SLF) is defined as the sum of the first-order sideband amplitudes about the fundamental dominant meshing component normalized by the standard deviation of the time signal average (Keller, 2003). Figure 4.45 shows the feature value as the crack grows:

$$SLF^e = \frac{RMC_{1,n_{\text{dominant}}-1}(x) + RMC_{1,n_{\text{dominant}}+1}(x)}{RMS(x)}$$

Feature 4: Energy Ratio. For an epicyclic gearbox, the *energy ratio* (ER) is defined as the standard deviation of the difference signal normalized by the standard deviation of the regular signal:

$$ER^e = \frac{RMS(d^e)}{RMS(r^e)}$$

where the regular signal is the inverse Fourier transform of all regular meshing components (RMCs) as defined in feature 1, harmonic ratio, that is,

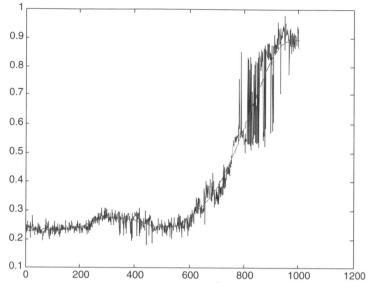

Figure 4.44   Sideband index as a function of the GAG cycles.

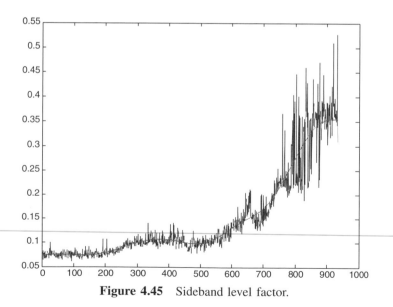

Figure 4.45   Sideband level factor.

$$r^e = F^{-1} \, (\text{RMC})$$

and the difference signal is defined as the time signal excluding the RMCs, as given by

$$d^e = x - r^e$$

Figure 4.46 shows the feature value of the energy ratio for the whole test duration.

Feature 5: Zero-Order Figure of Merit. The *zero-order figure of merit* (FM0) is defined as the peak-to-peak value of the time-average signal normalized by the sum of the amplitudes of the regular meshing frequencies for the whole spectrum (Fig. 4.47), as given by

$$\text{FMO}^e = \frac{\max(x) - \min(x)}{\text{RMC}}$$

Feature 6: Variance of the Energy Ratio (VER) during One Revolution. VER is defined to describe the inconsistency of the data characteristics during one shaft revolution, which is expected because the crack is in the gear plate at the foot of post 4, and the gear plate becomes asymmetric. Figure 4.39 shows

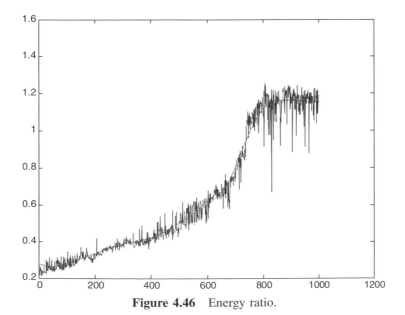

Figure 4.46 Energy ratio.

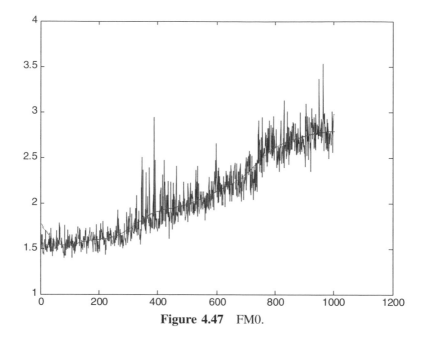

Figure 4.47 FM0.

that the TSA data segment (which is one revolution long) has nonuniform magnitude as the five planets pass by the accelerometer fixed on the gearbox's housing. This nonuniformity may increase as the crack size becomes bigger. To calculate VER, the regular signal and the difference signal are both divided into 100 segments, and the energy ratio for each segment is calculated. VER is the standard deviation of the 100 data points of the energy ratio. Figure 4.48 is the VER feature value from GAG 1 to GAG 1000 as the gear-plate crack grows from its initial value of 1.4 inch.

4.6 REAL-TIME IMAGE FEATURE EXTRACTION AND DEFECT/FAULT CLASSIFICATION

Feature extraction is also an important step in real-time product inspection and quality control based on machine vision technologies in a variety of application domains. We will introduce briefly a general scheme for image processing and defect/fault classification. Technical publications—books, journal articles, conference presentations, etc.—cover the topic extensively from both a generic and an application-specific standpoint (Boyer and Oz-guner, 2001; Corke, 2006; Liu et al., 1997; Torheim et al., 2001; Todorovic and Nechyba, 2004). Researchers at the Georgia Institute Technology have developed and applied machine vision methodologies successfully for online inspection and defect/fault recognition of industrial products (Ding et al., 2002). Of interest in the CBM/PHM application domain are, of course, those

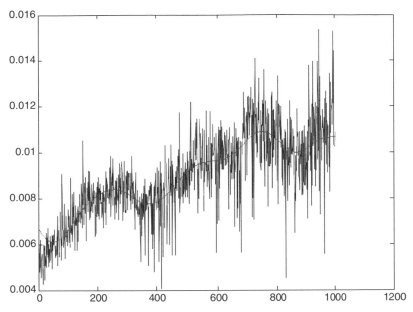

Figure 4.48   Variance of the energy ratio.

imaging methodologies that can produce reasonable classification results on-line in almost real time if an incipient failure (hot spots, corrosion precursors, etc.) is to be detected automatically and in a timely manner. Similar analysis approaches can be applied to offline inspection.

Figure 4.49 shows the proposed scheme. A color or gray-scale image is preprocessed and segmented to detect suspected defects or faults of targeted components. Then the features of each suspected target are extracted and fed into the classifier for identification. The segmentation and classification processes are repeated until a conclusion is reached with a prespecified level of confidence.

4.6.1   Image Preprocessing

The objective of this step is to find suspected defect targets by using image processing techniques. Image enhancement for further processing typically is required for images corrupted with noise and clutter. Preprocessing tools

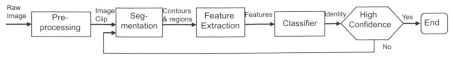

Figure 4.49   The scheme of the vision-based inspection and fault classification algorithm.

based on image segmentation have found great utility in a variety of applications (Charroux et al., 1996; Kass et al., 1988; Stewart et al., 2000). For example, a segmentation technique based on active contours ("snakes") offers the possibility for a fast and fully automated segmentation method that is implemented using level-set theory (Yezzi et al., 2000). It is highly optimized for real-time segmentation of images of reasonable size by using region-based curve-evolution equations and statistics in a natural way for bimodal images (Ding et al., 2002). Since the forces that drive the curve evolution depend on the statistical information of each region instead of absolute thresholded intensities, this approach is robust to intensity variations across the image. Moreover, it is fully global in that the evolution of each contour point depends on all the image data. To further expedite the process, measures can be taken to decrease the amount of data and provide a better initial estimate for the segmentation algorithm.

Figure 4.50 shows an example of the segmentation process using the coupled curve-evolution method on a fan-bone image for poultry product defect inspection. The scheme is applied online in a poultry processing plant and is intended to assist the operator in identifying those parts that do not meet strict bone-free requirements. This algorithm is efficient and accurate when the fan bone has a good contrast with respect to its neighboring tissue. The example also illustrates the utility of image-processing tools to online defect inspection of industrial and food-processing products. Similar techniques, of course, can be applied to machinery defects or faults.

4.6.2 Feature Extraction and Classification

The feature extraction and classification task identifies defect regions from the extracted binary data that result from the contours obtained by the segmentation routine. Candidate features that can be tested for the region classification task include global features, color features, and shape features:

| (a) | (b) | (c) |

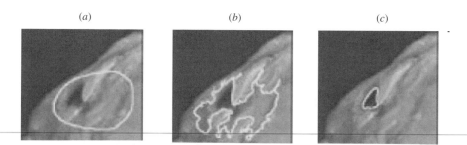

Figure 4.50 Evolution of the "snake" with an arbitrary initial contour: (a) initial contour; (b) $n = 150$; (c) $n = 850$.

- The global features of a region are computed from the position and contrast between the region and its neighbors.
- The color features of a region include the absolute intensities in red-green-blue (RGB) space, hue-intensity-saturation (HIS) space, and their relative relationships.
- The shape features can be boundary- or region-based, where among the former are the radius (mean and standard deviation), curvature, perimeter, and Fourier descriptors and among the latter are aspect ratio, area, and moments.

To select appropriate features for the classification task, metrics such as separability, robustness to segmentation inaccuracy, and robustness to color variations can be considered and employed. Using the poultry inspection as an example, the regions from the segmentation task can be classified into three classes: fan bone, shadow, and edge, which are the dark regions caused by fan bone, dark meat, and transition regions from meat to the background, respectively. Only when all the regions for one part are claimed as non-fan bone (shadow or edge), this part is claimed as fan bone-free.

4.6.3 Recursive Segmentation and Classification

Although the snake algorithm is robust and fast, it sometimes fails to capture the exact defect boundary when other features exist in the image. Figure 4.51 is a typical example of segmentation inaccuracy, where undersegmentation occurs around the defect region. Because the defect is very close to the transition region, whose color is dark and similar to that of the defect region, the defect region of interest is connected to the transition region. This behavior totally changes the shape of the binary region and therefore causes misclassification.

In order to improve the segmentation accuracy, a recursive segmentation and classification scheme is adopted. In this scheme, the confidence level of classification is used to evaluate the performance of the algorithm. This is

Figure 4.51 An unsuccessful example of the snake routine.

based on the fact that if the segmentation is accurate, the feature vectors in the feature space will be separable and, therefore, can be classified with high confidence. On the other hand, segmentation errors (under- or oversegmentation) tend to distort the shape of regions and change the intensity distribution, so the feature vector is moved further away from its cluster, and the identity of the region is difficult to be determined. In this case, further segmentation is needed to improve accuracy. This segmentation and classification process is repeated until a confident conclusion is reached.

Figure 4.52 is the result of reapplying the snake algorithm to the undersegmented region of Fig. 4.51 using a different initial contour estimate. By taking the undersegmented clip as a bimodal image, the snake algorithm effectively separates the defect from the transition region.

4.7 THE VIRTUAL SENSOR

It is often true that machine or component faults are not accessible directly for monitoring their growth behavioral patterns. The sensors that yield the required information directly simply may not exist, or they may be too expensive, or the desirable sensor location may be inaccessible. Still, the variables or states to be monitored are crucial to the diagnostic task. Consider, for example, the case of a bearing fault. No sensors exist that give direct measurement of the crack dimensions when a bearing is in an operational state. That is, there is no such device as a "fault meter" capable of providing direct measurements of fault evolution. Examples of a similar nature abound. In such situations, one must use some digital signal-processing or decision-making techniques that read the available sensor outputs and then provide information about the quantity or quality of interest. The combination of the

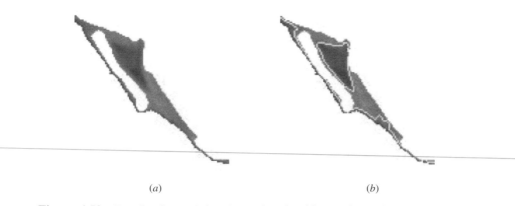

(a)   (b)

Figure 4.52 Result of reapplying the snake algorithm to the undersegmented region: (a) raw clip; (b) final contour.

available sensors and the DSP routines is known as a *virtual sensor* (or *soft sensor* or *analytic sensor*). There are many types of virtual sensors. A typical example of a virtual sensor is the Kalman filter, to be discussed later.

We will illustrate the notion of an analytic measurement via a simple example: Consider fluid flowing through an orifice, as shown in Fig. 4.53. The flow rate Q is a nonlinear function of the pressure drop $(P_1 - P_2)$, as shown below:

$$Q = K(P_1 - P_2)^{1/2}$$

where K is a constant. We can think of this equation as a "local" model describing the relation between two variables Q and $\Delta P = (P_1 - P_2)$. Consider further that ΔP is a measurable (and validated) quantity. If it is impractical to measure Q (i.e., a flowmeter cannot be inserted in the flow path), then Q may be estimated (inferred) from this "local" model. We call Q an *analytic measurement*.

The implication of this notional example, therefore, for diagnostic purposes is to use available sensor data to map known measurements to a "fault measurement." The virtual sensor can be realized as a neural network or any other mapping construct. Two potential problem areas must be considered. First, laboratory or constructed experiments are required to train the neural network; second, the physical measurement, that is, the pressure drop across the orifice in our example, must be a validated measurement.

Sensor-data validation presents a major challenge in terms of ensuring that the measurements are correct within bounds. Formal techniques, such as parity relations, have been proposed for sensor-data validation. A parity relation is formulated as the residual between measurements derived from different sources. A validated estimate of the measured variable is formed using a consistent subset of its input measurements (Fig. 4.54). Using a parity-space technique, we can determine failure(s) of input measurements by employing measurement inconsistency and excluding failed measurements from subsequent consideration. It is well known that three measurements are required to determine whether a sensor is faulty or not. A voting scheme typically is invoked to suggest which one of the three measurements is faulty.

As an example of a virtual sensor, Marko and colleagues (1996) report the development of a neural-network-based virtual or ideal sensor used to diag-

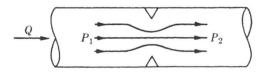

Figure 4.53 Fluid flow through an orifice.

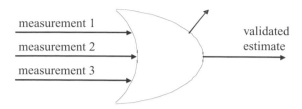

Figure 4.54   Sensor-data validation.

nose engine combustion failure, known as *misfire detection*. Their technique employs a recurrent neural net as the classifier that takes such measurable inputs as crankshaft acceleration, engine speed, engine load, and engine ID and produces a misfire diagnostic evaluation as the output. The same concept may be exploited to design a virtual sensor that takes as inputs measurable quantities or features and outputs the time evolution of the fault pattern. In general, a virtual sensor may be implemented as a neural network, a look-up table, a fuzzy-logic expert system, or other similar mapping tools. A schematic representation of a neural network as a virtual sensor is shown in Fig. 4.55.

4.8   FUSION OR INTEGRATION TECHNOLOGIES

Within an automated health management system, there are many areas where fusion technologies play a contributing role. These areas are shown in Fig. 4.56. At the lowest level, data fusion can be used to combine information from a multisensor data array to validate signals and create features. One

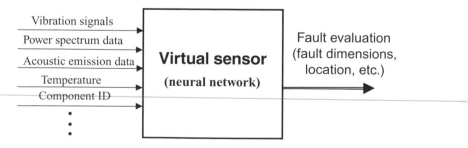

Figure 4.55   A schematic representation of a neural network as a virtual sensor.

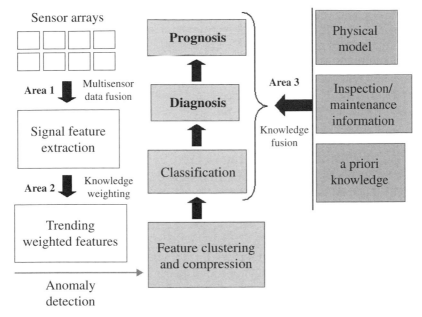

Figure 4.56 Fusion application areas.

example of data fusion is combining a speed signal and a vibration signal to achieve time-synchronous averaged vibration features.

At a higher level (area 2), fusion may be used to combine features in intelligent ways so as to obtain the best possible diagnostic information. This would be the case if a feature related to particle count and size in a bearing's lubrication oil were fused with a vibration feature such as kurtosis. The combined result would yield an improved level of confidence about the bearing's health. Finally, knowledge fusion (area 3) is used to incorporate experience-based information such as legacy failure rates or physical model predictions with signal-based information.

However, one of the main concerns in any fusion technique is the danger of producing a fused system result that actually performs worse than the best individual tool. This is so because poor estimates can serve to drag down the better ones. The solution to this concern is to weigh a priori the tools according to their capability and performance. The degree of a priori knowledge is a function of the inherent understanding of the physical system and the practical experience gained from its operating history. The ideal knowledge fusion process for a given application should be selected based on the characteristics of the a priori system information.

4.8.1 Fusion Architectures

Identifying the optimal fusion architecture and approach at each level is a vital factor in ensuring that the realized system truly enhances health monitoring capabilities. A brief explanation of some of the more common fusion architectures will be given next.

The centralized fusion architecture fuses multisensor data while it is still in its raw form, as shown in Fig. 4.57. At the center of this architecture, the data are aligned and correlated during the first stage. This means that the competitive or collaborative nature of the data is evaluated and acted on immediately. Theoretically, this is the most accurate way to fuse data; however, it has the disadvantage of forcing the fusion processor to manipulate a large amount of data. This is often impractical for real-time systems with a relatively large sensor network (Hall and Llinas, 1997).

The autonomous fusion architecture, shown in Fig. 4.58, quells most of the data management problems by placing feature extraction before the fusion process. The creation of features prior to the actual fusion process provides the significant advantage of reducing the dimensionality of the information to be processed. The main undesirable effect of pure autonomous fusion architectures is that the feature fusion may not be as accurate as in the case of raw data fusion because a significant portion of the raw signal has been eliminated.

Hybrid fusion architectures take the best of both approaches and are considered often the most practical because raw data and extracted features can be fused in addition to the ability to "tap" into the raw data if required by the fusion center (Fig. 4.59).

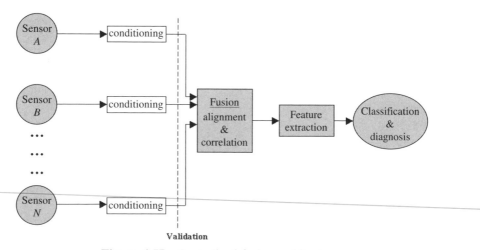

Figure 4.57 Centralized fusion architecture.

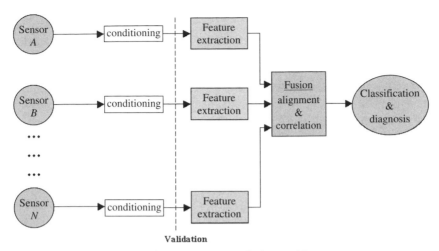

Figure 4.58 Autonomous fusion architecture.

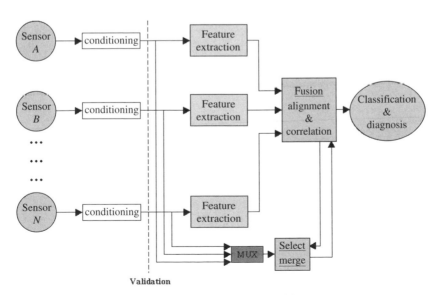

Figure 4.59 Hybrid fusion architecture.

4.8.2   Fusion Techniques

Various techniques are available for performing data, feature, or knowledge fusion. Discovering which technique is best suited for a particular application can be a daunting task. In addition, there are no hard and fast rules about what fusion technique or architecture works best for any particular application. The following subsections will describe some common fusion approaches such as Bayesian inference, Dempster-Shafer combination, weighting/voting, neural network fusion, and fuzzy-logic inference. In addition, a set of metrics will be discussed for independently judging the performance and effectiveness of the fusion techniques within a diagnostic system.

Bayesian Inference.   Bayesian inference can be used to determine the probability that a diagnosis is correct, given a piece of a priori information. Analytically, this process is described as follows:

$$P(f_1|O_n) = \frac{P(O_n|f_1) \cdot P(f_1)}{\displaystyle\sum_{j=1}^{n} P(O_n|f_j) \cdot P(f_j)}$$

where $P(f_1|O_n)$ is the probability of fault f given a diagnostic output O, $P(O_n|f_1)$ is the probability that a diagnostic output O is associated with a fault f, and $P(f_1)$ is the probability of the fault f occurring.

Bayes' theorem is only able to analyze discrete values of confidence from a diagnostic classifier (i.e., it observes it or it does not). Hence a modified method has been implemented that uses three different sources of information: an a priori probability of failure at time t [$P_{FO(t)}$], the probability of failure as determined from the diagnostic classifier data [$C_{D(i,t)}$], and feature reliability that is independent of time [$R_{D(i)}$]. Care must be taken to prevent division by zero. Thus

$$P_{f(t)} = \frac{\displaystyle\sum_{i=1}^{N} C_{D(i,t)}}{P_{FO(t)} \cdot \displaystyle\sum_{i=1}^{N} R_{D(i)}}$$

The Bayesian process is a common and well-established fusion technique, but it also has some disadvantages. The knowledge required to generate the a priori probability distributions may not always be available, and instabilities in the process can occur if conflicting data are presented or the number of unknown propositions is large compared with the known ones.

Dempster-Shafer Fusion.   The Dempster-Shafer method addresses some of the problems discussed earlier and specifically tackles the a priori probability issue by keeping track of an explicit probabilistic measure of the lack of information. The disadvantage of this method is that the process can become impractical for time-critical operations in large fusion problems. Hence the proper choice of a preferred method should be based on the specific diagnostic/prognostic issues that are to be addressed.

In the Dempster-Shafer approach, uncertainty in the conditional probability is considered. Each input feature has fuzzy membership functions associated with it representing the possibility of a fault mode. Each feature, therefore, represents an "expert" in this setting. These possibility values are converted to basic probability assignments for each feature. Dempster's rule of combination then is used to assimilate the evidence contained in the mass functions and to determine the resulting degree of certainty for detected fault modes. Dempster-Shafer theory is a formalism for reasoning under uncertainty (Shafer, 1976). It can be viewed as a generalization of probability theory with special advantages in its treatment of ambiguous data and the ignorance arising from them. Chapter 5 introduces a few basic concepts from Dempster-Shafer theory.

Weighting/Voting Fusion. Both the Bayesian and Dempster-Shafer techniques can be computationally intensive for real-time applications. A simple weighted-average or voting scheme is another approach to the fusion problem. In both these approaches, weights are assigned based on a priori knowledge of the accuracy of the diagnostic/prognostic techniques being used. The only condition is that the sum of the weights must be equal to 1. Each confidence value then is multiplied by its respective weight, and the results are summed for each moment in time. That is,

$$P(F) = \sum_{n=1}^{i} C_{(i,t)} \cdot W_{(i)}$$

where i is the number of features, C is the confidence value, and W is the weight value for that feature. Although simple in implementation, choosing proper weights is of critical importance to getting good results. Poor weighting often can result in a worse fused output than using the best individual outcome.

Fuzzy-Logic Inference Fusion. Fuzzy-logic inference is a fusion technique that uses the membership-function approach to scale and combine specific input quantities to yield a fused output. The basis for the combined output comes from scaling the membership functions based on a set of rules developed in a rule base. Once this scaling is accomplished, the membership functions are combined by one of various methodologies such as summation, maximum, or a "single best" technique. Finally, the scaled and combined membership functions are used to calculate the fused output by taking either the centroid, the maximum height, or the midpoint of the combined function.

An example of a feature fusion process using fuzzy logic is given in Fig. 4.60. In this example, features from an image are being combined to help determine if a "foreign" object is present in an original image. Image features such as tonal mean, midtones, kurtosis, and many others are combined to give a single output that ranks the probability of an anomalous feature being present in the image.

Neural-Network Fusion. A well-accepted application of artificial neural networks (ANNs) is data and feature fusion. The ability of ANNs to combine information in real time with the added capability of autonomous relearning (if necessary) makes them very attractive for many fusion applications. Section 5.3 in Chap. 5 gives a description of ANN fundamentals.

Training a neural network for a fusion application involves the process of adjusting the weights and evaluating the activation functions of the numerous interconnections between the input and output layers. Two fundamental types of learning methods are used for feature fusion applications: unsupervised and supervised. In the unsupervised method, learning is autonomous; that is, networks discover properties of the data set and learn to reflect those properties in their output. This method is used to group similar input patterns to facilitate processing of a large number of training data. In the supervised method, a "teacher" is present during the training phase that tells the network how well it is performing or what the correct behavior should be; that is, it is used for specifying what target outputs should result from an input pattern.

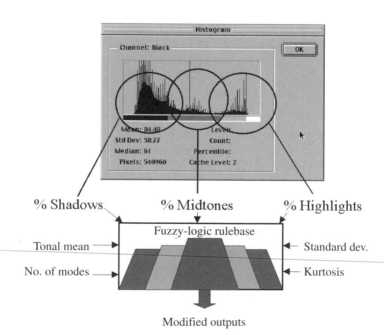

Figure 4.60 Example of fuzzy-logic inference.

A representative application of neural-network fusion would be to combine individual features from different feature-extraction algorithms to arrive at a single representative feature. An example of this type of neural network fusion will be given in the next subsection.

4.8.3 Example Fusion Results

The fusion techniques previously described have been implemented on various vibration data sets acquired during a series of transitional run-to-failure experiments on an industrial gearbox. In these tests, the torque was cycled from 100 to 300 percent load starting at approximately 93 hours. The drive gear experienced multiple broken teeth, and the test was stopped at approximately 114 hours. The data collected during the test were processed by many feature-extraction algorithmic techniques that resulted in 26 vibration features calculated from a single accelerometer attached to the gearbox housing. The features ranged in complexity from a simple RMS level to measures of the residual signal (gear mesh and sidebands removed) from the time-synchronous-averaged waveform. More information on these vibration features may be found in Byington and Kozlowski (1997). Figures 4.61 and 4.62 show plots of two of these features, kurtosis and NA4, respectively. (To get the feature of NA4, the residual signal r is first constructed. The quasi-normalized kurtosis of the residual signal then is computed by dividing the fourth moment of the residual signal by the square of its run-time average variance. The run-time average variance is the average of the residual signal over each time signal in the run ensemble up to the point at which NA4 is currently being calculated. NA4 is nondimensional and designed to have a nominal value of 3 if the residual signal is purely Gaussian.) The second line in each of these plots is the *ground-truth severity,* or the probability of failure as determined from visual inspections (discussed next).

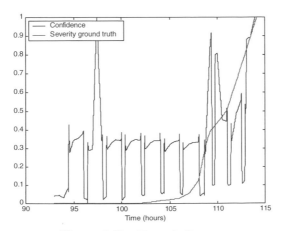

Figure 4.61   Kurtosis feature.

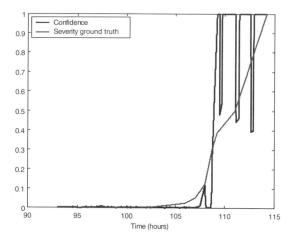

Figure 4.62 NA4 feature.

Borescopic inspections of the pinion and drive gear for this particular test run were performed to bound the time period in which the gear had experienced no damage and when a single tooth had failed. These inspection results, coupled with the evidence of which features were best at detecting tooth cracking prior to failure features (as determined from the diagnostic metrics discussed later), constituted the a priori information used to implement the Bayesian inference, weighting/voting, neural-network, and Dempster-Shafer fusion processes.

The seven best vibration features, as determined by a consistent set of metrics described in the next section, were assigned weights of 0.9, average performing features were weighted 0.7, and low performers 0.5 for use in the voting scheme. These weights are directly related to the feature reliability in the Baysian inference fusion. Similarly, the best features were assigned the uncertainty values of (0.05), average performers (0.10), and low performers (0.15) for the Dempster-Shafer combination. The prior probability of failure required for the neural-network, Bayesian inference, and Dempster-Shafer fusions was built on the experiential evidence that a drive-gear crack will form in a mean time of 108 hours with a variance of 2 hours.

The results of the Dempster-Shafer, Bayesian, and weighted fusion techniques on all 26 features are shown in Fig. 4.63. All three approaches exhibit an increase in their probability of failure estimates at around 108 hours (index 269). Clearly, the voting fusion is most susceptible to false alarms; the Bayesian inference suggests that the probability of failure increases early on but is not capable of producing a high confidence level. Finally, the Dempster-Shafer combination provides the same early detection, achieves a higher confidence level, but is more sensitive throughout the failure transition region.

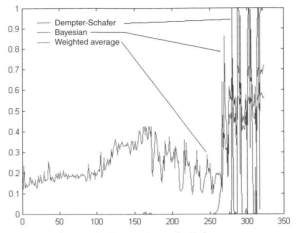

Figure 4.63 Fusion of all features.

Next, the same fusion algorithms were applied to just the best seven features, as defined by the metrics. It can be observed that the fusion of these seven features produces more accurate and stable results, as shown in Fig. 4.64. Note that the Dempster-Shafer combination can now retain a high confidence level with enhanced robustness throughout the critical failure transition region.

Finally, a simple backpropagation neural network was trained on four of the top seven features previously fused (RMS, kurtosis, NA4, and M8A). In order to train this supervised neural network, the probability of failure as defined by the ground truth was required as a priori information, as described

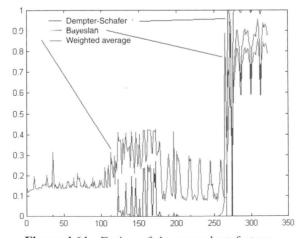

Figure 4.64 Fusion of the seven best features.

earlier. The network automatically adjusts its weights and thresholds (not to be confused with the feature weights) based on the relationships it sees between the probability-of-failure curve and the correlated feature magnitudes. Figure 4.65 shows the results of the neural network after being trained by these data sets. The difference between the neural-network output and the ground-truth probability-of-failure curve is due to errors that still exist after the network parameters have been optimized to minimize these errors. Once trained, the neural-network fusion architecture can be used to intelligently fuse these same features for a different test under similar operating conditions.

4.9   USAGE-PATTERN TRACKING

In addition to sensor monitoring, CBM/PHM systems for critical assets such as aircraft, vehicles, gas turbines, etc. must possess the capability to track their usage patterns. For example, in the aircraft case, regime recognition, including both flight maneuvering and automated gross weight calculations, is necessary to effectively estimate useful life remaining of life-limited components. In addition to individual usage tracking, the current life limits of components potentially can be extended by analysis of the severity of actual aircraft history versus the analytic model. Moreover, flight maneuvering must be identified with criteria consistent with load surveys. Gross weight (GW) and distribution of weight (such as center of gravity) are probably the most difficult usage variables to measure automatically. Schemes to deduce GW from flight parameters are available. For aerospace, transportation systems, and many industrial processes, we are interested in an estimate of loading

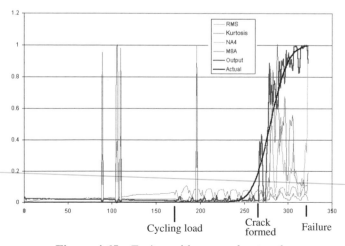

Figure 4.65   Fusion with a neural network.

profiles that become essential inputs to fatigue damage models for prognosis of the remaining useful life. Such loading profiles are expressed as functions of ground-air-ground (GAG) cycles or cycle time or simply time the system is in an operational state. Archived usage data may be used to derive probabilistic estimates of damage accumulation from the system's "birth" to its current state. The latter form the initial damage distribution for prognosis once the fault has been detected and isolated.

Archived usage data may be used to predict time to failure or product yield in manufacturing and industrial processes. Rietman and Beacky (1998) describe the application of intelligent data-processing tools to predict failures in a plasma reactor for etching semiconductor wafers. They used continuous streams of usage data on one reactor collected over a period of 3.5 years to monitor fluctuations in the standard deviation of the flow rate of a gas to the reactor in order to predict using a neural-network construct a pressure failure that could disable the reactor. They reported a prediction of 1 hour before the actual failure. In later work (Rietman et al., 2001), they scaled up their efforts for prediction of failures in manufacturing yield by monitoring the behavior of 23 processing steps (200 processing tools) simultaneously with independent and combined neural networks. Pareto charts were generated automatically and sent to the yield-analysis engineers, who were able to select the most likely processing step that would have adverse effect on yield.

4.10   DATABASE MANAGEMENT METHODS

Central to a reliable, efficient, and robust diagnostic and prognostic architecture is a well-designed database and a flexible database management schema. The OSA-CBM and MIMOSA efforts aimed at establishing database standards for the CBM/PHM arena are described in Sec. 5.1 of Chap. 5. The database must be capable of storing a variety of sensor data preferably in a client-server relational database configuration. The relational database management system facilitates data representation, data conversion, and data transfer for use with feature extraction and diagnostic and prognostic algorithms. Figure 4.66 depicts a conceptual schematic of the function of the data

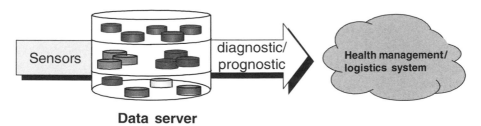

Data server

Figure 4.66   Data server as a component of the CBM/PHM architecture.

server in the CBM/PHM architecture, whereas Fig. 4.67 shows a sample core database scheme.

The data server must provide storage for setup information, collected dynamic (time domain, phase, and spectral) data, static (process) data, and analysis results. A secondary database may extend the functionality of the main database by including tables that define schedules and schedule components, the multiplexer input mapping, sensor calibration information, mapping to the main database, and mapping for reporting results to the graphical user interface (GUI).

- Data access interfaces allow applications to access persistent data from externally managed sources via Structured Query Language (SQL).
- Data Access Objects (DAO) are used to access desktop databases.
- Remote Data Objects (RDO) use Open Database Connectivity (ODBC) architecture for accessing client-server databases.
- ActiveX Data Objects (ADO) are designed as a high-performance, easy-to-use interface. It combines the bet feature of and replaces DAO and RDO

A relational database management system (RDMS) provides the following functions:

- Organizes large amounts of data in linked tables to make the design easy to understand
- Provides a relationally complete language for data definition, retrieval, and update

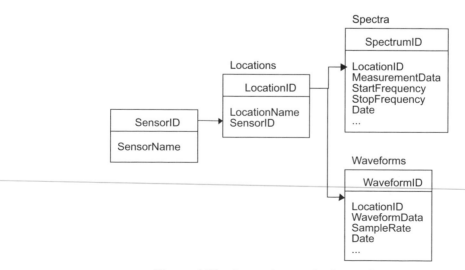

Figure 4.67   A sample core database scheme.

- Provides data integrity rules that define consistent database states to improve data reliability
- Examples of RDBMS include Oracle, Microsoft SQL Server, and desktop database applications, such as Microsoft Access

SQL is a well-defined, standard language used for query, updating, and managing relational databases. By using SQL, an application can ask the database to perform tasks rather than requiring application code to perform them. Effective use of SQL can minimize the amount of data sent across the network. Finally, the SQL SELECT statement returns information from the database as a set of selected rows. A typical example is

```
Select MeasurementsDate, MeasurementValue from
ProcessReadings
Where PointID = & PointID & and MeasurementDate > &
LastDate
```

4.11 REFERENCES

S. A. Adewusi and B. O. Al-Bedoor, "Wavelet Analysis of Vibration Signals of an Overhang Rotor with a Propagating Transverse Crack," *Journal of Sound and Vibration* **246**:777–793, 2001.

F. Akbaryan and P. R. Bishnoi, "Fault Diagnostics of Single-Variate Systems Using a Wavelet-Based Pattern Recognition Technique," *Industrial and Engineering Chemistry Research* **40**:3612–3622, 2001.

J. Altmann and J. Mathew, "Multiple Band-Pass Autoregressive Demodulation for Rolling-Element Bearing Fault Diagnostics," *Mechanical Systems and Signal Processing* **15**:963–977, 2001.

F. Aminian and M. Aminian, "Fault Diagnostics of Analog Circuits using Bayesian Neural Networks with Wavelet Transform as Preprocessor," *Journal of Electronic Testing: Theory and Applications (JETTA)* **17**:29–36, 2001.

C. Bieman, W. J. Staszewski, C. Boller, and G. R. Tomlinson, "Crack Detection in Metallic Structures Using Piezoceramic Sensors," *Key Engineering Materials* **167**: 112–121, 1999.

D. M. Blunt and W. J. Hardman, "Detection of SH-60 Helicopter Main Transmission Planet Gear Fault Using Vibration Analysis," Report No. NAWCADPAX/RTR-2000/123, Naval Air Warfare Center Aircraft Division, Patuxent River, MD, 2000.

K. L. Boyer and T. Ozguner, "Robust Online Detection of Pipeline Corrosion from Range Data," *Machine Vision and Applications* **12**:291–304, 2001.

C. S. Byington and J. D. Kozlowski, "Transitional Data for Estimation of Gearbox Remaining Life," *Proceedings of the 51st Meeting of the MFPT,* April 1997, Virginia Beach, VA.

B. Charroux, S. Phillipp, and J.-P. Cocquerez, "Image Analysis: Segmentation Operator Cooperation Led by the Interpretation," *Proceedings of the International Conference on Image Processing,* Vol. 3. 1996, pp. 939–942, Lausanne, Switzerland.

D. Chen and W. J. Wang, "Classification of Wavelet Map Patterns Using Multilayer Neural Networks for Gear Fault Detection," *Mechanical Systems and Signal Processing* **16**:695–704, 2002.

Z. W. Chen, "Research on Fault Diagnostics of the Electrohydraulic Servo-System," *China Mechanical Engineering* **11**:544–546, 2000 (in Chinese).

Z. W. Chen and Y. X. Lu, "Signal Singularity Detection and Its Application," *Journal of Vibration Engineering* **10**:148–155, 1997 (in Chinese).

P. Corke, "The Machine Vision Toolbox," *IEEE Robotics & Automation Magazine* (in press, 2006).

X. M. Deng, Q. Wang, and V. Giurgiutiu, "Structural Health Monitoring Using Active Sensors and Wavelet Transforms," *Proceedings of SPIE* **3668**:363–370, 1999.

Y. Ding, A. Yezzi, B. Heck, W. Daley, G. Vachtsevanos, and Y. Zhang, "An On-line Real-time Automatic Visual Inspection Algorithm for Surface Bone Detection in Poultry Products," *Second WSEAS International Conference on Signal, Speech, and Image Processing*, Koukounaries, Skiathos Island, Greece, September 25–28, 2002.

X. Z. Dong, Z. X. Geng, and Y. Z. Ge, "Application of Wavelet Transform in Power System Fault Signal Analysis," *Proceedings of the Chinese Society of Electrical Engineering* **17**:421–424, 1997 (in Chinese).

D. L. Donoho, "Denoising by Soft-Thresholding," *IEEE Transactions on Information Theory* **41**:613–627, 1995.

R. C. Eisenmann, Sr. and R. C. Eisenmann, Jr., *Machinery Malfunction Diagnosis and Correction.* Englewood Cliffs, NJ: Prentice-Hall, 1998.

S. J. Engel, B. J. Gilmartin, K. Bongort, and A. Hess, "Prognostics, the Real Issues Involved with Prediting Life Remaining," *Aerospace Conference Proceedings,* Vol. 6, March 2000, pp. 457–469, Big Sky, MT.

M. A. Essawy, S. Diwakar, S. Zein-Sabbato, and M. Bodruzzaman, "Helicopter Transmission Fault Diagnostics Using Neuro-Fuzzy Techniques," *Intelligent Engineering Systems Through Artificial Neural Networks* **7**:661–666, 1997.

D. Forrester, "A Method for the Separation of Epicyclic Planet Gear Vibration Signatures," *Proceedings of Acoustical and Vibratory Surveillance Methods and Diagnostic Techniques,* Senlis, France, 1998, pp. 539–548.

R. Frank, *Understanding Smart Sensors,* 2d ed. Norwood, MA: Artech House, 2000.

L. C. Freudinger, R. Lind, and M. J. Brenner, "Correlation Filtering of Modal Dynamics Using the Laplace Wavelet," *Proceedings of IMAC* **2**:868–877, 1998.

D. Hall and J. Llinas, "An Introduction to Multisensor Data Fusion," *Proceedings of the IEEE,* January 1997.

A. Hambaba and E. Huff, "Multiresolution Error Detection on Early Fatigue Cracks in Gears," *IEEE Aerospace Conference Proceedings* **6**:367–372, 2000.

Z. J. He and Y. Y. Zi, "Laplace Wavelet and Its Engineering Application," *Journal of Engineering Mathematics* **18**:87–92, 2001 (in Chinese).

S. J. Henkind and M. C. Harrison, "An Analysis of Four Uncertainty Calculi," *IEEE Transactions on Systems, Man, and Cybernetics* **18**(5): pp. 700–714, 1988.

I. M. Howard, "An Investigation of Vibration Signal Averaging of Individual Components in an Epicyclic Gearbox," Technical Report ARL-PROP-R-185, Australian Department of Defense, Aeronautical Research Laboratory, Melbourne, Australia, 1991.

S. S. Hu, C. Zhou, and W. L. Hu, "New Structure Fault Detection Method Based on Wavelet Singularity," *Journal of Applied Sciences* **18**:198–201, 2000 (in Chinese).

M. Kass, A. Witkin, and D. Terzopoulos, "Snakes: Active Contour Models," *Int. J. Computer Vision* **xx**:1, 4, 321–331, 1988.

J. Keller and P. Grabill, "Vibration Monitoring of a UH-60A Main Transmission Planetary Carrier Fault," American Helicopter Society 59th Annual Forum, Phoenix, Arizona, 2003.

S. Krishnan and R. M. Rangayyan, "Denoising Knee Joint Vibration Signals Using Adaptive Time-Frequency Representations," *Canadian Conference on Electrical and Computer Engineering* **3**:1495–1500, 1999.

F. L. Lewis, *Applied Optimal Control and Estimation: Digital Design and Implementation.* Englewood Cliffs, NJ: Prentice-Hall, Series, 1992.

J. Lin, H. X Liu, and L. S. Qu, "Singularity Detection Using Wavelet and Its Application in Mechanical Fault Diagnostics," *Signals Processing* **13**:182–188, 1997 (in Chinese).

J. Liu, Y. Y. Tang, and Y. C. Cao, "An Evolutionary Autonomous Agent's Approach to Image Feature Extraction," *IEEE Transactions on Evolutionary Computation* **1**(2):141-158, 1997.

J. K. Ma, Y. F. Liu, and Z. J. Zhu, "Study on Identification of Dynamic Characteristics of Oil-Bearings Based on Wavelet Analysis," *Chinese Journal of Mechanical Engineering* **36**:81–83, 2000 (in Chinese).

S. Mallat and W. L. Hwang, "Singularity Detection and Processing with Wavelets," *IEEE Transactions on Information Theory* **38**:617–643, 1992.

K. A. Marko, J. V. James, T. M. Feldkamp, C. V. Puskorius, J. A. Feldkamp, and D. Roller, "Applications of Neural Networks to the Construction of "Virtual" Sensors and Model-Based Diagnostics," *Proceedings of ISATA 29th International Symposium on Automotive Technology and Automation*, Florence, Italy, June 1996, pp. 133–138.

MATLAB Reference Guide. Cambridge, MA: MathWorks, Inc., 1994.

MATLAB Reference Guide. Cambridge, MA: MathWorks, Inc., 2004.

K. McClintic, M. Lebold, K. Maynard, C. Byington, and R. Campbell, "Residual and Difference Feature Analysis with Transitional Gearbox Data," *Proceedings of the 54th Meeting of the Society for Machinery Failure Prevention Technology,* Virginia Beach, VA, 2000, pp. 635–645.

P. D. McFadden, "Detecting Fatigue Cracks in Gears by Amplitude and Phase Demodulation of the Meshing Vibration," *Journal of Vibration, Acoustics, Stress, and Reliability in Design* **108**(2):165–170, 1986.

P. D. McFadden, "Examination of a Technique for the Early Detection of Failure in Gears by Signal Processing of the Time Domain Average of the Meshing Vibration," *Mechanical Systems and Signal Processing* **1**(2):173–183, 1987.

P. D. McFadden, "Interpolation Techniques for Time Domain Averaging of Gear Vibration," *Mechanical Systems and Signal Processing* **3**(1):87–97, 1989.

P. D. McFadden, "A Technique for Calculating the Time Domain Averages of the Vibration of the Individual Planet Gears and Sun Gear in an Epicyclic Gearbox," *Journal of Sound and Vibration* **144**(1):163–172, 1991.

P. D. McFadden, "Window Functions for the Calculation of the Time Domain Averages of the Vibration of the Individual Planet Gears and Sun Gear in an Epicyclic Gearbox," *Journal of Vibration and Acoustics* **116**:179–187, 1994.

P. D. McFadden and J. D. Smith, "An Explanation for the Asymmetry of the Modulation Sidebands about the Tooth Meshing Frequency in Epicyclic Gear Vibration," *Proceedings of the Institution of Mechanical Engineers, Part C: Mechanical Engineering Science,* **199**(1):65–70, 1985.

J. McNames, "Fourier Series Analysis of Epicyclic Gearbox Vibration," *Journal of Vibration and Acoustics* **124**(1):150–152, 2002.

S. Menon, J. N. Schoess, R. Hamza, and D. Busch, "Wavelet-Based Acoustic Emission Detection Method with Adaptive Thresholding," *Proceedings of SPIE* **3986**:71–77, 2000.

A. V. Oppenheim and R. W. Schafer, *Digital Signal Processing*. Englewood Cliffs, NJ: Prentice-Hall, 1975.

Z. K. Peng and F. L. Chu, "Application of the Wavelet Transformin Machine Condition Monitoring and Fault Diagnosis: A Review with Bibliography," *Mechanical Systems and Signal Processing* **18**:199–221, 2004.

Z. Peng, F. Chu, and Y. He, "Vibration Signal Analysis and Feature Extraction Based on Reassigned Wavelet Scalogram," *Journal of Sound and Vibration* **253**:1087–1100, 2002a.

Z. Peng, Y. He, Z. Chen, and F. Chu, "Identification of the Shaft Orbit for Rotating Machines Using Wavelet Modulus Maxima," *Mechanical Systems and Signal Processing* **16**:623–635, 2002b.

Z. Peng, Y. He, Q. Lu, and F. Chu, "Feature Extraction of the Rub-Impact Rotor System by Means of Wavelet," *Journal of Sound and Vibration* **259**:1000–1010, 2003.

J. Pineyro, A. Klempnow, and V. Lescano, "Effectiveness of New Spectral Tools in the Anomaly Detection of Rolling Element Bearings," *Journal of Alloys and Compounds* **310**:276–279, 2000.

V. V. Polyshchuk, F. K. Choy, and M. J. Braun, "Gear Fault Detection with Time-Frequency Based Parameter NP4," *International Journal of Rotating Machinery* **8**(1):57–70, 2002.

S. T. Quek, Q. Wang, L. Zhang, and K. K. Ang, "Sensitivity Analysis of Crack Detection in Beams by Wavelet Technique," *International Journal of Mechanical Sciences* **43**:2899–2910, 2001.

E. A. Rietman and M. Beachy, "A Study on Failure Prediction in a Plasma Reactor," *IEEE Transactions on Semiconductor Manufacture* **11**(4):670–680, 1998.

E. A. Rietman, S. A. Whitlock, M. Beachy, A. Roy, and T. L. Willingham, "A System Model for Feedback Control and Analysis of Yield: A Multistep Process Model of Effective Gate Length, Poly Line Width, and IV Parameters," *IEEE Transactions on Semiconductor Manufacture* **14**(1):32–47, 2001.

A. N. Robertson, K. C. Park, and K. F. Alvin, "Extraction of Impulse Response Data via Wavelet Transform for Structural System Identification," *American Society of Mechanical Engineers, Design Engineering Division (Publication)* **84**:1323–1334, 1995a.

A. N. Robertson, K. C. Park, and K. F. Alvin, "Identification of Structural Dynamics Models Using Wavelet-Generated Impulse Response Data," *American Society of*

Mechanical Engineers, Design Engineering Division (Publication) **84**:1335–1344, 1995b.

A. N. Robertson, K. C. Park, and K. F. Alvin, "Extraction of Impulse Response Data via Wavelet Transform for Structural System Identification," *Journal of Vibration and Acoustics, Transactions of the ASME* **120**:252–260, 1998a.

A. N. Robertson, K. C. Park, and K. F. Alvin, "Identification of Structural Dynamics Models using Wavelet-Generated Impulse Response Data," *Journal of Vibration and Acoustics, Transactions of the ASME* **120**:261–266, 1998b.

M. Ruzzene, A. Fasana, L. Garibaldi, and B. Piombo, "Natural Frequencies and Damping Identification Using Wavelet Transform: Application to Real Data," *Mechanical Systems and Signal Processing* **11**:207–218, 1997.

P. D. Samuel and D. J. Pines, "Vibration Separation and Diagnostics of Planetary GearTrains," *Proceedings of the American Helicopter Society 56th Annual Forum,* Virginia Beach, VA, 2000.

P. Samuel and D. Pines, "Planetary Gear Box Diagnostics using Adaptive Vibration Signal Representations: A Proposed Methodology," *Proceedings of the American Helicopter Society 57th Annual Forum,* Washington, DC, May 2001.

A. Saxena, B. Wu, and G. Vachtsevanos, "A Methodology for Analyzing Vibration Data from Planetary Gear Systems using Complex Morlet Wavelets," *Proceedings of American Control Conference (ACC 2005)*, Portland, OR, June 2005.

C. W. de Silva, Control Sensors and Actuators, Prentice Hall, Englewood Cliffs, New Jersey, 1989.

G. Shafer, *A Mathematical Theory of Evidence.* Princeton, NJ: Princeton University Press, 1976.

D. M. Shen, and W. Gao, "Research on Wavelet Multiresolution Analysis Based Vibration Signal Compress," *Turbine Technology* **42**:193–197, 2000 (in Chinese).

K. Shibata, A. Takahashi, and T. Shirai, "Fault Diagnostics of Rotating Machinery Through Visualization of Sound Signals," *Mechanical Systems and Signal Processing* **14**:229–241, 2000.

SKF, *Vibration Diagnostic Guide, www.skfreliability.com,* 2000.

W. J. Staszewski, "Vibration Data Compression with Optimal Wavelet Coefficients, Genetic Algorithms in Engineering Systems," *Innovations and Applications (Conference Publication)* **446**:186–189, 1997.

W. J. Staszewski, "Structural and Mechanical Damage Detection Using Wavelets," *Shock and Vibration* **30**:457–472, 1998a.

W. J. Staszewski, "Wavelet-Based Compression and Feature Selection for Vibration Analysis," *Journal of Sound and Vibration* **211**:735–760, 1998b.

W. J. Staszewski, S. G. Pierce, K. Worden, and B. Culshaw, "Cross-Wavelet Analysis for Lamb Wave Damage Detection in Composite Materials Using Optical Fibres," *Key Engineering Materials* **167**:373–380, 1999.

D. Stewart, D. Blacknell, A. Blake, R. Cook, and C. Oliver, "Optimal Approach to SAR Image Segmentation and Classification," *IEEE Proceedings Radar, Sonar Navigation* **147**(3):134–142, 2000.

M. Tanaka, M. Sakawa, and K. Kat, "Application of Wavelet Transform to Compression of Mechanical Vibration Data," *Cybernetics and Systems* **28**:225–244, 1997.

P. Tappert, A. von Flotow, and M. Mercadal, "Autonomous PHM with Blade-Tip-Sensors: Algorithms and Seeded Fault Experience," *IEEE Proceedings of Aerospace Conference,* Vol. 7. pp. 7–3295, 2001.

P. Tappert, *P&W DARPA LCF Project Summary Report.* Hood Technology Corporation, November 2003, Hood River, OR.

A. Teolis, *Computational Signal Processing with Wavelets.* Boston, MA: Birkhauser, 1998.

S. Todorovic and M. C. Nechyba, "A Vision System for Intelligent Mission Profiles of Micro Air Vehicles," *IEEE Transactions on Vehicular Technology* **53**(6):1713–1725, 2004.

G. Torheim, F. Godtliebsen, D. Axelson, K. A. Kvistad, O. Haraldseth, and P. A. Rinck, "Feature Extraction and Classification of Dynamic Contrast-Enhanced T2*-Weighted Breast Image Data," *IEEE Transactions on Medical Imaging* **20**(12): 1293–1301, 2001.

G. Vachtsevanos and P. Wang, "A Wavelet Network Framework for Diagnostics of Complex Engineered Systems," *Maintenance and Reliability Conference.* Knoxville, TN: MARCON, 1998.

D. F. Walnut, *An Introduction to Wavelet Analysis.* Boston, MA: Birkhauser, 2002.

W. J. Wang and P. D. McFadden, "Applications of Wavelets to Gearbox Vibration Signals for Fault Detection," *Journal of Sound and Vibration* **192**(5):927–939, 1996.

B. Wu, A. Saxena, R. Patricks, and G. Vachtsevanos, "Vibration Monitoring for Fault Diagnosis of Helicopter Planetary Gears," *16th IFAC World Congress,* Prague, 2005.

X. G. Xu, Y. C. Ji, and W. B. Yu, "The Application of Wavelet Singularity Detection in Fault Diagnostics of High-Voltage Breakers," *IECON Proceedings* **3**:490–494, 2001.

H. T. Yang and C. C. Liao, "A Denoising Scheme for Enhancing Wavelet-Based Power Quality Monitoring System," *IEEE Transactions on Power Delivery* **16**:353–360, 2001.

G. G. Yen and K. C. Lin, "Conditional Health Monitoring Using Vibration Signatures," *Proceedings of the IEEE Conference on Decision and Control* **5**:4493–4498, 1999.

G. G. Yen and K. C. Lin, "Wavelet Packet Feature Extraction for Vibration Monitoring," *IEEE Transactions on Industrial Electronics* **47**:650–667, 2000.

A. Yezzi, A. Tsai, and A. Willsky, "Medical Image Segmentation via Coupled Curve Evolution Equations with Global Constraints," *Proceedings IEEE Workshop on Mathematical Methods in Biomedical Image Analysis,* Hilton Head Island, South Carolina, 2000, pp. 12–19.

Z. Y. Zhang, J. L. Gu, and W. B. Wang, "Application of Continuous Wavelet Analysis to Singularity Detection and Lipeschitz Exponents Calculation of a Rub-Impact Signal," *Journal of Vibration, Measurement and Diagnostics* **21**:112–115, 2001 (in Chinese).

H. Zheng, Z. Li, and X. Chen, "Gear Fault Diagnosis Based on Continuous Wavelet Transform," *Mechanical Systems and Signal Processing* **16**:447–457, 2002.

Y. Zheng, D. H Tay, and L. Li, "Signal Extraction and Power Spectrum Estimation Using Wavelet Transform Scale Space Filtering and Bayes Shrinkage," *Signal Processing* **80**:1535–1549, 2000.

J. Zou, J. Chen, Y. P. Pu, and P. Zhong, "On the Wavelet Time–Frequency Analysis Algorithm in Identification of a Cracked Rotor," *Journal of Strain Analysis for Engineering Design* **37**:239–246, 2002.

CHAPTER 5

FAULT DIAGNOSIS

5.1 INTRODUCTION

Fault diagnosis, in contrast to the less well-understood prognosis, has been the subject of numerous investigations over the past decades. Researchers in such diverse disciplines as medicine, engineering, the sciences, business, and finance have been developing methodologies to detect fault (failure) or anomaly conditions, pinpoint or isolate which component/object in a system/ process is faulty, and decide on the potential impact of a failing or failed component on the health of the system. The diversity of application domains is matched by the plurality of enabling technologies that have surfaced over the years in attempts to diagnose such detrimental events while meeting certain performance specifications. Research and development (R&D) activities have increased more recently owing to the need to maintain legacy equipment, to improve their reliability, and to make them available when needed. We will focus here only on a few historical references that provided a significant impetus to this condition-based maintenance/prognostic, and health management (CBM/PHM) component and addressed primarily machine-related fault phenomena.

The U.S. Navy initiated an integrated diagnostic support system (IDSS), which includes adaptive diagnostics, feedback analysis, fault precursors, and some prognostic capability (Rosenberg, 1989). ONR's condition-based maintenance program was aimed at enhancing the safety and affordability of Navy and Marine Corps platforms and vehicles. An integrated system that processes sensor data from shipboard machinery, assesses machine condition, and provides early fault detection and prediction of time to failure was developed, implemented, and tested successfully on a shipboard chiller by an R&D research team (Hadden et al., 1999a, 1999b). The open systems architecture (OSA) CBM initiative suggests seven levels of activity to integrate transducing with data acquisition, signal processing, condition monitoring, health assessment, prognosis, decision support, and presentation. Figure 5.1 depicts the seven-layered OSA CBM architecture. An OSA CBM XML scheme is proposed for networking and communication capabilities. It is noteworthy that

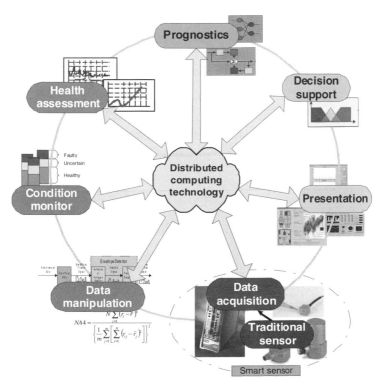

Figure 5.1 The seven-layered OSA CBM architecture.

the proposed decision-support level may integrate information to support a "decision to act" based on information from the prognostic layer, external constraints, mission requirements, financial incentives, etc. In parallel, International Standards Ogranization (ISO) standards have attempted to capture the conceptual framework for processing models aimed at providing the basic CBM/PHM modules from data acquisition and analysis to health assessment, prognostic assessment, and advisory generation. Figure 5.2 depicts the ISO 13374-1 processing model. R&D activities by several major companies (GE, Boeing, Lockheed, and Honeywell, among many others) have produced promising results focusing on specific application domains but lacking the widespread acceptance as technology standards. Concerning the enabling technologies, statistical distribution, Bayesian networks, stochastic autoregressive moving-average models, state-estimation techniques, knowledge-intensive expert systems, finite state machines, and Markov chain models and soft computing tools (neural nets, fuzzy logic, genetic algorithms) have been suggested as potential means for mechanical system diagnosis/prognosis. Fault diagnosis, or fault detection and identification (FDI), is a mature field with contributions ranging from model-based techniques to data-driven configurations that capitalize on soft computing and other "intelligent" tools (Konrad and Isevmann, 1996; Mylaraswamy and Venkatasubramanian, 1997).

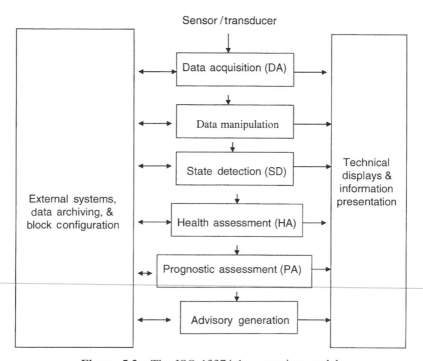

Figure 5.2   The ISO 13374-1 processing model.

Excellent surveys of design methods can be found in the classic papers by Willsky (1976) and Isermann (1984).

Modern diagnostic algorithms require massive databases that form a collection of data from multiple and diverse sensor suites. Information typically is extracted, in the form of features or condition indicators, from such databases and used as the input to diagnostic routines whose goal is to declare impending failure conditions. Machine diagnosis and prognosis for CBM involves an integrated system architecture with the following main modules:

1. A sensor suite and suitable sensing strategies to collect and process data of critical process variables and parameters
2. A failure modes and effects criticality analysis (FMECA) module that determines and prioritizes according to the frequency of occurrence and their severity possible failure modes and catalogs systematically effect–root-cause relationships
3. An operating-mode identification routine that determines the current operational status of the system and correlates fault-mode characteristics with operating conditions
4. A feature extractor that selects and extracts from raw data features or condition indicators to be used by the diagnostic module
5. A diagnostic module—the diagnostician—that assesses through online measurements the current state of critical machine components
6. A prognostic module—the prognosticator—that estimates the remaining useful life of a failing component/subsystem
7. The final module of the integrated architecture—the maintenance scheduler—whose task is to schedule maintenance operations without affecting adversely the overall system functionalities, of which the machine in question is only one of the constituent elements.

We focus in this chapter on fundamental technologies that enable the accurate and timely diagnosis of fault or incipient-failure conditions. Fault prognosis issues are addressed in Chap. 6. Elements of fault diagnosis and prognosis and their enabling technologies are, by necessity, interwoven and difficult to delineate from one another clearly. For this reason, some aspects of fault initiation, detection, and propagation will be covered in this chapter, although the propagation or time-evolution component may be more directly relevant to the prognosis module of the CBM/PHM architecture.

5.2 THE DIAGNOSTIC FRAMEWORK

5.2.1 Introduction and Definitions

In this section we review relevant definitions and basic notions in fault diagnosis. Over the past years, the meaning of defining terms has changed, conforming to the incorporation of machine fault diagnosis tools into the CBM/PHM framework. Agreement is still lacking within the CBM/PHM community on a standardized and accepted vocabulary. However, the military and other sectors are converging to the following definitions:

- *Fault diagnosis.* Detecting, isolating, and identifying an impending or incipient failure condition—the affected component (subsystem, system) is still operational even though at a degraded mode.
- *Failure diagnosis.* Detecting, isolating, and identifying a component (subsystem, system) that has ceased to operate.

The terms *fault detection, isolation,* and *identification* are typically employed to convey the following meaning:

- *Fault (failure) detection.* An abnormal operating condition is detected and reported.
- *Fault (failure) isolation.* Determining which component (subsystem, system) is failing or has failed.
- *Fault (failure) identification.* Estimating the nature and extent of the fault (failure).

In order to estimate the remaining useful lifetime or time to failure of a failing component, that is, to conduct fault prognosis, and maximize uptime through CBM, we seek to determine accurately and without false alarms impending or incipient failure conditions, that is, faults. Thus fault diagnosis is aimed at determining accurately and without false alarms impending or incipient failure conditions. The designer's objective is to achieve the best possible performance of a given diagnostic routine by minimizing false positives and false negatives while reducing the time delay between the initiation and detection/isolation of a fault event. A detailed discussion of performance metrics for fault diagnosis is deferred until Chap. 7. We will summarize in the next subsection only those significant elements of fault diagnosis requirements and performance metrics that will set the stage for a better understanding of the failure mechanisms we seek to detect and identify as well the enabling technologies needed to realize this essential module of the CBM/ PHM architecture.

5.2.2   Fault Diagnosis Requirements and Performance Metrics

CBM/PHM system designers must adhere to design requirements imposed by the system user and conforming to overall system objectives. Although it may be possible to define a generic set of requirements associated with diagnostic routines, specific quantitative requirements definitions are system- and application-dependent. In general, the CBM/PHM system must

- Ensure enhanced maintainability and safety while reducing operational and support (O&S) cost
- Be designed as an open systems architecture
- Closely control PHM weight
- Meet reliability, availability, maintainability, and durability (RAM-D) requirements
- Meet monitoring, structural, cost, scalability, power, compatibility, and environmental requirements

More specifically, fault diagnosis requirements include the following:

- The PHM system shall detect no less than x percent of all system or subsystem (component) failures being monitored and no less than y percent of system failures specifically identified through reliability centered maintenance (RCM) and FMECA studies.
- The PHM system shall demonstrate a fault-isolation rate to one (or x) organizational removable assembly (ORA) units for all faults that are detected by PHM.
- Fault-identification requirements may be defined as to the impact of a particular fault on the system's operating characteristics.
- For false alarms, the PHM must have a demonstrated mean operating period between false alarms of x hours.
- If built-in testing (BIT) capability is available, the PHM diagnosis BIT shall complete an end-to-end test within an x-minute test period.
- Additional requirements may be defined in terms of allowable false positives, false negatives, and the time window between fault initiation and fault detection.

Specific numerical quantities must be prescribed by the end user. It should be pointed out that there are no accepted standards for CBM/PHM requirements. The community is currently debating these issues as the technological basis advances and systems are developed and implemented.

Fault-diagnostic algorithms must have the ability to detect system performance, degradation levels, and faults (failures) based on physical property changes through detectable phenomena. The faults detected are typically a subset of those identified through FMECA analyses. Furthermore, such algorithms must have the ability to identify the specific subsystem and/or component within the system being monitored that is failing, as well as the specific failure mechanism that has occurred. The potential impact of the fault on the operational integrity of the system is also a desirable outcome of diagnosis. Since the ultimate objective of a CBM/PHM system is to predict the time progression of the fault and estimate the time to failure of the failing component, it is essential that faults be detected and isolated as early as possible after their initiation stage.

Reducing false alarms, false positives and false negatives, is a major objective of a diagnostic system. Several means are available that may facilitate this important task:

- Data validation
- Optimal feature selection/extraction
- Optimal choice of fault detection/identification algorithms
- Combining (fusing) evidence (Dempster-Shafer theory)
  - Conflict resolution and ignorance representation
  - Fusion of fault-classification algorithms
  - Declaring a fault only when confidence (certainty) exceeds a set threshold

Enabling technologies for each one of these means are discussed in sections of this and other chapters of this book. Performance metrics for fault diagnostic algorithms are detailed in Chap. 7.

5.2.3 Fault Monitoring and Diagnosis Framework

Fault detection and identification (FDI) strategies have been developed in recent years and have found extensive utility in a variety of application domains. We are employing the term FDI here to imply primarily fault detection and the determination of the specific fault component/subsystem because it has been prevalent in the technical literature in recent years. A general FDI structure is depicted in Fig. 5.3. In this figure, a residual signal that represents a deviation from standard operating conditions is generated by comparing, for example, a model output with the actual system output. Based on this signal, one makes decisions about the operating condition of the machinery. Methods to generate the residual are given in this chapter and can be based on time signal statistics, system identification, frequency domain techniques, etc.

The enabling technologies typically fall into two major categories: model-based and data-driven, as shown in Fig. 5.4. Model-based techniques rely on an accurate dynamic model of the system and are capable of detecting even

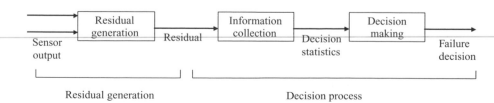

Figure 5.3   General FDI structure.

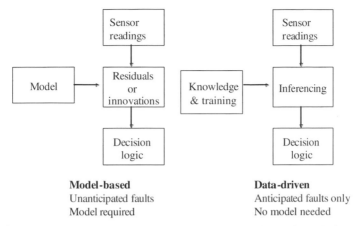

Figure 5.4   Model-based and data-driven fault diagnosis techniques.

unanticipated faults. They take advantage of the actual system and model outputs to generate a "discrepancy" or residual (innovation), as it is known, between the two outputs that is indicative of a potential fault condition. In one method, a bank of filters is used to pinpoint the identity of the actual fault component. On the other hand, data-driven techniques often address only anticipated fault conditions, where a fault "model" now is a construct or a collection of constructs such as neural networks, expert systems, etc. that must be trained first with known prototype fault patterns (data) and then employed online to detect and determine the faulty component's identity.

A general scheme for online fault monitoring and diagnosis is shown in Fig. 5.5. This shows the entire online procedure required for monitoring and

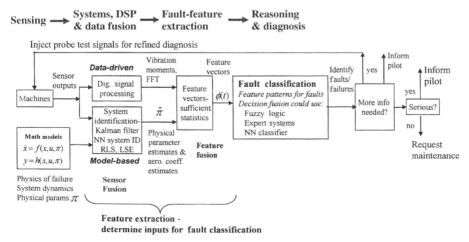

Figure 5.5   Online fault monitoring and diagnosis.

diagnosis and serves as a focus for the discussion in this chapter. The major components of this procedure are data collection through sensor outputs and then building of a *feature vector* that contains enough information about the current machine operating condition to allow for fault identification and classification. The information contained in the feature vector can be obtained by three means: (1) model-based methods (e.g., identify physical model parameters using Kalman filtering or recursive least squares), (2) data-driven methods (e.g., vibration signal statistical moments, vibration frequency spectrum information), and (3) statistical regression and clustering techniques on existing historical legacy data. Once the current feature vector has been obtained, it is used as input to a fault-classification block that can contain various types of decision-making algorithms. If more information is needed before a decision can be reached, probing signals can be sent back to the machine (e.g., inject test vibration signal into a machinery mount to test for loose fittings). We now discuss the components in Fig. 5.5.

Fault-Feature Vector. The fault-feature vector is a set of data that is useful for determining the current fault status of the machinery. Feature vectors may contain a variety of information about the system. This includes both system parameters relating to fault conditions (bulk modulus, leakage coefficient, temperatures, pressures) and vibration and other signal analysis data (energy, kurtosis, discrete Fourier transform, etc.). Feature vector components are selected using physical models and/or historical legacy data such as that available in the USN SH-60 HIDS study (Hardman et al., 1999). Physical models might show that components such as bulk modulus and leakage coefficient should be included, and historical failure data might show the importance of vibration signature energy, kurtosis, etc. Different feature vectors will be needed to diagnose different subsystems.

In probabilistic terms, the feature vector $\varphi[x(t)]$ is a *sufficient statistic* for the full state and history $x(t)$ of the system that determines the fault/failure condition $z(t)$. That is, the fault likelihood function probability density function (PDF) can be decomposed as

$$f[z(t)/x(t)] = g[x(t)]h\{\varphi[x(t)], z(t)\} \tag{5.1}$$

so that knowledge of $\varphi[x(t)]$ is sufficient to diagnose the fault state $z(t)$. The data component factor $g[x(t)]$ depends only on $x(t)$ and is not needed for diagnosis. It is not necessary to know the full system state and history $x(t)$. Note that throughout this chapter boldface lettering is used to designate vector or matrix quantities.

The feature vectors are extracted using system identification and digital signal-processing techniques, as shown in examples that follow. $\varphi[x(t)]$ is monitored continuously and is a time-varying vector. At each time, the fault classification may be determined by comparing $\varphi[x(t)]$ to a library of stored fault patterns or to predetermined alarm conditions. $\varphi[x(t)]$ contains the *symptoms* of the fault status, whereas the fault pattern library (FPL) contains *di-*

agnostic information in terms of *known fault patterns*. Many techniques have been used to create the FPL for fault classification, as we shall elaborate in the rest of the chapter, including physical modeling, statistical techniques, historical fault-test data, rule-based expert systems, reasoning systems, fuzzy logic, neural networks, etc.

5.3 HISTORICAL DATA DIAGNOSTIC METHODS

For many industrial and military machines, there is a large amount of historical data from previous failures. For instance, a U.S. Navy strategy is being followed to develop and demonstrate diagnosis and prognosis tolls for helicopter drive trains. The SH-60 Helicopter Integrated Diagnostic System (HIDS) Program was initiated to develop and integrate advanced mechanical diagnostic techniques for propulsion and power drive systems. Extensive tests have been carried out on deliberately seeded faults on SB 2205 roller bearings (Engel et al., 2000; Hardman et al., 1999). A gear tooth is weakened by implanting an electronic discharge machine notch in the tooth root. This creates a localized stress concentration that leads to formation of a crack that grows and eventually causes tooth failure. Using collocated sensors, vibration signals were recorded at all gearboxes, shafting, and hangar bearings, oil monitoring was performed, and in-flight rotor track and balance capability was monitored. This program has led to a large amount of stored historical data for use in diagnosis and prognosis in currently operating machinery. Comparison of measured signals in such machinery with suitable stored failure data can indicate the presence of fault conditions.

Historical data may include vibration measurements or moment information (energy, RMS value, kurtosis, etc.) on machines prior to, leading up to, and during failures. Time signals measured on operating machinery may be compared with these historical data to determine if a fault is in progression. The challenge is that quantity of the data is simply overwhelming, and it must be turned into a useful form of information for diagnostic purposes. Many methods exist for analyzing these data and turning them into useful forms for decision making, including statistical correlation/regression methods, fuzzy-logic classification, and neural-network clustering techniques. In this section we can only hint at some methods of analyzing these data and turning them into useful information.

The next subsection briefly introduces recent attempts at standardizing data archiving and data management tasks under the OSA-CBM/MIMOSA initiative as a preliminary step leading to our discussion of diagnostic methods that exploit historical databases. Detailed descriptions can be found at the MIMOSA Web site (*www.mimosa.org*).

OSA-CBM/MIMOSA. The focus of the OSA-CBM program has been the development of an open-system standard for distributed software components. *Open system* refers to a systems engineering approach that facilitates the integration and interchangeability of components from a variety of sources. An open-system standard should consist of publicly available descriptions of component interfaces, functions, and behavior. The architecture should not be exclusive to any specific hardware implementations, operating systems, or software technology. OSA-CBM is a specification for transactions between components within a CBM system. The core of the standard is the object-oriented data model, defined using Unified Modeling Language (UML) syntax. The OSA-CBM UML data model is a mapping of key concepts for diagnosis, prognosis, and data transactions. MIMOSA (Machinery Information Management Open Systems Alliance, *www.mimosa.org*) is a standard for data exchange between asset management systems. Figure 5.6 depicts the MIMOSA repository architecture, highlighting the data archival management path. In addition to MIMOSA, other existing or emerging standards include AI-ESTATE and IEEE 1451.2.

5.3.1 Alarm Bounds

Figure 5.7 shows the concept of alarm bounds or fault tolerance limits. These are prespecified boundaries on measured sensor signals that indicate when an anomaly situation may be occurring. If the measured signal or one of its moments passes outside these bounds, an alarm state is signaled. Industrial software packages such as the LabVIEW Sound and Vibration Toolkit by National Instruments provide highly effective means for measuring desired sensor signals, data logging, user specification of alarm bounds, and alarm functions when the bounds are violated. Note that the signal itself has confidence bounds, provided, for instance, by the Kalman filter error covariance, that can be taken into account in advanced decision making for diagnosis. Methods of prediction to estimate remaining useful life are given in Chap. 6.

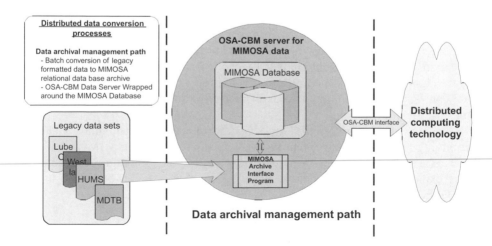

Figure 5.6 Virtual OSA-CBM data server.

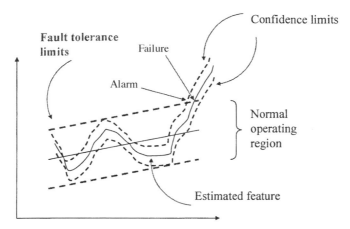

Figure 5.7   Alarm bounds on measured sensor signals.

The alarm bounds can be determined by two basic methods. The first method involves computing the bounds based directly on the current measured signal. An example is the Bollinger bands (*www.bollingerbands.com*) computed as the moving average plus the moving variance and the moving average minus the moving variance. When signal variations exceed the Bollinger bands, it has been observed that significant trend changes may be imminent. The advantage of computing the alarm bounds from the actual measured signals is that the bounds are generated automatically as time passes, and no pretraining or historical data are needed.

However, in some situations, historical data are available that contain valuable information. In this case, alarm bounds can be determined based on historical data and past observed failure events.

5.3.2 Statistical Legacy Data Classification and Clustering Methods

In many cases, one has available statistical legacy fault data such as that shown in Fig. 5.8. In such figures, each point (x, z) represents one previously measured anomaly instance. In the data shown, one observes that as vibration magnitude increased in previously recorded tests, drive tooth gear wear was measured to increase. Such data are highly useful for diagnosis; however, the sheer magnitude of recorded data is overwhelming, and mathematical techniques are required to render them into a useful form.

Statistical Correlation and Regression Methods. Statistical correlation and regression methods, such as Pearson's correlation given in Chap. 3 on sensor and sensing strategies, are useful for extracting information from archived historical data. Statistical clustering methods can reveal the structure inherent in data records, as shown in Fig. 5.9, where the data are shown to be clustered into three fault conditions.

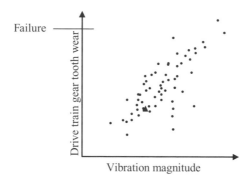

Figure 5.8 Sample of statistical legacy fault data.

These figures have only two dimensions. However, historical data can have many dimensions or components (e.g., vibration magnitude, kurtosis, frequency spectrum content), which adds to the complexity of data interpretation. To incorporate statistical legacy data into a diagnostic system, one may determine which data components to include in the feature vector $\phi(t)$. Components that are significantly correlated with the fault of interest should be included in the feature vector. Bootstrapping techniques can be used to determine confidence levels for the correlation. The MATLAB Statistics Toolbox (MATLAB, 2002) has many tools for performing such statistical analysis.

Conditional expected values are an alternative means for extracting useful information. Given N statistical sample values of vectors $\mathbf{x}^i \in \mathbf{R}^n$ of measurable data and their associated faults z^i, it is known that a consistent estimator for the joint PDF is given by the Parzen estimator:

$$
P(x, z) = \frac{1}{(2\pi)^{(n+1)/2}\, \sigma^{n+1}}
$$

$$
\frac{1}{N} \sum_{i=1}^{N} \exp\left[-\frac{(x - x^i)^T (x - x^i)}{2\sigma^2} \right] \exp\left[-\frac{(z - z^i)^2}{2\sigma^2} \right] \quad (5.2)
$$

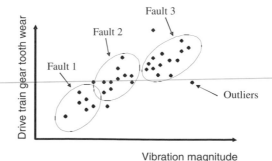

Figure 5.9 Clustering methods reveal the structure in statistical legacy data.

Substituting this into the formula for the conditional expected value, that is,

$$E(z/x) = \frac{\int zp(x,\,z)\,dz}{\int p(x,\,z)\,dz} \tag{5.3}$$

one obtains the expected value of the fault state z given the actual observed data x as

$$E(z/x) = \frac{\displaystyle\sum_{i=1}^{N} z^i \exp\left[-\dfrac{(x - x^i)^T(x - x^i)}{2\sigma^2}\right]}{\displaystyle\sum_{i=1}^{N} \exp\left[-\dfrac{(x - x^i)^T(x - x^i)}{2\sigma^2}\right]} \tag{5.4}$$

The conditional expected value gives the value of the z variable (which could be a fault condition) given the measured data points x^i. This allows the statistical fault data to be incorporated into a diagnostic decision system.

This development holds for non-Gaussian as well as Gaussian statistical data. The smoothing parameter σ is selected to give a good compromise between accuracy of fault diagnosis and the ability to generalize to new data sets. More discussion is given about Parzen PDF estimation in Chap. 6.

Fuzzy-Logic Classification. Fuzzy-logic techniques can be used to perform data analysis by interactive computer-aided inspection and manipulation techniques. Figure 5.10 shows a sample of this approach. Fuzzy-logic clustering leads directly to rules of the form

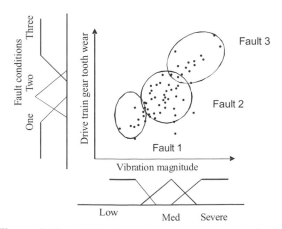

Figure 5.10   FLCT fuzzy-logic clustering techniques.

If (vibration magnitude is medium), then (condition is fault 2)

which can be incorporated directly into diagnostic decision rule bases for fault classification. The *MATLAB Fuzzy Logic Toolbox* (2002) is useful for such applications.

Neural-Network Classification and Clustering. Several techniques for neural-network (NN) classification and clustering are available. Neural networks (Haykin, 1994; Lippmann, 1987; Lewis et al., 1999) are highly useful for classification and clustering owing to their formal mathematical structure, their intuitive appeal as being based on biologic nervous systems for data processing, and their learning features. The simplest NNs have only one layer; more complex NNs have two or three layers. A two-layer NN is shown in Fig. 5.11. This NN has two layers of weights (shown by arrows), one input layer of neurons, a hidden layer of neurons (circles), and an output layer of neurons (circles). At each neuron in the hidden and output layers, one can place a nonlinear *activation function,* which is responsible for the powerful processing power of the NN.

NNs have a rich mathematical structure. Mathematically, a multi-input, single-output two-layer NN is described by the equation

$$u(x) = \sum_{i=1}^{L} z_i \, \sigma\left(\sum_{j=1}^{n} v_{ij}x_j + v_{i0}\right) + z_0 \tag{5.5}$$

with v_{ij}, v_{i0} the first-layer weights and thresholds, z_i, z_0 the second-layer weights and thresholds, n the number of inputs, and L the number of hidden-layer neurons. The activation functions $\sigma(\cdot)$ may be selected as hard limiters, sigmoids, tanh, etc.

However, for purposes of engineering CBM, a few basic concepts suffice, and much of what one needs is contained in the *MATLAB Neural Network*

Neural network (NN)

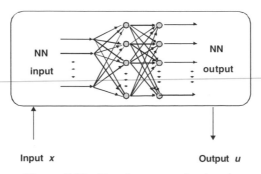

Figure 5.11   Two-layer neural network.

Toolbox (1994). NNs are especially useful for prediction and prognosis, and more mathematical information about them is presented in Chap. 6. A novel NN design based on wavelet theory and called *wavelet neural* network (WNN) is discussed in the next section and is used to classify fault conditions when the fault indicators are high-frequency signals such as those originating from vibration emission sensors.

It has been shown that a two-layer NN with suitably chosen structure can approximate a large class of functions that appear in practical applications. On the other hand, a two-layer NN can only partition data into convex regions, as shown in Fig. 5.12. To solve more general sorts of classification problems, a three-layer NN is needed. Historically speaking, the acceptance of NNs was hindered until it was shown that the two-layer NN can solve the exclusive-or problem, which is depicted in the middle band of the figure.

The next examples show how to perform classification and clustering on available data points. They essentially use NN techniques to turn data into information in a form that is useful for diagnosis.

Example 5.1   Neural Network Classification of Data. Figure 5.13 shows a measured data set of eight points. It is desired to classify these data into the four groups

    Group 1: o (1,1), (1,2)
    Group 2: x (2,-1), (2,-2)
    Group 3: + (-1,2), (-2,1)
    Group 4: # (-1,-1), (-2,-2)

This may be accomplished easily using the *MATLAB Neural Networks Toolbox* (1994) as follows:

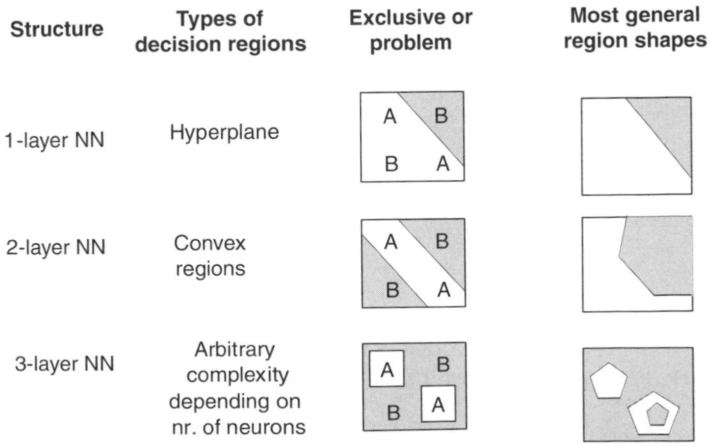

Structure	Types of decision regions	Exclusive or problem	Most general region shapes
1-layer NN	Hyperplane		
2-layer NN	Convex regions		
3-layer NN	Arbitrary complexity depending on nr. of neurons		

Figure 5.12   Types of neural-network decision regions.

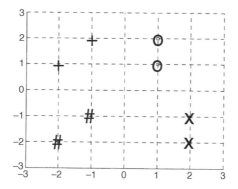

Figure 5.13   Measured data requiring classification.

Training the NN. Denote the four groups by (0,0), (0,1), (1,0), and (1,1), respectively, and place the data points and their corresponding desired grouping into two vectors as

$$\mathbf{X} = \begin{bmatrix} 1 & 1 & 2 & 2 & -1 & -2 & -1 & -2 \\ 1 & 2 & -1 & -2 & 2 & 1 & -1 & -2 \end{bmatrix} \quad \text{data input}$$

$$\mathbf{Y} = \begin{bmatrix} 0 & 0 & 0 & 0 & 1 & 1 & 1 & 1 \\ 0 & 0 & 1 & 1 & 0 & 0 & 1 & 1 \end{bmatrix} \quad \text{desired target vector}$$

It is desired to construct an NN that classifies the data $\mathbf{X}$ into the four groups as encoded in the target vector $\mathbf{Y}$. Now the following MATLAB code is used to configure and train a two-layer NN to classify the data into the desired groups:

```
R=[-2 2;-2 2]; % define 2-D input space
netp=newp(R,2); % define 2-neuron NN
p1=[1 1]'; p2=[1 2]'; p3=[2 -1]'; p4=[2 -2]'; p5=[-1 2]';
p6=[-2 1]'; p7=[-1 -1]'; p8=[-2 -2]';
t1=[0 0]'; t2=[0 0]'; t3=[0 1]'; t4=[0 1]'; t5=[1 0]';
t6=[1 0]'; t7=[1 1]'; t8=[1 1]';
P=[p1 p2 p3 p4 p5 p6 p7 p8];
T=[t1 t2 t3 t4 t5 t6 t7 t8];
netp.trainParam.epochs = 20; % train for max 20 epochs
netp = train(netp,P,T);
```

The data must be presented to the NN enough times to allow full training. In this example, we selected 20 epochs. The result is shown in Fig. 5.14, where two lines have been computed by the NN to divide the data into the required four groups.

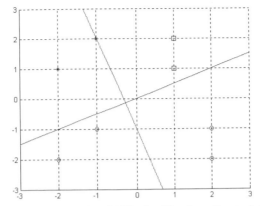

Figure 5.14 Result of NN classification after training.

Classifying Current Faults. This classification has been performed on the eight available historical data points. The importance of this approach for diagnosis is as follows. Suppose that one now measures using available sensors a new data point $P1 = (1, -1)$. This is presented to the NN using the MATLAB command

```
Y1=sim(netp,P1).
```

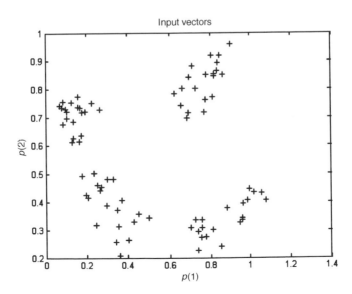

Figure 5.15 Historical data from 80 measured previous runs.

The result will be that the data are classified as belonging to group 2, which could be a certain fault condition.

Example 5.2 Clustering of Data Using Neural Networks. One might have available historical legacy data of the type shown in Fig. 5.15, which show 80 data points that might correspond to previously measured fault conditions in a class of machinery. To interpret such data and turn them into useful information, one may cluster them using a structure known as a *competitive NN with Kohonen training* (Haykin, 1994). This yields information about groupings of fault conditions.

Training the NN. To cluster these data using MATLAB (1994), first, one forms a 2×80 matrix **P** containing the (x, y) coordinates of the 80 data points (see previous example). Then, the following MATLAB code is run:

```
% make new competitive NN with 8 neurons
net = newc([0 1;0 1],8,.1);
  % train NN with Kohonen learning
net.trainParam.epochs = 7;
net = train(net,P);
w = net.IW{1};
  %plot
plot(P(1,:),P(2,:),'+r');
xlabel('p(1)');
ylabel('p(2)');
hold on;
circles = plot(w(:,1),w(:,2),'ob');
```

The result is shown in Fig. 5.16, where the circles plotted were generated by the trained NN and are centered in eight clusters identified by the NN. We selected the number of cluster neurons as eight, as seen in the MATLAB code. The number of clusters may be varied by the user. In this example, six cluster neurons probably would have been a better choice for the given data.

Classifying the Current Fault. Now that the historical data have been clustered into eight fault conditions, one applies this to fault diagnosis in the following manner. Suppose that a new data point (0, 0.2) is measured from a sensor reading. By using the MATLAB commands

```
p = [0; 0.2];
a = sim(net,p)
```

the NN is run, and the new data point is associated into the fault cluster marked by neuron number 1.

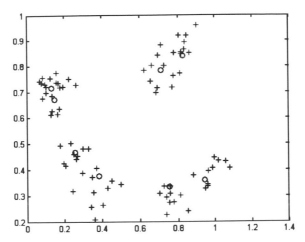

Figure 5.16 Eight clusters identified by the NN. Circles are shown at the centers of the clusters.

5.4 DATA-DRIVEN FAULT CLASSIFICATION AND DECISION MAKING

Referring to Fig. 5.5, we have discussed sensor selection, selecting the fault-feature vector, and model-based methods, data-driven methods, and historical legacy data. A number of techniques have been given for extracting useful information from measured data. Once the feature vector parameters and signals have been measured or computed for the running machinery, one must analyze the result to determine the current fault operating condition. This may be accomplished using fault classification decision making, as shown in Fig. 5.17. A fault-pattern library can be formed using previously observed faults or expert information about the machinery. Then the fault-feature vector

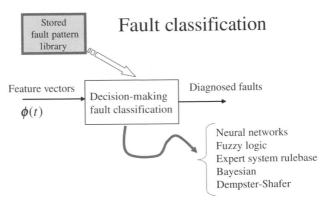

Figure 5.17 Fault classification decision making using a fault-pattern library.

can be compared with the stored patterns of anomalous situations. Several decision-making tools can be used, including rule-based, fuzzy-logic, Bayesian, Dempster-Shafer, neural networks, etc. Instead of using a fault-pattern library, alternative approaches are multimodel and filter-bank methods, where a separate model or filter is used for each known fault condition. All these topics are now introduced.

Example 5.3   A Data-Driven Diagnostic Decision-Making System. In contrast to model-based approaches, data-driven fault diagnostic techniques rely primarily on process and PHM data (data from sensors specifically designed to respond to fault signals) to model a relationship between fault features or fault characteristic indicators and fault classes. Such "models" may be cast as expert systems or artificial neural networks or a combination of these computational intelligence tools. They require a sufficient database (both baseline and fault conditions) to train and validate such diagnostic algorithms before their final online implementation within the CBM/PHM architecture. They lack the insight that model-based techniques provide regarding the physics of failure mechanisms, but they do not require accurate dynamic models of the physical system under study. They respond only to anticipated fault conditions that have been identified and prioritized in advance in terms of their severity and frequency of occurrence, whereas model-based methods may be deployed to detect even unanticipated faults because they rely on a discrepancy or residual between the actual system and model outputs. Over the past few years, a number of diagnostic reasoners have been reported based on well-known computational intelligence tools such as fuzzy logic and neural networks (Propes and Vachtsevanos, 2003; Vachtsevanos and Wang, 1998). Moreover, specialized neural-network and neurofuzzy constructs have been suggested to address the idiosyncratic character of certain fault modes and their characteristic signatures in the time, frequency, and wavelet domains.

The diagnostic system may include additional modules that are intended to fuse the results of multiple classifiers (when more than one fault diagnostic algorithm is employed) and provide confidence or uncertainty bounds for true fault declarations. Classifier fusion techniques have been reported, and certainty factors use notions from Dempster-Shafer evidential theory to suggest if the fault evidence exceeds a specified threshold so that false alarms may be limited. We will illustrate the data-driven approach to diagnosis via an example of an integrated architecture developed at the Georgia Institute of Technology (Zhang et al., 2002; Propes et al., 2002).

Figure 5.18 is a schematic representation of a data-driven fault diagnostic system. The main modules of the architecture address issues of sensor data processing, operational mode identification, fault-feature extraction, diagnosis via multiple classifiers, and confirmation through a certainty factor based on Dempster-Shafer theory.

The reasons to use multiple classifiers as the diagnostic algorithms are as follows: It is generally possible to break down the sensor data (and, correspondingly, the symptomatic evidence) into two broad categories. The first

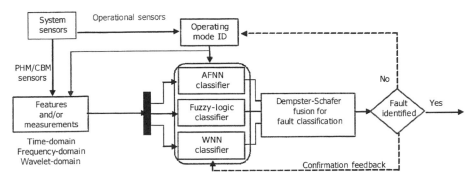

Figure 5.18   A data-driven fault diagnostic system.

one is concerned with low-bandwidth measurements, such as those originating from process variables, temperature, pressure, flow, etc., whereas the second exemplifies high-bandwidth measurements, such as vibrations, current spikes, etc. Failure modes associated with the first category may develop slowly, and data are sampled at slow rates without loss of trending patterns. High-frequency phenomena, though, such as those accompanying a bearing failure, require a much faster sampling rate in order to permit a reasonable capture and characterization of the failure signature. Moreover, process-related measurements and associated failure-mode signatures are numerous and may overlap, thus presenting serious challenges in resolving conflicts and accounting for uncertainty. This dichotomy suggests an obvious integrated approach to the fault diagnosis problem; that is, process-related faults may be treated with a fuzzy-rule-base set as an expert system, whereas high-bandwidth faults are better diagnosed via a feature extractor/neural-network classifier topology.

5.4.1   Fault-Pattern Library

In preceding examples we have seen how to obtain rules such as those shown in Table 5.1. This is the fault-pattern library (FPL). The FPL maps the observed conditions (as contained in the fault-feature vector) into fault-condition classes. Note that this is a *rule-based system* with antecedents (the left-hand column) that can be physical parameters, sensor data and moments, frequency content information, etc. The consequent (right-hand column) contains the resulting fault condition.

5.4.2   Decision-Making Methods

Based on the FPL or other approaches, several decision-making techniques may be used, as mentioned previously. Some of these are now discussed.

Fuzzy-Logic Decision Making.   In Table 5.1 there appear categorizations such as "IF (leakage coeff. *L* is large)" that invite a clear characterization of

TABLE 5.1   Fault-Pattern Library

Condition	Fault Mode
IF (leakage coeff. L is large)	THEN (fault is hydraulic system leakage)
IF (motor damping coeff. B is large) AND (piston damping coeff. B_p is large)	THEN (fault is excess cylinder friction)
IF (actuator stiffness K is small) AND (piston damping coeff. B_p is small)	THEN (fault is air in hydraulic system)
IF (base mount vibration energy is large)	THEN (fault is unbalance)
IF (shaft vibration second mode is large) AND (motor vibration RMS value is large)	THEN (fault is gear tooth wear)
IF (third harmonic of shaft speed is present) AND (kurtosis of load vibration is large)	THEN (fault is worn outer ball bearing)
Etc.	Etc.

the term *large*. Table 5.2 shows that if the leakage coefficient is 10 percent or more above nominal value, then the fault condition exists. Such information may be obtained by a combination of physical modeling, expert opinion, and historical legacy fault data.

An alternative method of characterizing valuation in the fault-feature vector entries is provided by fuzzy logic (Kosko, 1997; Ross, 1995). Fuzzy logic was developed by Lotfi Zadeh and has a rich and fascinating mathematical structure (Zedeh, 1973). However, for purposes of engineering CBM, a few

TABLE 5.2   Parameter Deviations That Indicate Fault Conditions

Fault/Failure Description	Perturbations of the Physical Model Parameters						
	J	B	C	K	L	M_p	B_p
Hydraulic system leakage					+10%		
Control surface loss	−10%					−10%	
Excessive hydrolic cylinder friction		+10%					+10%
Rotor mechanical damage		−10%		−10%			
Motor magnetism loss		−10%	−10%				
Air in hydraulic system				−10%			−10%

Source: Skormin et al. (1994).

basic concepts suffice, and much of what one needs is contained in the MATLAB Fuzzy Logic Toolbox (2002).

We have indicated in Fig. 5.10 how to use fuzzy-logic methods for classifying historical data. This leads directly to rules of the form

If (vibration magnitude is medium), then (condition is fault 2)

which may be entered into the FPL. Another example is given in Fig. 5.19, which characterizes the number of broken bars in motor drives in terms of magnitudes of the first and second sideband frequency component (Filippetti, 2000).

A fuzzy-logic system is shown in Fig. 5.20 that has three components. First, the sensed data are classified into linguistic terms, such as *large, medium, small, too hot,* etc. This is accomplished using prespecified *membership functions.* These terms are used next in a look-up table rule base of the form shown in Table 5.1. In usual practice, the consequents give the *representative values* of the outputs, which are decoded mathematically using the *output membership functions* into numerical values. Mathematically, the numerical output of a fuzzy-logic system is given by

$$u(x) = \frac{\sum_{i=1}^{N} z^i \prod_{j=1}^{n} \mu_{ij}(x_j)}{\sum_{i=1}^{N} \prod_{j=1}^{n} \mu_{ij}(x_j)} \tag{5.6}$$

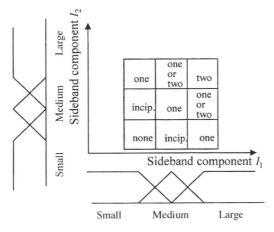

Figure 5.19 FPL rulebase to diagnose broken bars in motor drives using sideband components of vibration signature FFT. Number of broken bars = none, one, two; Incip. = incipient fault. (Filippetti, 2000).

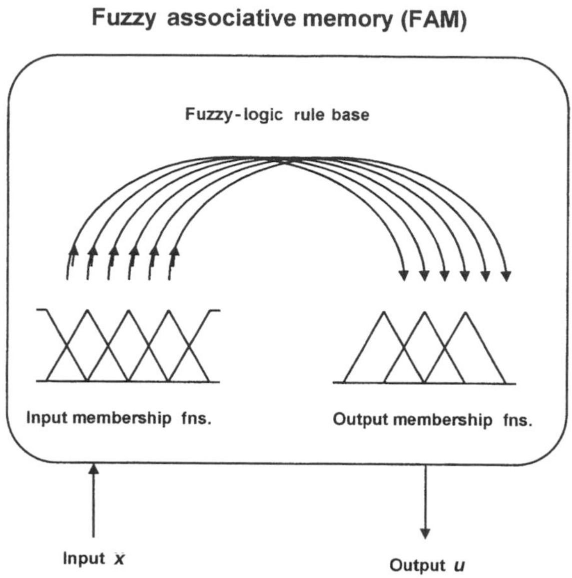

Fuzzy associative memory (FAM)

Fuzzy-logic rule base

Input membership fns.

Output membership fns.

Input *x*

Output *u*

Figure 5.20 Fuzzy-logic system.

where x_j are the observed data values (rule antecedents), z^i are the consequents associated with them by the rule base, termed *control representative values,* and $\mu_{ij}(x_j)$ are the membership functions. This formula assumes product inferencing and centroid defuzzification using singleton control membership functions.

However, in CBM diagnostic applications, one does not need this full machinery. In fact, the third defuzzification step is not needed for the FPL because all we are interested in is the consequences of the rule base, which yield the current fault conditions.

A representative set of membership functions for a sensor output on a machine is shown in Fig. 5.21. The numerical values for the center and end points of the membership functions may be determined from historical data, expert opinion, or physical modeling.

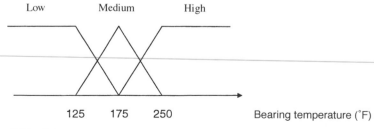

Low   Medium   High

125   175   250   Bearing temperature (°F)

Figure 5.21 Membership functions "low," "medium," and "high" for bearing temperature.

Note that for an actual observed value of temperature, one may have more than one linguistic value. For instance, the temperature reading of 235°F can be both medium and high, as shown in Fig. 5.22. In this instance, the temperature is high to the degree of 0.6 and medium to the degree of 0.4. Suppose that the look-up table FPL has the two rules:

IF (Bearing Temp is medium) THEN (no fault)
IF (Bearing Temp is high) THEN (fault is clogged oil filter)

Then one might say that there is a 60 percent chance of a clogged oil filter given the sensor reading of 235°F.

5.4.3 Dempster-Shafer Evidential Theory

The Dempster-Shafer theory of evidence module is incorporated into the system for uncertainty management purposes. Each input feature has membership functions associated with it representing the possibility of a fault mode. Each feature, therefore, represents an expert in this setting. These possibility values are converted to basic probability assignments for each feature. Dempster's rule of combination then is used to assimilate the evidence contained in the basic probability assignments and to determine the resulting degree of certainty for detected fault modes. Dempster-Shafer theory is a formalism for reasoning under uncertainty (Shafer, 1976). It can be viewed as a generalization of probability theory with special advantages in its treatment of ambiguous data and the ignorance arising from it. We summarize a few basic concepts from Dempster-Shafer theory.

Let X denote a set of fault modes that are mutually exclusive and exhaustive. In the Dempster-Shafer theory, the set X is called a *frame of discernment*, and each fault mode in X is called a *singleton*. The power set of X (the set of all subsets of X) will be denoted as $P(X)$. Any subset of X (including X itself, the empty set $\emptyset$, and each of the singletons) is called a *hypothesis*.

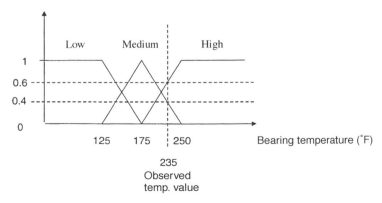

Figure 5.22 Observed temperature value of 235°F.

In diagnosis, each piece of evidence (sign or symptom) is considered to be either present or absent. Associated with each piece of evidence is a basic probability assignment (BPA). A BPA is a function that assigns to each hypothesis a basic probability number (BPN). The BPN for a hypothesis $A \in P(X)$ is denoted by $m(A)$. The BPNs are in the range [0 1] and must sum to 1 over all hypotheses. If the sum of the BPNs for all of X's proper subsets equals k, then the BPN for X is defined to be $1 - k$. Furthermore, the BPN for $\emptyset$ is defined to be zero.

The *belief* in A (sometimes also called the *lower probability* of A) is the sum of the BPN of A plus the BPNs of all of A's proper subsets. The belief function is denoted by Bel(A). A belief in a hypothesis is also in the range of [0 1], and intuitively, 1 means total belief in the hypothesis, and 0 means lack of belief (not to be confused with disbelief). The distinction between $m(A)$ and Bel(A) is crucial: $m(A)$ is the amount of belief committed exactly to A, whereas Bel(A) is the total amount of belief committed to A. The two numbers are not necessarily the same [$m(A) \leq$ Bel(A)].

As noted earlier, a belief of 0 for a hypothesis does not mean disbelief; rather, it means a lack of belief. The disbelief in a hypothesis A is given by Bel(A^c), where A^c is used to denote the set-theoretic complement of A. The *plausibility* of A (sometimes also called the *upper probability* of A) is defined to be PI(A) = $1 -$ Bel(A^c). Plausibility can be thought of as the extent to which the evidence does not support the negation of a hypothesis. Note that Bel(A) must be no greater than PI(A) for all A[Bel(A) $\leq$ PI(A)]. The interval between Bel(A) and PI(A) is known as the *belief interval*. Intuitively, the smaller the magnitude of the belief interval, the more conclusive the piece of evidence is with respect to the hypothesis.

A major feature of Dempster-Shafer theory is Dempster's rule of combination, which allows one to combine or integrate effectively evidence derived from multiple sources. Its impact on fault diagnosis is obvious because it provides a means to combine fault-feature or fault-classifier outputs while accounting for incomplete or even conflicting information. If m_1 and m_2 are two basic probability assignments (BPAs) or mass functions induced by two independent sources, then the BPA induced by the conjunction of evidence is calculated by

$$m_1 \otimes m_2 \ (C) = \frac{\displaystyle\sum_{A_i \cap B_j = C} m_1(A_i)m_2(B_j)}{1 - \displaystyle\sum_{A_i \cap B_j = \emptyset} m_1(A_i)m_2(B_j)} \tag{5.7}$$

when $A \neq \emptyset$. By definition, $\emptyset$ must be assigned BPN zero. The division by $1 - \displaystyle\sum_{A_i \cap B_j = \emptyset} m_1(A_i)m_2(B_j)$ is a normalization to force the BPNs to sum to 1.

Intuitively, the combination rule narrows the hypothesis set by distributing the BPNs (hence belief) into smaller and smaller subsets. The resulting BPNs then can be used to compute Bel(*C*) and PI(*C*) for the hypothesis *C*.

An important advantage of the Dempster-Shafer theory is its ability to express degrees of ignorance. Commitment of belief to a subset does not force the remaining belief to be committed to its complement, that is, Bel(*B*) + Bel(*B^c*) ≤ 1. Therefore, disbelief can be distinguished from a lack of evidence for belief. Another advantage is the several aspects of the theory that are intuitively pleasing. For example, the notion that evidence can be relevant to subsets, rather than just to singletons, nicely captures the idea of narrowing in on a diagnosis.

The theory assumes that pieces of evidence are independent. However, it is not always reasonable to assume independent evidence. Also, there is no theoretical justification for the combination rule (Henkind and Harrison, 1988).

Example 5.4   A Fuzzy Diagnostic System with Dempster-Shafer Combination Rule Applied to the Navy Centrifugal Chiller Problem. A combined fuzzy-logic/Dempster-Shafer approach is used to determine if a failure (or impending failure) has occurred and to assign a degree of certainty or confidence to this declaration. Figure 5.23 depicts the essential elements of the fuzzy-logic-based diagnostic process.

The fuzzy diagnostic system takes features as inputs and then outputs any indications that a failure mode may have occurred in the plant. The fuzzy-logic system structure is composed of four blocks: fuzzification, the fuzzy

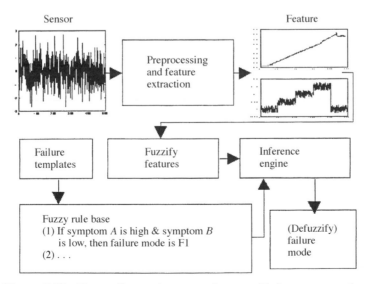

Figure 5.23   Fuzzy diagnostic system layout with feature extraction.

inference engine, the fuzzy rule base, and defuzzification, as shown in Fig. 5.24. The fuzzification block converts features to degrees of membership in a linguistic label set such as low, high, etc. The fuzzy rule base is constructed from symptoms that indicate a potential failure mode. And it can be developed directly from user experience, simulated models, or experimental data.

Suppose that for a "refrigerant charge low" fault mode, the following symptoms are observed:

- Low evaporator liquid temperature
- Low evaporator suction pressure
- Increasing difference between evaporator liquid temperature and chilled water discharge temperature (D-ELT-CWDT)

Three sensors are used to monitor these symptoms:

- Evaporator liquid temperature (ELT)
- Evaporator suction pressure (ESP)
- Chilled water discharge temperature (CWDT)

A graphic representation of two typical fuzzy rules used in diagnosing the refrigerant charge low fault is shown Fig. 5.25. These fuzzy rules are applied to the sensor data, and the crisp fault diagnosis result can be obtained using fuzzy min-max operations (to aggregate the results from multiple fuzzy rules), as illustrated in Fig. 5.26. The last step defuzzifies the resulting output, using the centroid method, to a number between 0 and 100. This is finally compared with a threshold to determine whether or not a failure mode should be reported.

Taking advantage of Dempster's rule of combination, we proceed to estimate the degree of certainty in a particular fault mode declaration. First, we must answer the question: Who are the experts? In this case, the sensors, membership functions, and the associated fuzzy rule base constitute the expert pool. Then, suppose that the following pieces of evidence are available:

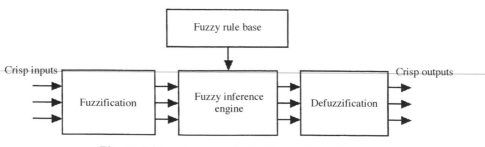

Figure 5.24 The fuzzy-logic diagnostic system.

1. If ELT* is Low   AND   Slope of D-ELT-CWDT* THEN Failure is High

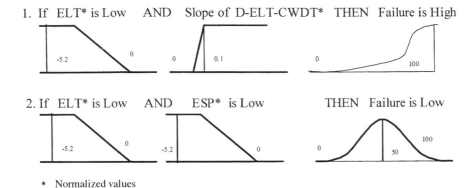

2. If ELT* is Low   AND   ESP* is Low   THEN Failure is Low

* Normalized values

Figure 5.25 Fuzzy rules for the fault diagnosis of refrigerant charge low.

EVAPORATOR LIQUID TEMPERATURE SENSOR (EVALUATED FROM INPUT
MEMBERSHIP FUNCTIONS)

High	0.2
Medium	0.3
Low	0.5

EVAPORATOR SUCTION PRESSURE SENSOR (EVALUATED FROM INPUT
MEMBERSHIP FUNCTIONS)

High	0.1
Medium	0.2
Low	0.7

1. Apply fuzzy operation and implication method

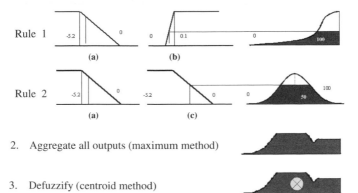

Rule 1

(a)   (b)

Rule 2

(a)   (c)

2. Aggregate all outputs (maximum method)

3. Defuzzify (centroid method)

Figure 5.26 Inference and defuzzification.

We construct a mass function for each sensor using this evidence and then combine them using Dempster's rule of combination, which is give by

$$m_{12}(A) = \frac{\sum_{B \cap C = A} m_1(B) m_2(C)}{1 - K} \qquad K = \sum_{B \cap C = \emptyset} m_1(B) m_2(C)$$

We compare the resulting mass function with the fuzzy rule base. IF evaporator liquid temperature is low AND evaporator suction pressure is low, THEN failure mode is refrigerant charge is low. From the resulting mass function, we might obtain

$m(\{\text{failure mode is refrigerant charge low}\}) = 0.62$
$m(\{\text{failure mode is NOT refrigerant charge low}\}) = 0.14$
$m(\{\text{failure mode is refrigerant charge low}\},$
$\{\text{failure mode is NOT refrigerant charge low}\}) = 0.24$

We therefore can calculate a measure of the degree of certainty from the mass function according to

$DOC(\{\text{Failure Mode is Refrigerant Charge Low}\})$
$= m(\{\text{Failure Mode is Refrigerant Charge Low}\})$
$- Belief(\{\text{Failure Mode is NOT Refrigerant Charge Low}\})$
$= 0.62 - 0.14 = 0.48$
Normalizing this in the range [0, 1], we have a measure of *Belief* = 0.73.

Figure 5.27 depicts the time evolution of the degree of certainty for the refrigerant charge low as additional evidence accumulates. A threshold level is set by the designer aiming at reducing false alarms while achieving an accurate fault mode assessment but still declaring the diagnostic outcome as early as possible.

The same procedure can be applied to combine evidence from multiple sensors in the diagnosis of evaporator tubes freezing fault, as shown below:

Temperature sensor tells us (expert 1)

$M(\{\text{evaporator tubes are freezing}\}) = 0.5$
$M(\{\text{evaporator tubes are not freezing}\}) = 0.3$
$M(\{\text{evaporator tubes are freezing, evaporator tubes are not freezing}\})$
$= 0.2$

Pressure sensor tells us (expert 2)

$M(\{\text{evaporator tubes are freezing}\}) = 0.2$
$M(\{\text{evaporator tubes are not freezing}\}) = 0.1$
$M(\{\text{evaporator tubes are freezing, evaporator tubes are not freezing}\})$
$= 0.7$

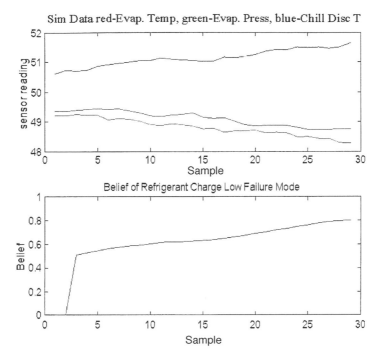

Figure 5.27 Degree of certainty calculation for refrigerant charge low.

Resulting combined evidence:

$M(\{\text{evaporator tubes are freezing}\}) = 0.551$
$M(\{\text{evaporator tubes are not freezing}\}) = 0.292$
$M(\{\text{evaporator tubes are freezing, evaporator tubes are not freezing}\})$
$= 0.157$

5.4.4 The Wavelet Neural Network

The wavelet neural network (WNN) is also used as one component of the classifier. Considering a typical process, the fault modes may include leakage, corrosion, crack, etc. It is generally possible to break down the sensor data into two categories: low-bandwidth data, such as temperature data, flow rate data, etc., and high-bandwidth data, such as vibration data, acoustic data, etc. The data from the first category may vary slowly, whereas those from the second category may change rapidly. The fuzzy-logic classifier operates on the incoming low-bandwidth sensor data, whereas the WNN classifies the fault modes with high-bandwidth data.

The WNN is a tool that maps complex features or fault indicators to their respective fault modes. Over the past few years, in addition to the WNN, polynomial neural-network (PNN), probabilistic neural-network, radial basis function neural-network, and neurofuzzy constructs, among others, have been applied successfully to fault diagnosis and prognosis. Many investigators also have adopted commercial off-the-shelf (COTS) type neural-network routines based on the classic backpropagation principle (MATLAB, 1994). Examples of such data-driven methods abound in the technical literature (Ivakhnenko et al., 1993). PNN is a network representation of $R^n \rightarrow R$ functions. The network nodes (*neurons*) are low-order polynomials of a few inputs. The PNN design problem involves the synthesis of the $R^n \rightarrow R$ mapping from a database of input/output pairs. The objective is to find a net structure (size and topology) and its parameters (weights) so as to represent the database but also preserve the network's generalization properties. The first PNN was synthesized with the group method of data handling (GMDH) (*www.gmdh.net*).

The WNNs found application in fault diagnosis and prognosis because of their multiresolution property and other attractive attributes. Wavelets are a family of transforms with windows of variable width that allow the resolution of a signal at various levels of the Heisenberg uncertainty tradeoff ($\Delta\omega\Delta x$). The resulting neural network is a universal approximator, and the time-frequency localization property of wavelets leads to reduced networks at a given level of performance (Zhang and Benveniste, 1992).

In a neural network, a finite-duration function $f : R \rightarrow R$ can be represented as the weighted sum of local and orthogonal basis functions given by

$$f(x) = \sum_{p=-\infty}^{\infty} \sum_{q=-\infty}^{\infty} c_{p,q}\psi_{p,q}(x)$$

where the coefficients $c_{p,q}$ are the projections of $F(x)$ on the basis functions $\psi_{p,q}(x)$. Wavelets are a kind of local and orthogonal basis function with the time (or space) and frequency atoms of the form

$$\psi_{p,q}(x) = \sqrt{a_q}\psi[a_p(x - b_p)]$$

where a_p is called the *dilation parameter,* b_p is the *translation parameter,* and ψ is the mother wavelet of a particular choice. Figure 5.28 shows typical wavelets. A multidimensional wavelet may be expressed as $\psi_{a,b}(x) = \sqrt{a^n}\psi(a\|x - b\|)$. Figure 5.29 shows a "Mexican hat" type of wavelet in this category.

Figure 5.28 Typical wavelets.

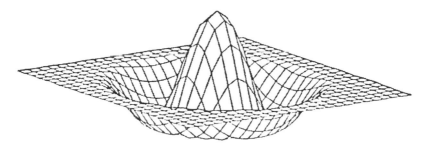

Figure 5.29   A multidimensional wavelet.

A family of wavelets derived from translations and dilations of a single (mother) function is used to synthesize a WNN. Figure 5.30 shows a typical WNN. The output of each node is a nonlinear function of all its inputs. The network represents an expansion of the nonlinear relationship between inputs x and outputs $f(x)$ or y into a space spanned by the functions represented by the activation functions of the network's nodes.

The WNN design task involves identification of its structure and parameters. The *network structure* is determined by the number of inputs and outputs, the number of wavelet nodes, and the connection topology. The structure identification problem is reduced to determination of the number of wavelet nodes because the number of inputs and outputs is determined by the problem definition, and the internal topology is defined to allow for full connectivity. Training data are clustered via k-means clustering in the input-output product space, and each cluster represents a potential WNN node. The WNN parameters are determined via a Levenberg-Marquardt algorithm (Gill et al., 1981), which is a compromise between the steepest descent and the Newton methods. For a fixed network structure, training of the WNN requires a minimization of the arithmetic average squared modeling error with respect to the network parameters. Figure 5.31 depicts a static and a dynamic WNN. The dynamic WNN version is used in fault prognosis, as discussed in Chap. 6.

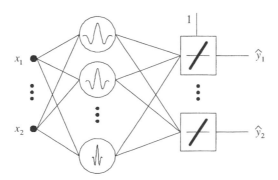

Figure 5.30   A typical wavelet neural network.

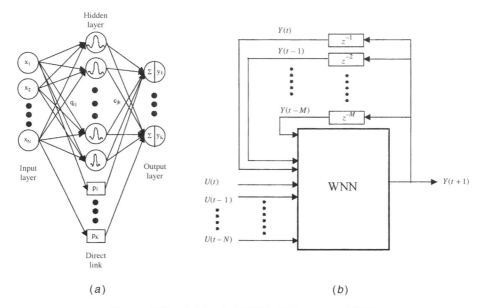

Figure 5.31 (*a*) Static WNN. (*b*) Dynamic WNN.

A basic fault classification scheme employing a WNN is shown in Fig. 5.32. The WNN is trained offline and then implemented online. A schematic representation of these two processes is shown in Fig. 5.33. Figure 5.34 shows flowcharts for the offline training routine and the online implementation, respectively.

5.4.5 Wavelet Fuzzy Fault Classifier

The wavelet functions can be used as feature extractors only. In conjunction with a classifier—fuzzy-rule set or neural network—they form a diagnostic tool that takes advantage of the multiresolution property and provide a trade-off between the time and frequency domains. Thus more frequency scales are employed when the fault information resides primarily in the frequency domain. Figure 5.35 shows a schematic representation of such an architecture. An information matrix is derived whose elements are coefficients of the wavelet signal expansion. A fuzzy version of the wavelet coefficients is entered as inputs to the fuzzy inferencing engine. The latter's output are the fault modes corresponding to the fault indicators (Mufti and Vachtsevanos, 1995).

5.4.6 Multimodel and Filter-Bank Fault Classification

An alternative approach to fault classification can be obtained by using the fault-detection filter theory that creates an estimated output based on a base-

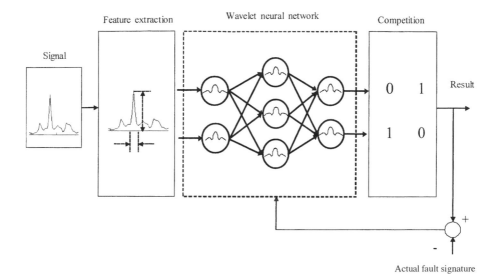

Figure 5.32 The WNN fault classification paradigm.

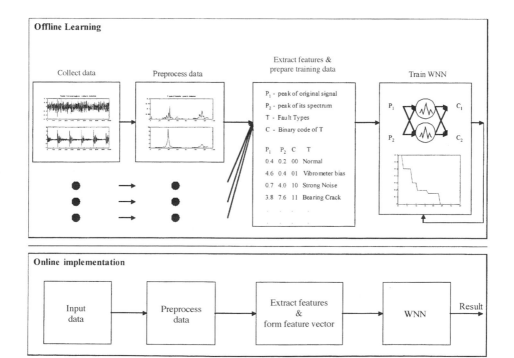

Figure 5.33 The WNN offline learning and online implementation.

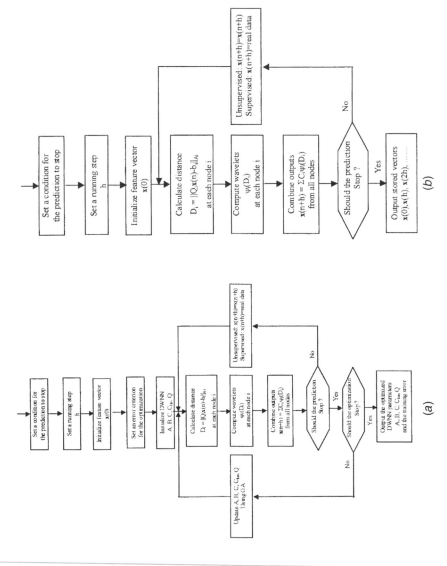

Figure 5.34 (*a*) Offline WNN flowchart. (*b*) Online implementation.

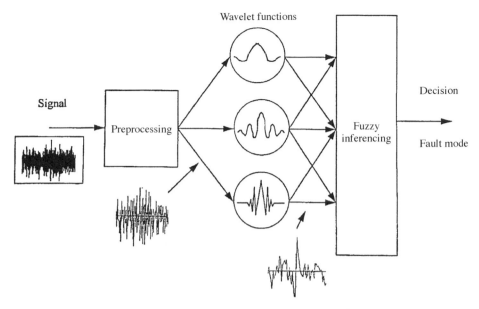

Figure 5.35   A wavelet-fuzzy fault classifier.

line model. The filter residual, composed of the difference between the estimated output and the measured one, is constructed to have an invariant direction in the presence of an element from a set of a priori faults that allows both detection and identification. Figure 5.36 depicts a failure-detection system involving a failure-sensitive primary filter. The variable ξ denotes information concerning detected failures.

Failure-sensitive filters are designed to be sensitive to new abrupt changes in the data stream and are applicable to a wide variety of failures. Most designs are limited to time-invariant systems. Classic examples of failure-sensitive filters include the Beard-Jones detection filter (Jones, 1973), Will-

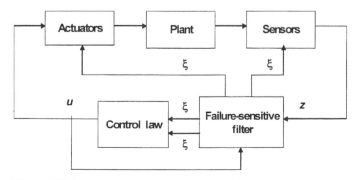

Figure 5.36   General configuration of failure detection system.

sky's multiple-hypothesis filter design method (Willsky, 1976), and Chen's jump-process formulation (Chen et al., 2001), among others.

Innovations-based detection systems address a wide range of actuator and sensor failures and are capable of isolating different failures by using knowledge of the individual effects. They employ a linearly growing bank of matched filters to compute the best failure estimates with a consequent increase in computational complexity. Qualification factors of such failure monitoring systems include

- *Isolability.* A measure of the model's ability to distinguish between certain specific failure modes. Enabling technologies include incidence matrices involving both deterministic (zero-threshold) and statistical (high-threshold) isolability.
- *Sensitivity.* A qualitative measure characteristic of the size of failures. This factor depends on the size of the respective elements in the system's matrices, noise properties, and the time to failure. Filtering typically is used to improve sensitivity, but it is rather difficult to construct a straightforward framework.
- *Robustness.* This factor refers to the model's ability to isolate a failure in the presence of modeling errors. Improvements in robustness rely on algebraic cancelation that desensitizes residuals according to certain modeling errors.

An example of an innovation-based detection system is Willsky's generalized likelihood ratio test (GLRT) (Willsky, 1976). The system dynamic model is customized to account for a specific failure mode, that is, actuator or sensor failure. The dynamic model, therefore, is provided as follows:

Dynamic Jump

$$x(k + 1) = \Phi(k)x(k) + B(k)u(k) + w(k) + v\delta_{k+1,\theta} \qquad (5.8)$$

where θ is unknown time of failure, and

$$\delta_{ij} = \begin{cases} 1 & i = j \\ 0 & \text{otherwise} \end{cases}$$

Dynamic Step

$$x(k + 1) = \Phi(k)x(k) + B(k)u(k) + w(k) + v\sigma_{k+1,\theta} \qquad (5.9)$$

where

$$\sigma_{ij} = \begin{cases} 1 & i \geq j \\ 0 & i < j \end{cases}$$

Sensor Jump

$$z(k) = Hx(k) + ju(k) + v(t) + v\delta_{k,\theta} \tag{5.10}$$

Sensor Step

$$z(k) = Hx(k) + ju(k) + v(t) + v\sigma_{k,\theta} \tag{5.11}$$

Innovations are then given by

$$\gamma(k) = G(k; \theta)v + \tilde{\gamma}(k) \tag{5.12}$$

where G is a failure signature matrix, and γ is the no-failure innovation. Using a chi-squared random variable defined by

$$I(k) = \sum_{j=k-N+1}^{k} \gamma'(j)V^{-1}(j)\gamma(j) \tag{5.13}$$

we can determine if there is a failure based on

$$l(k) > \varepsilon \Rightarrow \text{failure}$$

$$l(k) \leq \varepsilon \Rightarrow \text{no failure} \tag{5.14}$$

5.5 DYNAMIC SYSTEMS MODELING

Modeling in terms of ordinary differential equations (ODEs) provides a powerful method for extracting physical models of processes and can be performed in terms of Lagrangian or Hamiltonian dynamics, approximation methods applied to partial differential equations, distributed models, and other techniques. ODEs can be further expressed in the state-variable form, which opens the door for exploiting a host of modern techniques for signal and system processing based on firmly established numerical computational techniques.

Example 5.5 Physical Model Determination of Fault Modes, Feature Vector, and Sensors for an Electrohydraulic Flight Actuator. This example is taken from Skormin et al., 1994. We will use physical modeling principles to determine the feature vector for the electrohydraulic flight actuator shown in Fig. 5.37.

There are several main subsystems, and their associated fault modes are

Control surface loss
Excessive bearing friction
Hydraulic system leakage
Air in hydraulic system
Excessive cylinder friction
Malfunction of pump control valve
Rotor mechanical damage
Motor magnetism loss

Using physical modeling principles, differential equations can be used to model the various subsystems, and the resulting transfer functions are

Motor dynamics

$$\frac{\omega(s)}{T(s)} = \frac{1}{Js + B}$$

Pump/piston dynamics

$$\frac{X(s)}{F(s)} = \frac{1}{(M_p s + B_p)s}$$

Actuator system dynamics

$$\frac{P(s)}{R(s)} = \frac{1}{(A^2/K)s + L}$$

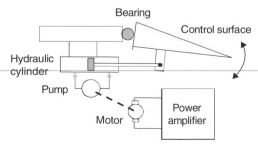

Figure 5.37   Electrohydraulic flight actuator.

where the physical parameters are motor inertia J and damping B, piston mass M_p and damping B_p, and hydraulic actuator area A and stiffness K. For a specific actuator, these parameters will have specific numerical values.

Fault Feature Vector. For fault diagnosis purposes, it is desirable to relate the physical model parameters suggested earlier to the fault modes. This parameter to fault mode mapping may be achieved by polling domain experts, that is, operators, equipment manufactures, flight technicians, etc. or through simulations. Table 5.3 shows a typical actuator fault condition in relation to fault-mode mapping. Therefore, the feature vector may be selected as the parameter vector:

$$\varphi(t) = [J \ B \ M_p \ B_p \ K \ L]^T$$

Sensor Selection. The feature-vector entries must be measured next through an appropriate sensor suite. In most cases, the elements of the feature vector are physical parameters that, unfortunately, cannot be measured directly. A recursive-least-squares (RLS) parameter-estimation approach may be called on to resolve this dilemma. It is instructive to digress at this point and present some fundamental concepts of system identification before returning to our actuator example and employing an RLS technique to estimate the feature-vector entries given measurable signals from an available suite of sensors. Chapter 6 gives a summary of system dynamic models based on ARMA, ARMAX, and ARIMA, as well as parameter identification.

5.5.1 Kalman Filter

The Kalman filter (Lewis, 1986) is a dynamic systems tool for estimating unknown states by combining current measurements with the most recent state

TABLE 5.3 Fault Condition for the Actuator Based on the Physical Parameters

Condition	Fault Mode
IF (leakage coeff. L is large)	THEN (fault is hydraulic system leakage)
IF (motor damping coefficient B is large) AND (piston damping coefficient B_p is large)	THEN (fault is excess cylinder friction)
IF (actuator stiffness K is small) AND (piston damping coefficient B_p is small)	THEN (fault is air in hydraulic system)
Etc.	Etc.

estimate. It provides estimates of quantities of interest that may in themselves not be directly measurable. In the case of linear systems and Gaussian noise, these estimates are optimal in the sense of minimizing an estimation-error covariance. Knowledge of noise processes is used to minimize this estimation-error covariance via the optimal determination of the so-called Kalman gain. The Kalman filter also provides *confidence bounds* for the estimates in terms of the error covariance matrix. These are important in prognosis applications, as we will see in Chap. 6. There are well-developed techniques for implementing the Kalman filter, using MATLAB and other software packages.

A discrete-time system with internal state $\mathbf{x}_k$ and sensor measurements $\mathbf{z}_k$ may be described in terms of the recursive difference equation, that is,

$$\mathbf{x}_{k+1} = \mathbf{A}_k\mathbf{x}_k + \mathbf{B}_k\mathbf{u}_k + \mathbf{G}_k\mathbf{w}_k$$

$$\mathbf{z}_k = \mathbf{H}_k\mathbf{x}_k + \mathbf{v}_k \tag{5.15}$$

where $\mathbf{u}_k$ is a control input, $\mathbf{w}_k$ is a *process noise* that captures uncertainties in the process dynamics, such as modeling errors and unknown disturbances (e.g., wind gusts in aircraft), and $\mathbf{v}_k$ is a *measurement noise*. Both these noises are assumed to be white Gaussian noises, independent of each other and of the initial state $\mathbf{x}_0$, which is unknown. Let

$$\mathbf{Q}_k = \mathbf{E}(\mathbf{w}_k\mathbf{w}_k^T) = \text{noise covariance of } \mathbf{w}_k \tag{5.16}$$

$$\mathbf{R}_k = \mathbf{E}(\mathbf{v}_k\mathbf{v}_k^T) = \text{noise covariance of } \mathbf{v}_k \tag{5.17}$$

with $\mathbf{R}_k \rangle 0$.

The Kalman filter is a digital signal-processing routine that combines the current sensor measurements and the current state estimates to produce the optimal state estimate in such a manner that the error covariance is minimized, assuming that all processes are Gaussian. The Kalman filter is given in terms of the *state estimate* at time k, denoted by $\hat{\mathbf{x}}_k$, and the *error covariance* $\mathbf{P}_k = \mathbf{E}[(\mathbf{x}_k - \hat{\mathbf{x}}_k)(\mathbf{x}_k - \hat{\mathbf{x}}_k)^T]$. The Kalman filtering algorithm consists of the following steps:

Kalman Filter Algorithm

Initialization: $\mathbf{P}_0 = \mathbf{P}_{\mathbf{x}_0}$, $\hat{\mathbf{x}}_0 = \bar{\mathbf{x}}_0$; set the estimate and covariance of the unknown initial state.

Time update (propagation effect of system dynamics):
   Error covariance update: $\mathbf{P}_{k+1}^- = \mathbf{A}_k\mathbf{P}_k\mathbf{A}_k^T + \mathbf{G}_k\mathbf{Q}_k\mathbf{G}_k^T$

Estimate update: $\hat{\mathbf{x}}_{k+1}^- = \mathbf{A}_k\hat{\mathbf{x}}_k + \mathbf{B}_k\mathbf{u}_k$
Measurement update at time $k + 1$ (effect of measurement z_{k+1}):
   Error covariance update: $\mathbf{P}_{k+1} + [(\mathbf{P}_{k+1}^-)^{-1} + \mathbf{H}_{k+1}^T\mathbf{R}_{k+1}^{-1}\mathbf{H}_{k+1}]^{-1}$
   Estimate update: $\hat{\mathbf{x}}_{k+1} = \hat{\mathbf{x}}_{k+1}^- + \mathbf{P}_{k+1}\mathbf{H}_{k+1}^T\mathbf{R}_{k+1}^{-1}(\mathbf{z}_{k+1} - \mathbf{H}_{k+1}\hat{\mathbf{x}}_{k+1}^-)$

In this algorithm, the notation $\hat{\mathbf{x}}_{k+1}^-$, $\mathbf{P}_{k+1}^-$ denotes the *predicted* estimate and error covariance at time $k + 1$ before the sensor measurement z_{k+1} at time $k + 1$ has been taken into account.

Figure 5.38 depicts schematically the error covariance update timing diagram. The time update generally increases the error covariance owing to the injection of uncertainty by the process noise $\mathbf{w}_k$, whereas the measurement update generally decreases the error covariance owing to the injection of information by the measurement z_k. The Kalman filter iteratively provides the optimal state estimate that minimizes the error covariance at each time step. The error covariance, in turn, gives confidence bounds on the estimates. A smaller error covariance means more accurate estimates.

To simplify the discussion and implementation of the filter, one defines the *Kalman gain* as

$$\mathbf{K}_k = \mathbf{P}_k \mathbf{H}_k^T \mathbf{R}_k^{-1} = \mathbf{P}_k^- \mathbf{H}_k^T (\mathbf{H}_k \mathbf{P}_k^- \mathbf{H}_k^T + \mathbf{R}_k)^{-1} \qquad (5.18)$$

where the second equality takes some steps to show. Then one may write the measurement update as

$$\hat{\mathbf{x}}_{k+1} = \hat{\mathbf{x}}_{k+1}^- + \mathbf{K}_{k+1}(\mathbf{z}_{k+1} - \mathbf{H}_{k+1}\hat{\mathbf{x}}_{k+1}^-) \qquad (5.19)$$

The estimate of the measured output is $\hat{\mathbf{z}}_k = \mathbf{H}_k\hat{\mathbf{x}}_k$, and the *predicted output* at time $k + 1$ is given as $\hat{\mathbf{z}}_{k+1}^- = \mathbf{H}_{k+1}\mathbf{x}_{k+1}^-$. According to the measurement update equation, if the predicted output is equal to the actual sensor reading z_{k+1} at time $k + 1$, then the state estimate does not need to be updated, and one simply takes $\hat{\mathbf{x}}_{k+1} = \hat{\mathbf{x}}_{k+1}^-$, the predicted value.

A diagram showing the Kalman filter appears in Fig. 5.39. Actual software implementation of the Kalman filter relies on several routines with excellent numerical stability based on the Riccati equation, including the square-root

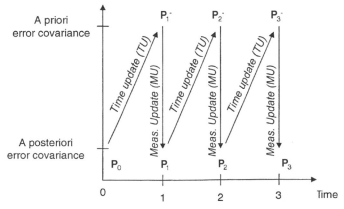

Figure 5.38 Error covariance update timing diagram.

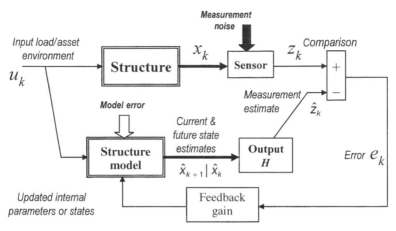

Figure 5.39 The Kalman filter.

formulations (Lewis, 1986). We will illustrate the utility of Kalman filtering in fault diagnosis via an example.

Example 5.6 Kalman Filter State Estimation. MATLAB software is given in the Appendix to simulate a state-variable system

$$\mathbf{x}_{k+1} = \mathbf{A}\mathbf{x}_k + \mathbf{B}\mathbf{u}_k + \mathbf{G}\mathbf{w}_k$$

$$\mathbf{z}_k = \mathbf{H}\mathbf{x}_k + \mathbf{v}_k$$

In this example, we consider an actuator known to have the state variable model

$$\mathbf{x}_{k+1} = \begin{bmatrix} 0 & 1 \\ -0.9425 & 1.9 \end{bmatrix} \mathbf{x}_k + \begin{bmatrix} 0 \\ 1 \end{bmatrix} u_k + \mathbf{w}_k$$

$$z_k = [1 \quad 0]\mathbf{x}_k + \mathbf{v}_k$$

with the process noise $\mathbf{w}_k$ as a normal 2-vector with each component having zero mean and variance 0.1, denoted as normal (0, 0.1). The measurement noise v_k is known to be normal (0, 0.1).

The output response z_k of the actuator with a unit step test input $\mathbf{u}_k$ is measured and is shown in Fig. 5.40. From the measured input and output signals, it is desired to estimate the two state components $x_1(k)$ and $x_2(k)$, which are not directly measurable. As shown in Fig. 5.41, this is a complex problem involving the estimation of two quantities of interest given available sensor measurements of only one output data stream.

The Kalman filter code in the Appendix allows us to reconstruct the states using the measured signals u_k and z_k. The actual states and their estimated

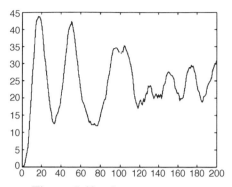

Figure 5.40 System output z_k.

values are plotted in Fig. 5.42. Part (a) shows a plot of the first state $x_1(k)$ and its estimate $\hat{x}_1(k)$, and part (b) shows the second state $x_2(k)$ and its estimate $\hat{x}_2(k)$. The estimates differ by a small amount from the actual values owing to the unknown noise, but the agreement is excellent, and it is difficult to see that there are actually two signals in each plot.

Confidence Bounds on the Estimates. It is important to note that the Kalman filter also gives confidence bounds on the estimates, which are important (especially in prognosis applications) in knowing how accurate the estimates are. These are obtained from the error covariance matrix $\mathbf{P}_k$. Writing the error covariance as

$$\mathbf{P}_k = \begin{bmatrix} p_{11} & p_{12} \\ p_{21} & p_{22} \end{bmatrix}$$

The covariance entries for this example, as computed by the software in the Appendix during the Kalman filtering run, are shown in Fig. 5.43. They in-

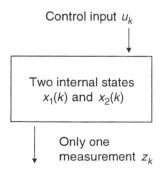

Figure 5.41 Estimating unknown internal states from measured input/output data.

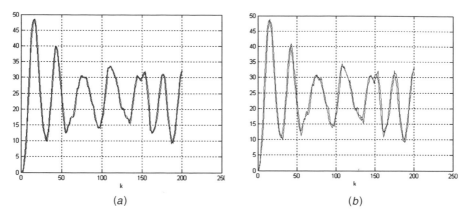

(a)   (b)

Figure 5.42 Actual and estimated values of states: (*a*) state component 1; (*b*) state component 2 (the plots are so close that they are in effect on top of each other).

dicate that the estimate for $x_1(k)$ is better than the estimate for $x_2(k)$ because its covariance is smaller.

5.5.2 Recursive Least-Squares (RLS) Parameter Estimation

A special case of the Kalman filter is known as *recursive least-squares* (*RLS*) *parameter identification.* It is useful for a broad class of parameter identification problems that occur in diagnosis applications when the dynamic model is not known. A SISO system can be described in state-space form as

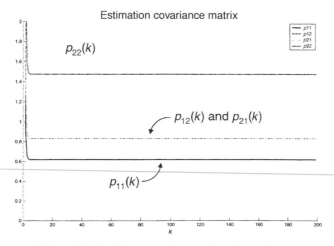

Figure 5.43 Error covariances for the estimates.

$$\mathbf{x}_{k+1} = \mathbf{A}\mathbf{x}_k + \mathbf{B}u_k, \qquad \mathbf{x}_0$$

$$y_k = \mathbf{H}\mathbf{x}_k$$

The input-output form is given by

$$Y(z) = H(z\mathbf{I} - \mathbf{A})^{-1}\mathbf{B}U(z) = H(z)U(z)$$

where $H(z)$ is the transfer function. In terms of the numerator and denominator polynomials, one has

$$Y(z) = H(z)U(z) = \frac{b_0 z^m + b_1 z^{m-1} + \cdots + b_{m-1} z + b_m}{z^n + a_1 z^{n-1} + \cdots + a_{n-1} z + a_n} U(z)$$

where the denominator degree is n, the numerator degree is $m \le n$, and the relative degree is $n - m$. This is written in autoregressive moving-average (ARMA) form as

$$y_k = -a_1 y_{k-1} - \cdots - a_n y_{k-n} + b_0 u_{k-d} + b_1 u_{k-d-1} + \cdots + b_m u_{k-d-m}$$

$$y_k = -\sum_{i=1}^{n} a_i y_{k-i} + \sum_{i=0}^{m} b_i u_{k-i-d}$$

The system delay is $d = n - m$.

It is often the case that the system model is unknown. Then, under certain general conditions, the unknown system parameters a_i, b_i can be *identified* from observations of its input u_k and output y_k using RLS. This is the same as identifying the unknown system matrices $\mathbf{A}$, $\mathbf{B}$, $\mathbf{H}$.

Do not confuse the *system identification problem* with the *state estimation problem,* for example, as solved by the Kalman filter. In the latter, one requires that the system model $\mathbf{A}$, $\mathbf{B}$, $\mathbf{H}$ be known, and the internal state x_k is estimated using input and output measurements.

To determine the unknown parameters, the ARMA form is written as

$$y_k = [-y_{k-1} \ -y_{k-2} \cdots -y_{k-n} \ u_{k-d} \cdots u_{k-d-m}] \begin{bmatrix} a_1 \\ a_2 \\ \vdots \\ a_n \\ b_0 \\ \vdots \\ b_m \end{bmatrix} + v_k$$

or $\qquad y_k = \mathbf{h}_k^T \boldsymbol{\theta} + v_k$

where the unknown system *parameter vector* is

$$\boldsymbol{\theta} = [a_1 \cdots a_n \; b_0 \cdots b_m]^T$$

and the known *regression matrix* is given in terms of previous outputs and inputs by

$$\mathbf{h}_k^T = [-y_{k-1} - y_{k-2} \cdots -y_{k-n} \; u_{k-d} \cdots u_{k-d-m}]$$

Since there are $n + m + 1$ parameters to estimate, one needs n previous output values and $m + 1$ previous input values. The unknown parameter vector can be found from the regression vector given the following recursive technique:

RLS Algorithm

$$\mathbf{P}_{k+1} = \mathbf{P}_k - \mathbf{P}_k \mathbf{h}_{k+1}(\mathbf{h}_{k+1}^T \mathbf{P}_k \mathbf{h}_{k+1} + \sigma_{k+1}^2)^{-1} \mathbf{h}_{k+1}^T \mathbf{P}_k \qquad (5.20)$$

$$\hat{\boldsymbol{\theta}}_{k+1} = \hat{\boldsymbol{\theta}}_k + \mathbf{P}_{k+1} \frac{\mathbf{h}_{k+1}}{\sigma_{k+1}^2}(y_{k+1} - \mathbf{h}_{k+1}^T \hat{\boldsymbol{\theta}}_k)$$

In the RLS algorithm, $\mathbf{P}_k$ is the error covariance, and $\hat{\boldsymbol{\theta}}_k$ is the estimate at time k of the system parameters $\boldsymbol{\theta}$. Thus the entries of $\mathbf{P}_k$ give an idea of how accurate the estimates are. We shall see that this is especially important in Chap. 6.

One must know the system degree n and the delay d for RLS identification because they are needed to construct the regression vector $\mathbf{h}_{k+1}$. There are various techniques for estimating these numbers.

To initialize the RLS algorithm, one may select $\hat{\boldsymbol{\theta}}_0 = \mathbf{0}$, $\mathbf{P}_0 = \gamma \mathbf{I}$, with γ a large positive number. This reflects the fact that initially nothing is known about the unknown parameters.

In the case of scalar outputs, the terms $(\mathbf{h}_{k+1}^T \mathbf{P}_k \mathbf{h}_{k+1} + \sigma_{k+1}^2)$ is a scalar, so the RLS algorithm requires no matrix inversions. Equation (5.20) is known as the *Riccati equation* (RE).

Note that the RLS algorithm can be derived by applying the Kalman filter to the system

$$\boldsymbol{\theta}_{k+1} = \boldsymbol{\theta}_k$$

$$y_k = \mathbf{h}_k^T \boldsymbol{\theta} + v_k$$

In fact, it can be shown that the RLS algorithm also can be written as

$$\mathbf{K}_{k+1} = \mathbf{P}_k \mathbf{h}_{k+1}(\mathbf{h}_{k+1}^T \mathbf{P}_k \mathbf{h}_{k+1} + \sigma_{k+1}^2)^{-1}$$

$$\hat{\boldsymbol{\theta}}_{k+1} = \hat{\boldsymbol{\theta}}_k + \mathbf{K}_{k+1}(y_{k+1} - \mathbf{h}_{k+1}^T \hat{\boldsymbol{\theta}}_k)$$

$$\mathbf{P}_{k+1} = (\mathbf{I} - \mathbf{K}_{k+1} \mathbf{h}_{k+1}^T) \mathbf{P}_k$$

where $\mathbf{K}_k$ is the Kalman gain.

In diagnostic applications, the system can be slowly time-varying (e.g., as parameters change with wear). A method for handling slowly time-varying system parameters is to employ *exponential data weighting*. This uses $Q = 0$ and time-varying measurement noise covariance of the form

$$\sigma_k^2 = R\alpha^{-2(k+1)}$$

with $R > 0$, so that the noise covariance decreases with time. This suggests that the measurements improve with time and so gives more weight to more recent measurements. That is, as data become older, they are exponentially discounted. The RLS algorithm with exponential data weighting is given as

RLS Algorithm with Exponential Data Weighting

$$\mathbf{P}_{k+1} = \alpha^2\left[\mathbf{P}_k - \mathbf{P}_k\mathbf{h}_{k+1}\left(\mathbf{h}_{k+1}^T\mathbf{P}_k\mathbf{h}_{k+1} + \frac{R}{\alpha^2}\right)^{-1}\mathbf{h}_{k+1}^T\mathbf{P}_k\right]$$

$$\mathbf{K}_{k+1} = \mathbf{P}_k\mathbf{h}_{k+1}\left(\mathbf{h}_{k+1}^T\mathbf{P}_k\mathbf{h}_{k+1} + \frac{R}{\alpha^2}\right)^{-1}$$

$$\hat{\boldsymbol{\theta}}_{k+1} = \hat{\boldsymbol{\theta}}_k + \mathbf{K}_{k+1}(y_{k+1} - \mathbf{h}_{k+1}^T\hat{\boldsymbol{\theta}}_k)$$

In order for RLS to identify the unknown parameters successfully, the input must excite all the modes of the unknown system. A time sequence u_k is said to be persistently exciting (PE) of order N if

$$\sum_{i=1}^{K}\begin{bmatrix} u_i \\ u_{i-1} \\ \vdots \\ u_{i-(N-1)} \end{bmatrix}[u_i \quad u_{i-1} \quad \cdots \quad u_{i-(N-1)}] > \rho I$$

for some K, where $\rho > 0$. It can be shown that one can solve for the $n + m + 1$ unknown ARMA parameters if u_k is PE of order $n + m + 1$. A signal u_k is PE of order N if its two-sided spectrum is nonzero at N points (or more). For example, a unit step function is PE of order 1, and a sine wave is PE of order 2.

In practical applications, if the input is not PE, one can inject a small *probing signal input* to excite all the system modes and allow for RLS parameter identification. To identify, for instance, six parameters, the probing signal must have at least three sine-wave components of different frequencies.

Example 5.7 RLS System Identification. The measured input u_k and output signals y_k of an unknown system are given in Fig. 5.44. It is known that the system is of second order with a delay $d = 2$. Therefore,

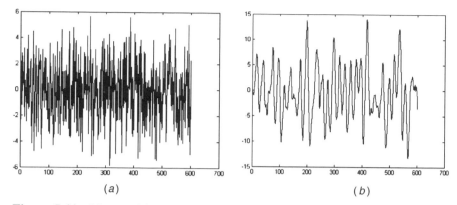

Figure 5.44 Measured input (a) and output (b) signals from an unknown system.

$$y_k = -a_1 y_{k-1} - a_2 y_{k-2} + b_0 u_{k-2}$$

for some unknown system parameters a_1, a_2, b_0. RLS can be used to find the parameters from the known input and output signals.

The RLS algorithm for this system is easy to write and is given in MATLAB code in the Appendix. As a result of the RLS algorithm, the unknown parameters are identified, and the system is found to be

$$y_k = 1.9 y_{k-1} - 0.95 y_{k-2} + 0.2 u_{k-2}$$

If one simulates this system using MATLAB with the input signal in Fig. 5.44a, the result is identical to the output signal given in Fig. 5.44b.

Confidence Bounds on the Estimates. It is important to note that RLS also gives confidence bounds on the estimates, which are given in the covariance matrix P_k.

Example 5.8 Sensor Selection and Feature Vector Estimation Using RLS for Electrohydraulic Flight Actuator. This example is from Skormin and colleagues (1994). In Example 5.5, we showed how to use physical modeling principles to determine a feature vector that is useful in diagnosis of an electrohydraulic flight actuator. If the feature vector is known, the current operating condition of the actuator can be determined, including any potential faults, by using a rule base such as Table 5.3. The feature vector was selected as

$$\varphi(t) = [J\ B\ M_p\ B_p\ K\ L]^T$$

in terms of the physical parameters, which are motor inertia J and damping B, pump mass M_p and damping B_p, and actuator area A, and stiffness K.

Unfortunately, these physical parameters cannot be measured directly via sensors.

Sensor Selection. The inputs and outputs of the dynamic blocks can be measured. For the motor dynamic block, for instance, one can measure armature current, pressure difference, and motor speed. This corresponds to the inputs and outputs of the subsystem in Fig. 5.45. (Note that from the armature current and pressure difference one can compute torque $T(t)$, which is the input to the motor transfer function.) This guides one to place sensors to measure these input/output quantities.

Given these measurable inputs and outputs, one can use system identification techniques such as recursive least squares to determine the parameters from the measured signals. The procedure of using measurable quantities to estimate the signals or parameters of interest is a virtual sensor (VS), as shown in Fig. 5.46, which is a composition of actual physical sensors plus some digital signal processing (DSP) that yields finally as the outputs the parameter features of interest.

To use RLS, one first must convert the continuous-time dynamic model, which is obtained from physical modeling principles, into a discrete-time model, as required for RLS.

Conversion of Continuous-Time Models into Discrete-Time Models. Procedures for converting continuous-time dynamics (i.e., differential equations) into discrete-time dynamics (i.e., difference equations) are given in Lewis (1992). According to the transfer function given in Fig. 5.45, the continuous-time differential equation model of the system is given by

$$J\dot{\omega} + B\omega = T \qquad (5.20)$$

because multiplication by s in the frequency domain corresponds to differentiation in the time domain. Using Euler's backward approximation for differentiation yields,

$$\frac{d\omega}{dt} \approx \frac{\omega_k - \omega_{k-1}}{\Delta} \qquad (5.21)$$

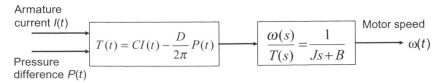

Armature current $I(t)$

$$T(t) = CI(t) - \frac{D}{2\pi} P(t)$$

Pressure difference $P(t)$

$$\frac{\omega(s)}{T(s)} = \frac{1}{Js+B}$$

Motor speed $\omega(t)$

Figure 5.45 Measurable signals in motor subsystem.

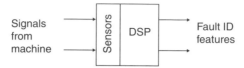

Figure 5.46   The virtual sensor consists of actual sensors plus additional DSP.

with k the time index and Δ the sample period. (One also could use Euler's forward approximation $\dot{\omega} \approx (\omega_{k+1} - \omega_k)/\Delta$, which would replace T_k by T_{k-1}.) The sample period should be chosen to be at least five times smaller than the smallest system time constant. Using this approximation in Eq. (5.20) yields the difference equation:

$$J\omega_k - J\omega_{k-1} + \Delta B\omega_k = \Delta T_k \tag{5.22}$$

or

$$\omega_k = \frac{J}{J + \Delta B}\,\omega_{k-1} + \frac{\Delta}{J + \Delta B}\,T_k \tag{5.23}$$

By defining

$$a_1 = -\frac{J}{J + \Delta B} \qquad b_0 = \frac{\Delta}{J + \Delta B} \tag{5.24}$$

one can write this difference equation as

$$\omega_k = -a_1\omega_{k-1} + b_0 T_k \tag{5.25}$$

This is a system in ARMA form with degree $n = 1$ and delay $d = 0$. Now RLS can be run given the input T_k (which is computed from the measured values of armature current I_k and pressure difference P_k) and the measured output ω_k. The result is the estimated parameters a_1, b_0. From these, the desired system parameters can be computed by solving Eq. (5.24) to obtain

$$J = \frac{-\Delta a_1}{b_0} \qquad B = \frac{1 + a_1}{b_0} \tag{5.26}$$

Once these system parameters have been identified, one can go to the rule base in Table 5.3 to determine whether a fault is present.

5.5.3   The Particle Filter Framework for Fault Detection/Diagnosis

Particle filtering is an emerging and powerful methodology for sequential signal processing with a wide range of applications in science and engineering

(Arulampalam, et al., 2002). It has captured the attention of many researchers in various communities, including those of signal processing, statistics, and econometrics.

Founded on the concept of sequential importance sampling (SIS) and the use of Bayesian theory, particle filtering is very suitable in the case where the system is nonlinear or in the presence of non-Gaussian process/observation noises, as in the case of engines, gas turbines, gearboxes, etc., where the nonlinear nature and ambiguity of the rotating machinery world is significant when operating under fault conditions. Furthermore, particle filtering allows information from multiple measurement sources to be fused in a principled manner, which is an attribute of decisive significance for fault-detection/diagnosis purposes.

The underlying principle of the methodology is the approximation of the conditional state probability distribution $p(z_k/x_k)$ using a swarm of points called *particles* (samples from the space of the unknowns) and a set of weights associated with them representing the discrete probability masses. Particles can be generated and recursively updated easily (Arulampalam et al., 2002) given a nonlinear process model (which describes the evolution in time of the system under analysis), a measurement model, a set of available measurements $z_{1:k} = (z_1, \ldots, z_k)$, and an initial estimation for the state PDF $p(x_o)$, as shown in the following equations.

$$x_k = f_k(x_{k-1}, \omega_k) \quad \leftrightarrow \quad p(x_k|x_{k-1})$$

$$z_k = h_k(x_k, v_k) \quad \leftrightarrow \quad p(z_k|x_k)$$

As in every Bayesian estimation problem, the estimation process can be achieved in two main steps, namely, prediction and filtering. On the one hand, prediction uses both knowledge of the previous state estimation and the process model to generate the a priori state PDF estimation for the next time instant, as is shown in the following expression (Doucet et al., 2001):

$$p(x_k|z_{1:k-1}) = \int p(x_k|x_{k-1})p(x_{k-1}|z_{1:k-1}) \, dx_{k-1}$$

On the other hand, the filtering step considers the current observation z_k and the a priori state PDF to generate the a posteriori state PDF by using Bayes' formula:

$$p(x_k|z_{1:k}) = \frac{p(z_k|x_k)p(x_k|z_{1:k-1})}{p(z_k|z_{1:k-1})}$$

The actual distributions then would be approximated by a set of samples and the corresponding normalized importance weights $\tilde{w}_k^i = \tilde{w}_k(x_{0:k}^i)$ for the ith sample:

$$p(x_k|z_{1:k}) \approx \sum_{i=1}^{N} \tilde{w}_k \, (x_{0:k}^i)\delta(x_{0:k} - x_{0:k}^i)$$

where the update for the importance weights is given by

$$w_k = w_{k-1} \frac{p(z_k|x_k)P(x_k|x_{k-1})}{q(x_k|x_{0:k-1}, z_{1:k})}$$

Fault detection/diagnosis implies the fusion of the inner information present in the feature vector (observations) in order to determine the operational condition (state) of a system. In this sense, a particle filter–based module for fault detection/diagnosis can be implemented by considering the nonlinear state model

$$\begin{cases} x_d(k+1) = x_d(k) + n(k) \\ x_c(k+1) = f_k[x_d\,(k),\, x_c\,(k),\, w(k)] \\ \text{Features}(k) = h_k[x_d\,(k),\, x_c\,(k),\, v(k)] \end{cases}$$

where f_k and h_k are nonlinear mappings, $x_d(k)$ is a collection of Boolean states associated with the presence of a particular operational condition in the system (normal operation, fault type 1, 2, etc.), $x_c(k)$ is a set of continuous-valued states that describe the evolution of the system given certain operational conditions, $n(k)$ is zero-mean independent and identically distributed (I.I.D.) white noise, and $w(k)$ and $v(k)$ are non-Gaussian noises that characterize the process and feature noise PDF, respectively.

The above-mentioned implementation allows the algorithm to modify the probability masses associated with each particle as new feature information is been received. Furthermore, the output of the fault detection/diagnosis module, defined as the percentage of the particle population that activates each Boolean state, gives a recursively updated estimation of the probability for any given fault condition.

Once the probabilities for every fault condition are well established via the particle filter module, any statistical test is valid to define a threshold to be used for alarm signals or prognosis module activation purposes.

Consider the following example where the proposed methodology has been applied with the purpose of detecting the existence of a crack in blades of a turbine engine. For this particular case, after an exhaustive analysis of the structural model of the turbine, it has been determined that the model is precise enough to describe the crack growth in the blade under nominal load conditions, that is,

$$N = P \cdot L^6 + f(L^5) + \omega(N)$$

where L is the length of the crack (in inches), N is the number of stress cycles applied to the material, P is a random variable with known expectation, f is a fifth-order polynomial function known from a FRANC-3D model, and $\omega(N)$ is I.I.D. white noise. The measurement equation is considered to be linear with additive Gaussian noise $v(N)$, that is, $y(N) = L(N) + v(N)$. On the other hand, the structure of the state equation is nonlinear; moreover, the noise $\omega(N)$ is nonnegative and therefore non-Gaussian.

Since the only available information in this application problem is the time of arrival (TOA) for each blade, some preprocessing techniques were needed in order to generate a feature that could be used for detection purposes. A schematic that illustrates the preprocessing steps is shown in Fig. 5.47.

The feature information, together with the crack-growth model and an appropriate feature-crack-size nonlinear mapping, has been used to run the particle filter module and discriminate between two Boolean states: absence of crack (normal operation) and presence of crack (fault condition).

The results for the detection module for the case of one particular blade are shown in Fig. 5.48. Although it is possible to observe some changes in the probability of failure around the two-hundredth cycle of operation, it is clear that only after the two-hundred and seventy-fifth cycle is the evidence in the feature strong enough to ensure the existence of a crack.

5.6 PHYSICAL MODEL–BASED METHODS

Model-based methods rely on the availability of a good dynamic model of the system. Such models can be derived using physical modeling principles,

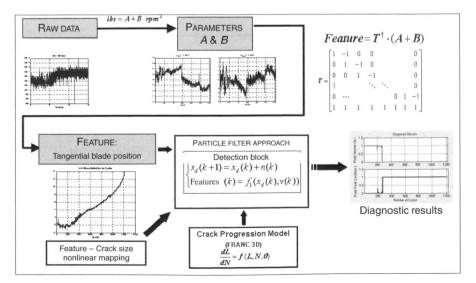

Figure 5.47   Particle filter diagnosis module for crack detection in a blade.

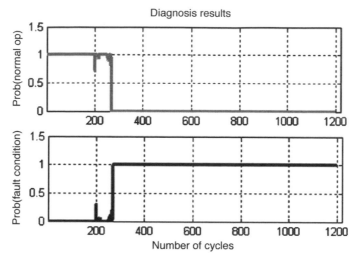

Figure 5.48   Results for crack detection in a blade.

or the model can be determined using online parameter estimation and system identification methods. Often the two approaches can be combined, as shown in the following examples.

The structural integrity of critical systems, such as aircraft, has been attracting major interest from the CBM/PHM community in recent years. Attention has focused primarily on crack initiation and propagation models. Crack initiation models, used as fault diagnostic tools, have been studied extensively by materials engineering experts, and an excellent review of the crack initiation constitutive equations can be found in Sehitoglu (1997). Three-dimensional finite-element models, such as ANSYS, FRANC-3D, and FASTRAN (Newman, 1992), are making inroads in the quest for effective and efficient modeling tools that capture robustly the fault behavior of structural components under stress or loading conditions. Corrosion phenomena also have been targeted.

5.6.1 Fatigue and Crack Propagation Models

Crack initiation models based on physical modeling principles are by now well understood and can be used to account for deviations from standard operating conditions. Crack models must account for load profiles and stress ratio, variations in temperature, sustained load hold time, and interactions of damage mechanisms such as fatigue, environment, and creep. Crack initiation and propagation models need to account for a plethora of uncertainties associated with material properties, stress factors, environmental conditions, etc. They are expressed, therefore, as stochastic equations, that is, probability density functions with all parameters considered as random quantities. Parameter distributions are estimated via Monte Carlo simulations, probabilistic neural networks, and others.

In such critical systems as aircraft, industrial and manufacturing processes, etc., defect (e.g., cracks and anomalies) initiation and propagation must be estimated for effective fault prognosis. Dynamic models of fatigue and fault progression can be derived by using principles of physics. Recent studies in materials research have focused on microstructural characterization of materials and modeling frameworks that describe robustly and reliably the anomaly initiation and propagation as a function of time (Friedman, 1990). Such models addressing the constitutive material properties down to the single-particle level generally are not suitable for real-time treatment owing to their immense computational complexity. Reduced-order models, called *metamodels,* usually are derived from the microstructural characterizations so that they may be used to estimate in almost real time the crack initiation and propagation with conventional computing platforms (Newman, 1992). Crack initiation models must account for variations in temperature, cycle frequency, stress ratio, sustained load hold time, and interactions of damage mechanisms such as fatigue, environment, and creep. A thorough description of the crack initiation constitutive equations can be found in Chestnut (1965).

Crack propagation models, clearly of interest from the prognosis viewpoint, are distinguished into two major categories: deterministic and stochastic models. Fatigue crack growth in such typical machinery components as bearings, gears, shafts, and aircraft wings is affected by a variety of factors, including stress states, material properties, temperature, lubrication, and other environmental effects. Variations of available empirical and deterministic fatigue crack propagation models are based on Paris' formula

$$\frac{d\alpha}{dN} = C_0(\Delta K)^n$$

where α = instaneous length of dominant crack
N = running cycles
C_0, n = material-dependent constants
ΔK = range of stress intensity factor over one loading cycle

Model parameters typically are estimated via a nonlinear recursive least-squares scheme with a forgetting factor (Ljung, 1999).

Stochastic crack growth models consider all parameters as random quantities. Thus the resulting growth equation is a stochastic differential equation.

Example 5.9 A Model of Corrosion Initiation and Propagation/Evolution. Structural-integrity concerns regarding military vehicles, sea vessels and aircraft, and civil structures such as bridges and pipelines, among many others, in corrosive environments have taken central stage in the CBM/PHM community because of their severe impact on the health of such legacy assets and the consequent cost associated with maintenance, repair, and overhaul. Figure 5.49 depicts the damage due to corrosion on aircraft structures. In the weakened state brought on by corrosion, materials become even more susceptible to the fatigue damage caused by the repeated cycling of even moderate stresses.

Current corrosion-fatigue models (Kondo, 1989; Goto and Nisitani, 1992) do not have the capability of integrating the new data that can be obtained using fiberoptic and other sensors placed at critical locations susceptible to corrosion. Predictive models will be more accurate when this new information is used directly in the predictive algorithms. In modeling crack initiation and the formation and growth of corrosion pits, a physics-based predictive model must account for the geometry of the structure and joints; loading as a function of time; level of moisture; chemistry, including pH and concentration of chloride ions (Cl^-) and perhaps other metal ions; temperature; residual stress; and microstructure of the material. Advances in new sensors enable detection of parameters critical to corrosion. This new sensor information should be coupled with data from traditional sensors such as accelerometers and strain gauges, as well as flight data. With the advances in sensor technology, commensurate advancements are also needed for predictive models that explicitly use this new sensor information.

Corrosion-fatigue damage is often generated along surfaces, at fastener holes, and between layers of multilayer joints. Usually this damage is in the form of small cracks that nucleate from corrosion-assisted fatigue or fretting fatigue. Small cracks under corrosion-fatigue can occur at the stress concentration generated by corrosion pits or through an embrittlement mechanism that negatively alters the local mechanical properties.

Crack growth rate and associated acceleration mechanisms are quite dependent on the environment. For example, the mechanism for environmentally

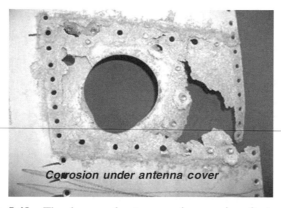

Corrosion under antenna cover

Figure 5.49   The damage due to corrosion on aircraft structures.

assisted cracking (EAC) is hydrogen embrittlement for lower concentrations of Cl^-, and for high concentrations of Cl^-, an electrochemical process (i.e., anodic dissolution at grain boundaries) contributes or dominates. EAC is a thermally activated process following an exponential Arrhenius relation. Increasing temperature to 150°F increases the rate of stress corrosion cracking (SCC) in some aluminum alloys by an order of magnitude compared with room temperature. Hence temperature is important to consider in the prediction model because the location of cracking often can be in an area that could get hot on a sunny summer day.

Pillowing from the entrapment of corrosion products in lap joints leads to the acceleration of fretting fatigue owing to the change in contact pressures between mating plates. Fretting that occurs near fasteners and between layers in multilayer joints can destroy the protective system, increasing the susceptibility of the high-strength aluminum alloy to corrosion damage. This leads to the need for a model algorithm involving multiple damage modules because more than one potential damage mechanism or mode of failure may operate. Fracture mechanics methods generally are well developed in predicting growth once damage has developed using an estimated initial flaw size (EIFS). However, the rate of crack growth is sensitive to the environment at the crack tip. Simply using information from new sensors will improve prediction of crack growth.

There are several possible reasons for the accelerated rate of crack propagation in a corrosive environment:

1. EAC coupled with small-amplitude ΔK (dithering effect)—perhaps at levels above K_{ISCC}
2. EAC coupled with periodic larger amplitude ΔK (or "conventional" corrosion-fatigue when operating at levels below K_{ISCC})
3. Increased thermal activation as temperature increases (room temperature versus 150°F), which shifts K_{ISCC} to the left and lowers stage II crack growth rate

The effect of frequency is important in predicting the kinetics of damage nucleation in a corrosion-fatigue environment. Typical flight loads consist of high mean stress and high-frequency cycles combined with low-frequency cycles associated with the pressurization cycle. In addition, two-step tests where the specimen is subjected to corrosion and then subjected to fatigue without corrosion can be used to determine damage summation under variable loading conditions and working environments.

The modeling effort is crucial toward a better understanding of the corrosion damage mechanisms but also in providing a fundamental platform for diagnostic and prognostic reasoners. In combination with actual measurements, corrosion models form the foundation for accurate and reliable assessments of the current and future status of critical airframe and other components.

5.6.2 The Utility of Finite-Element Analysis (FEA) in Model-Based Diagnosis

The Finite-Element Method. The *finite-element technique* is a methodology that serves to study the behavior of systems in many engineering disciplines, including thermal engineering, fluid mechanics, electromagnetism, and mechanics of materials. The method derives its name from the idea that a complex system or structure can be analyzed by dividing it in small parts called *elements,* each of which is fully defined in its geometry and/or properties, that is, is discrete (Moaveni, 2003) and *finite,* as represented in Fig. 5.50. Elements can be studied more easily than the complete structure they are a part of, and each can be considered as a continuous structural member (Rao, 1990). To solve a finite element problem, the method calls for a search of simultaneous solutions to the reaction problem of the elements due to applied loads or disturbances (i.e., forces, fields, heat, etc.), constraints (i.e., motion, position, temperature, etc.), and the interaction of adjacent elements (i.e., those connected by means of *nodes*). The problem is set up so that the solution converges toward the behavior of the total structure.

Introduction of the finite-element method generally is attributed to Courant (1943), who was then studying the vibration of systems. Important steps in the development of the method were taken later by Turner and colleagues (1956) and Clough (1960), leading to a formal establishment of the method in a book by Strang and Fix (1973). Through the years, the finite-element method has been applied successfully to a wide variety of problems, and in the present day, literature on the subject is abundant.

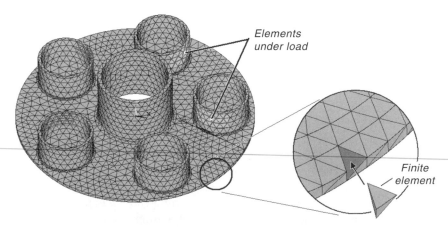

Figure 5.50 Computer model of a mechanical component. The design has been divided into finite elements through a process known as *meshing*.

Finite-element models allow a designer to analyze the static or dynamic response of a system to the application of loads or the presence of disturbances. In the case of structural studies, a model for finite-element analysis can be as simple as a single truss experiencing a point load or as complex as a complete engine with moving parts. Today's software tools based on FEA usually are available for a very specific engineering discipline, or else they are very large and complex program packages useful in a wide range of applications. For the first case, we find programs such as FASTRAN (Newman, 1992) and FRANC-3D (Cornell Fracture Group, 2003) that use finite elements or results from finite-element studies to simulate mechanical fracture growth. For the other case, we find powerful packages such as ANSYS (DeSalvo and Gorman, 1987) and ABAQUS (*www.abaqus.com*).

FEA for Fault Diagnosis. FEA simulations can be applied to many different situations to approximate the behavior of systems when operating conditions vary or when faults appear and progress. Thus the utility of FEAs for fault diagnosis is that they offer the potential to determine the way in which a condition of the system changes when a fault or anomaly is present. The idea is that this condition of the system should be measured in the form of a variable value, formally known as *state estimate,* or should be *extracted* from a system's signal in the form of a *feature*. The assumption is, of course, that the state or feature under consideration behaves similarly in both the FEA model and the actual working system.

Depending on the behavior of one or several of these states or features, it may be possible not just to diagnose the presence of a fault but also to assess its magnitude and development stage. In addition, the results from FEA simulations under different operating conditions can be used to compare the relative performance of different states or features under these varying situations. Thus they allow us to create a list, or table, that indicates what features best describe the system's health under specific circumstances. For example, the table might specify that one feature is better for detecting a fault at low-torque operating conditions of an engine, whereas another feature is better when working at high torques. Note that the *better* qualifier can refer to a variety of things depending on our interest, such as response sensitivity, noise immunity, statistical variance, etc. This process of choosing features depending on the operating conditions has come to be known as *feature selection.*

Example 5.10 A Case Study of a Crack on a Component of a Helicopter Gearbox. Structural FEA techniques are being used in conjunction with physics-based models of mechanical systems to simulate changes in system behavior and sensor measurements under specific operating and fault conditions. Consider, for example, the following case study of a helicopter gearbox with a crack.

The main transmission of a UH-60A Blackhawk helicopter uses a planetary gear system. For the study in question, the component known as the *planetary carrier plate* developed a crack on a region of high stress (Wu et al., 2004).

A physics-based vibration model for planetary gear systems originally developed by McFadden and Smith (1985) and refined by Keller and Grabill (2003) was used to simulate expected changes in the vibration of the system at different operating conditions and with different crack sizes. The planetary transmission in question is represented in Fig. 5.51.

The vibration model used was developed originally to describe the vibration of a healthy system, and hence it did not explicitly include a parameter to represent anomalies such as cracks on the system's components. Nevertheless, the crack was expected to cause deformation of the plate, which, in turn, changed the angular position of the planet gears. The angular position of the gears is an important parameter in the model, thus affecting the vibration of the system.

The FEA program known as ANSYS was used to perform an analysis of deformation of the planetary carrier plate owing to specific crack lengths. The results from this deformation translated into new relative positions for each of the five planet gears, which, in turn, could become the input to the vibration model to generate simulated vibration signals that represented the characteristics that would be observed as the crack grew. This is shown in Fig. 5.52.

The information gathered from this analysis may be useful to diagnose a specific crack length in similar systems because feature values tied to the general behavior of the design can be extracted from the simulation signals that represent the vibration at different crack lengths.

The application just discussed presents an example of how information gathered from FEAs can be used to develop diagnostic systems based on static analyses. However, the performance of this kind of diagnostic systems can be further enhanced by considering the dynamic aspect of the system's loads. Furthermore, FEA models also offer the potential to do fault prognosis studies, but the treatment of this subject is deferred to Chap. 6.

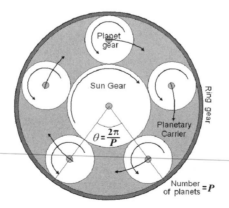

Figure 5.51   Representation of planetary gear transmission with five planet gears.

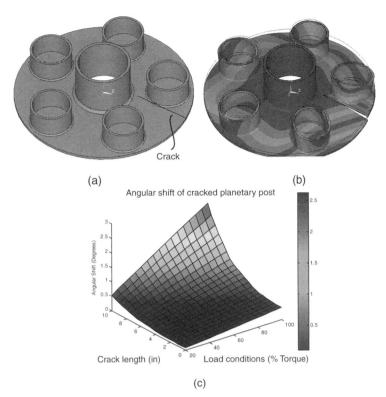

(a)   (b)

(c)

Figure 5.52 Results from FEA on a cracked planetary carrier plate. (*a*) Part design. (*b*) Deformation caused by one specific instance of operational loading in the presence of a crack. The shadings represent ranges of strain. (*c*) Combined results, from several simulations, for the angular position of one planet.

5.7 MODEL–BASED REASONING

Model-based reasoning (MBR) techniques have much to offer in addressing system-based fault diagnosis issues for complex systems. A higher-order reasoning capability, such as MBR, can be used to delineate the true cause of an anomalous condition and its potential impact on other healthy system components, as well as the identity of the fault component. MBR technologies take advantage of *reasoning* paradigms and build on qualitative notions and/ or graphic dependencies to arrive at an expedient yet not strictly analytic solution to the fault-detection and fault-propagation problem for systems consisting of interconnected subsystems/components. The most popular diagnostic MBR paradigm operates on a structural and functional description of the system to be diagnosed. The MBR approach used by DeKleer and Williams (1987) employs constraint propagation where known (measured) terminal values are propagated to other terminals; that is, value at other terminals are deduced using given transfer functions. Each propagated value is valid if those components are operational. This component dependency is recorded using a dependency-tracking mechanism. When two inconsistent values are propagated to a terminal in the system, a conflict set is formed from the union of the dependent component sets of each value. A *conflict set* simply means that one or more of the components in that set are faulty. As new measurements are made, the constraint-propagation mechanism continues to generate new conflict and fault candidates derived from a given set of conflicts. If a specific component appears in all known conflict sets, then that component would be a single-fault candidate.

A multibranched diagnostic tree then is generated using internal propagation. The constraint-propagation function is applied to numerical *intervals* instead of single numbers; that is, 2.65 maps into [1, 4]. The algorithm starts with known system inputs and determines through simulation the expected range for the output. Next, a qualitative value of OK is assigned to these calculated ranges, assuming that they belong to a normal or unfaulted state. The ranges above and below OK are defined to be HIGH and LOW, respectively. The tree generation begins by inputting one of the three possible qualitative values at the system output, representing one of the three branches from the top-level (root) node of the tree. The node at the other end of the branch represents the next measurement. From this new node, one of the three possible branches is again chosen, and the corresponding range is entered to initiate another cycle of propagation and next measurement selection. When a particular component is identified as faulty, one path of the diagnostic tree is completed. Iteration through all possible paths starting at the root then will generate the complete tree.

Example 5.11   Diagnostic Tree Design with Model-Based Reasoning.
To illustrate these concepts, let us look at a simple adder-multiplier example (Tong et al., 1989). The simple adder-multiplier circuit shown in Fig. 5.53 consists of three analog multipliers, labeled *M*1, *M*2, and *M*3, and two analog adders, labeled *A*1 and *A*2. For this simple circuit, a ternary diagnostic tree is constructed as shown in Fig. 5.54. The root and internal nodes represent measurements, leaf nodes represent failed components, empty circles denote

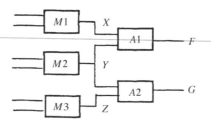

Figure 5.53   Analog adder-multiplier circuit.

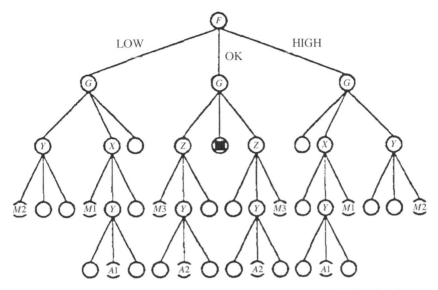

Figure 5.54   A ternary diagnostic tree for the adder multiplier circuit.

multiple simultaneous faults, while black squares denote either fault not detected or not isolatable. Fault diagnosis proceeds as follows: If terminal F is measured LOW (BAD) and terminal X is measured HIGH (BAD), then the ternary tree shown in Fig. 5.54 correctly indicates a multiple fault (represented by an open circle).

Model-Based Reasoning Approach to Fault Diagnosis. Critical components that are prone to failure usually are constituent elements of a larger subsystem or system. For example, a gearbox, found on most land, sea, and aerial vehicles, is an assembly of various components such as shafts, bearings, gears, etc. Such subsystems are instrumented with accelerometers mounted on their frame, which respond to vibration signals that might originate from any one of the gearbox components. It is crucial, therefore, to devise methodologies that are capable of determining which component is defective and, additionally, how specific component faults/failures may propagate from faulty to healthy components (the domino effect!), thus causing a catastrophic failure.

MBR belongs to this methodologic category. It uses all the system sensors to enhance and confirm fault isolation. It is an intuitive multilevel modeling approach with inherent cross-checking for false-alarm mitigation and multilevel correlation for accurate failure isolation. A glossary of terms associated with MBR is given below:

- *Model.* An executable, declarative structure representing an objective system (usually physical).

- *Model-based reasoning.* The use of explicit models to aid intelligent reasoning processes in achieving set goals within such domains as diagnosis, explanation, and training.
- *Structural model.* A model representation that shows what components are present in the system and how they are connected together.
- *Functional model.* An abstract model composed from representations of the fundamental physical processes manifested in the system. A functional model uses qualitative representation of real-world signals for reasoning about the interactions among these processes.

In the MBR framework, chains of functions indicate functional flows and components link to the functions they support, whereas sensors link to the functions they monitor and conditions/constraints link to the function they control. Figure 5.55 depicts the basic structure of a generic functional model.

For a specific example, we look at the fuel-supply system of the F35 Joint Strike Fighter (JSF) (Gandy and Line, 2000). Figure 5.56 shows the structural model of the fuel-supply system, whereas Fig. 5.57 depicts its functional model.

Most MBR approaches find their roots in the artificial intelligence domain and, more specifically, in qualitative simulation techniques (Shen and Leitch, 1991, 2005; Fouche and Kuipers, 1992; Davis, 1984). The model-based reasoning approach of De Kleer and Williams (1987) is a classic example. Their model-based diagnostic reasoner operates on a structural and functional description of the system to be diagnosed, as suggested earlier.

A qualitative measure of MBR's efficiency is the expected number of measurements needed to isolate a fault. The accuracy of the diagnosis is measured in terms of the minimum number of paths that cannot completely isolate the faulty component. If a component can be detected as being faulty, it should appear as a leaf node. Finally, the resolving power of the diagnostic tree is related to the number of multiple faults that can be discriminated.

Example 5.12   The Helicopter Intermediate Gearbox. The intermediate gearbox (IGB) of a helicopter is an assembly consisting of a pair of meshed spiral pinion gears whose shafts are supported at either end by bearings. It is located at the end of the tail boom of the helicopter and serves the purpose

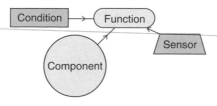

Figure 5.55   A generic functional model.

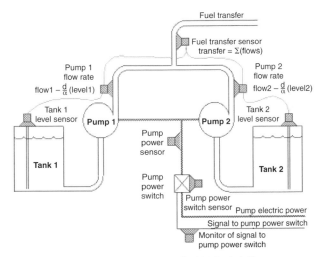

Figure 5.56   Structural model of JSF fuel delivery system.

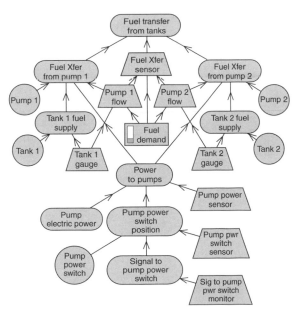

Figure 5.57   Functional model of JSF fuel delivery system.

of changing the direction of the tail rotor drive shaft up toward the tail rotor gearbox. In our example, we shall discuss the intermediate gearbox of the H-60 (Blackhawk) helicopter. A schematic view of the H-60 IGB is shown in Fig. 5.58, followed by the structural and functional diagrams in Figs. 5.59 and 5.60, respectively.

The IGB has multiple fault modes depending on which components go bad. For example, a bearing may develop a defect in the inner/outer race, a shaft or a gear pinion might develop a crack, or even a gear tooth may get worn out or chipped. All these fault modes give rise to characteristic vibration signatures. By looking at the corresponding frequency bands in the accelerometer readings, we can classify the fault modes present. Some common IGB fault classification heuristics are given below:

- Rotor imbalance—1 × shaft speed
- Shaft misalignment—2 × shaft speed, high axial vibration
- Mechanical looseness—higher harmonics of shaft speed
- Excessive bearing clearance—subsynchronous whirl instability
- Bearing race defect—$(n/2)$ × shaft speed, n is the number of balls
- Gear bevel defect—Gear natural freq., sidebands spaced at the running speed of the bad gear.

The MBR Approach to Fault Detection in the IGB.   For fatigue crack analysis an H-60 IGB pinion gear made of 9310 steel was used in a cyclic crack growth test. It was seeded with faults to initiate cracks. These faults consisted

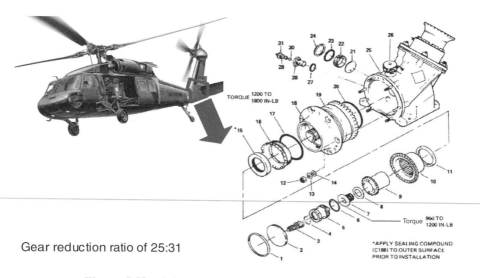

Gear reduction ratio of 25:31

Figure 5.58   Schematic of the H-60 intermediate gearbox.

To data acquisition module

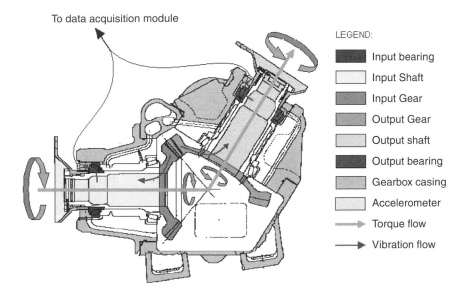

Figure 5.59 Structural diagram of the H-60 intermediate gearbox.

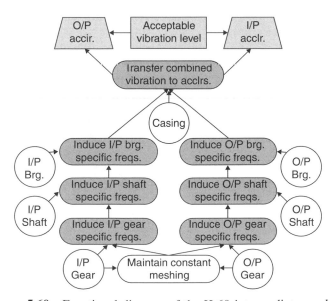

Figure 5.60 Functional diagram of the H-60 intermediate gearbox.

of notches made by an electric discharge machine (EDM) and were located at and parallel to the root of one of the gear teeth, as shown in Fig. 5.61. The crack growth test consisted of rotation in a spin pit at a constant high speed with a varying load cycle to simulate flight conditions. Data collection was done using two stud-mounted accelerometers at the input and the output of the IGB.

The data consist of 36 sets of accelerometer readings from both the input and the output ends of the gearbox. A number of features were extracted both in the frequency domain and in the wavelet domain for crack growth analysis. The observed behavior from the data was compared with the predicted behavior from the functional model of the system. Any discrepancy in the two patterns was used to detect a fault condition, whereas the heuristics presented earlier were used to classify the fault. A schematic of this approach is presented in Fig. 5.62.

Performance Metrics of MBR Methods. Performance metrics for MBR methods address such questions as How efficient is the tree? How accurate is the diagnosis? A quantitative measure of the efficiency is the expected number of measurements needed to isolate a fault. Diagnostic accuracy, on the other hand, is associated with the minimum number of paths that cannot completely isolate a faulty component. If a component can be detected as being faulty, it should appear as a leaf node. Finally, the resolving power of a tree is related to the number of multiple faults that can be discriminated. Qualitative simulation methods that employ a qualitative model of the given system may produce spurious behaviors. The task at hand, therefore, is not only to predict the system's fault behaviors as they propagate through the tree dependencies but also to decide on the *true* faulty component(s) among a number of possible candidates. Pruning spurious behaviors can be achieved

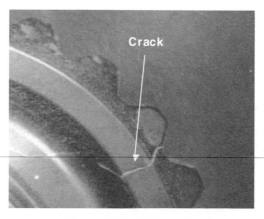

Figure 5.61 IGB crack.

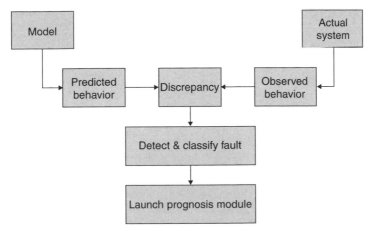

Figure 5.62 An MBR approach to fault detection and classification in the IGB.

by a number of techniques. Among them, the DeKleer and William's approach offers a solution to this problem. Another technique relies on stability considerations, that is, the definition of an energy-looking or Lyapunov function. The time rate of this function is used to decide on the true fault behavior and reject all spurious responses. This formulation fits well into the qualitative reasoning framework.

A few concluding comments are in order regarding model-based failure-detection methods. First, the design of such models depends, of course, on the application itself. Detection filters are analytically based and cover a wide range of system failures. While suboptimal, they are applicable primarily to time-invariant systems. Multiple-hypotheses techniques offer the best performance with full implementation. They are especially applicable to the detection of abrupt changes, but they lost their popularity because of their complex configuration. Same comments hold true for jump-process formulations. Innovations-based detection systems involve a wide range of procedures and offer various tradeoffs between performance and complexity. They constitute the most reliable failure detection and isolation systems with the considerations of isolability, sensitivity, and robustness.

5.8 CASE-BASED REASONING (CBR)

A case-based reasoner solves new problems by using or adapting solutions that were employed to solve old problems (Reisbeck and Shank, 1989). CBR offers a reasoning paradigm that is similar to the way people routinely solve problems. Aamodt and Plaza (1994) have described CBR as a cyclic process (Fig. 5.63) consisting of the four RE's:

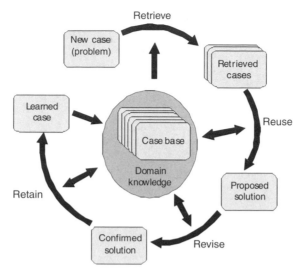

Figure 5.63 The CBR cycle. (Adapted from Aamodt & Plaza, 1994.)

· *Retrieve* the most similar case(s).
· *Reuse* the case(s) to attempt to solve the problem.
· *Revise* the proposed solution if necessary.
· *Retain* the new solution as a part of a new case.

A new problem is matched against cases in the case base, and one or more similar cases are retrieved. A solution suggested by the matching cases then is reused and tested for success. Unless the retrieved case is a close match, the solution probably will have to be revised, producing a new case that can be retained. Currently, this cycle rarely occurs without human intervention and most CBR systems are used mainly as case retrieval and reuse systems (Watson and Marir, 1994). Case revision (i.e., adaptation) is often carried out by the managers of the case base. With the current state of the art, the artificial intelligence (AI) systems are not completely autonomous yet; however, they have proven to be very successful as decision-support systems. Several applications use machine learning techniques for data mining and learning associations within CBR to propose an approximate solution and make it much simpler for the users to come up with a good solution. Table 5.4 shows a comparison between CBR and other technologies that are intended to store, retrieve, and reason about data, information, and fault cases.

Induction, as a logical decision process, forms the foundational basis for building a case history, identifying patterns among cases, ordering decision criteria by their discriminatory power, and partitioning cases into families, allowing the user to search the stored case library efficiently and extract knowledge from old cases. Figure 5.64 shows an ordered induction process, suggesting the steps required for its design and eventual use.

TABLE 5.4 Comparison of CBR with Other Technologies

Technology	When to Use	When Not to Use
Databases	Well-structured, standardized data and simple, precise queries possible	Complex, poorly structured data and fuzzy queries required
Information retrieval	Large volumes of textual data	Nontextual complex data types, background knowledge available
Statistics	Large volumes of well-understood data with a well-formed hypothesis	Exploratory analysis of data with dependent variables
Rule-based systems	Well-understood, stable, narrow problem area and justification by rule trace acceptable	Poorly understood problem area that constantly changes
Machine learning	Generalized rules required from a large training set and justification by rule trace acceptable	Rules not required, and justification by rule trace unacceptable
Neural networks	Noisy numerical data for pattern recognition or signal processing	Complex symbolic data or when justification required
CBR	Poorly understood problem area with complex structured data changing slowly with time and justification required	When case data not available, or if a complex adaptation required, or if an exact optimal answer is required

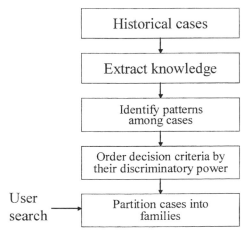

Figure 5.64 The induction process as a decision-support tool.

Entry points in an induction scheme for equipment reliability and maintainability include failure symptoms (e.g., abnormal noise, vibrations, thrust deficiency, high oil consumption, etc.), whereas target slots may be aiming at faulty equipment (e.g., failed component or replacement part), and the information extraction module is intended to extract symptom-specific technical failures. A CBR approach to prognosis and health management builds on the induction paradigm and entails additional modules for coding, adaptation, and other PHM specific functions. Figure 5.65 shows the CBR approach to fault management.

The CBR designer is faced with two major challenges: *coding* of cases to be stored into the case library or case base and *adaptation*, that is, how to reason about new cases so as to maximize the chances of success while minimizing the uncertainty about the outcomes or actions. Additional issues may relate to the types of information to be coded in a case, the type of database to be used, and questions relating to the programming language to be adopted.

Challenge 1: Coding. The information to be coded must achieve a desired level of abstraction as well as the capability to describe specific situations. Moreover, the coding scheme must be capable of addressing both quantitative and qualitative information. Information about fault conditions and corrective actions typically is expressed in a dual mode: in terms of such quantitative

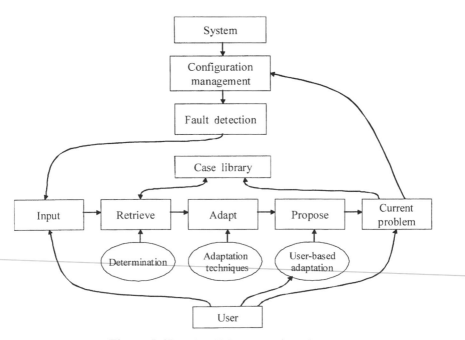

Figure 5.65 The CBR approach to PHM.

statements as "The temperature is exceeding 100°F" but also in qualitative terms such as "The vibration levels are very high." A case must have as a minimum the following fields:

- *Condition*. Describes the existing conditions when the case is logged.
- *Measurement*. Qualitative and quantitative measurements or features.
- *Symptom* (optional). Other unusual observations. Can be all textural information.
- *Component*. Describes the component affected and the corresponding fault.
- *Action*. Describes what actions/decisions were taken in the past.
- *Result*. Describes the success and failures associated with the preceding actions.
- *Explanation*. Includes an explanation why a particular decision was taken.

A relational database management system offers unique advantages to the CBR designer because (1) it organizes large amounts of data in linked tables to make the case base easy to understand, (2) it provides a relationally complete language for data definition, retrieval, and update, and (3) it provides data integrity rules that define consistent database states to improve data reliability. Figure 5.66 is an example of a relational database. A simple example of a possible case structure is shown in Fig. 5.67 for illustration purposes. Figure 5.68 depicts a case as an element of the case base.

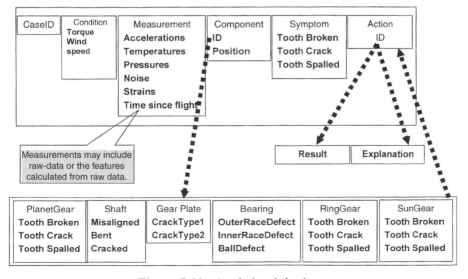

Figure 5.66 A relational database.

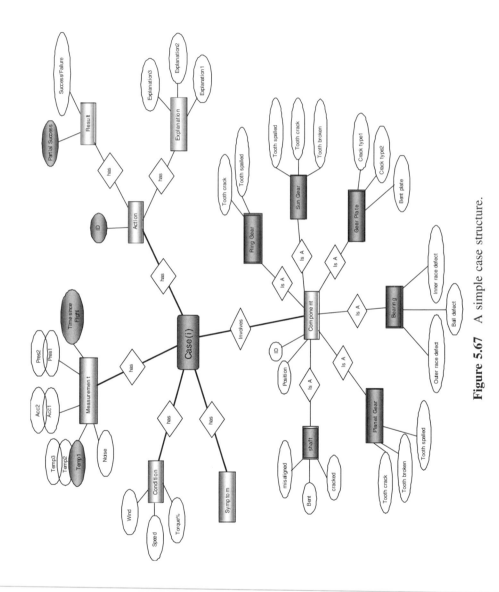

Figure 5.67 A simple case structure.

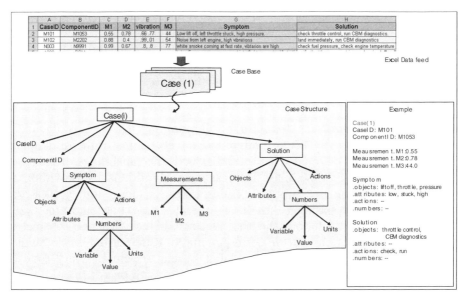

	A	B	C	D	E	F	G	H
1	CaseID	ComponentID	M1	M2	vibration	M3	Symptom	Solution
2	M101	M1053	0.55	0.78	.66 .77	44	Low lift off, left throttle stuck, high pressure.	check throttle control, run CBM diagnostics.
3	M102	M2202	0.88	0.4	.99, .01	54	Noise from left engine, high vibrations	land immediately, run CBM diagnostics
4	N003	N9991	0.99	0.67	.8, .8	77	white smoke coming at fast rate, vibtarion are high	check fuel pressure, check engine temperature

Figure 5.68 An illustration of the case base and one of its cases.

Challenge 2: Adaptation. In many situations, a case returned from the case base might not be the exact solution sought. An adapted solution, therefore, must be generated in the solution space that will match, as closely as possible, the input problem description, as shown schematically in Fig. 5.69.

Conventional methods for creating the adapted solution include interpolation and extrapolation. Learning algorithms and other computational intelligence-based tools, such as fuzzy inferencing, neural networks, etc., also may be used. A learning adaptation algorithm compares similar attributes and

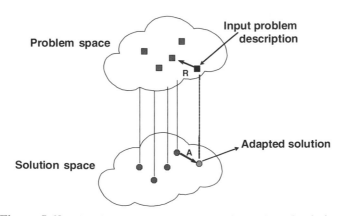

Figure 5.69 A schematic representation of an adapted solution.

attempts to find two data items that differ only in one attribute. The solution difference is the value of the change in the attribute. This approach offers an advantage over other methods because it does not require the addition, re-training, or reinduction of new rules. Figure 5.70 shows a detailed CBR schematic illustrating the indexing, similarity metrics, and adaptation rules, as well as their interdependencies.

Example 5.13   Dynamic Case Based Reasoning (DCBR). A dynamic case-based reasoning (DCBR) paradigm circumvents some of the difficulties associated with conventional schemes and extends the capabilities of the case base, the stored fault cases, and the reasoning algorithms by providing a continuous evolution over time of existing cases as more similar ones are encountered and emphasizing those case attributes that contribute more significantly to a correct solution. The basic idea behind the "dynamic" character of the CBR is to store a *statistic* corresponding to different fields in a single case instead of storing several cases of a similar structure. Every time a problem case is presented, it can be either grouped with one of the old cases based on its similarity or considered as a new case.

- If a new case is detected, the case library is updated by including this case.
- If this case matches an already existing one, the statistic vector of the corresponding representative case is updated.

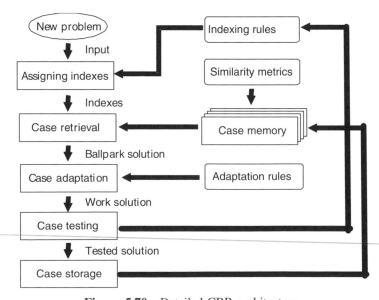

Figure 5.70   Detailed CBR architecture.

Figure 5.71 is a schematic of the DCBR engine. Figure 5.72 summarizes the salient features of the DCBR system, and Fig. 5.73 is an illustration of a dynamic case.

An attractive feature of the DCBR is the dynamic reasoning tool it employs for adaptation. An inferencing mechanism suggests a solution based on the best match with retrieved cases. A schematic representation of the inferencing mechanism is shown in Fig. 5.74.

The following factors enhance its effectiveness:

- Retrieval can be made more precise based on a statistical or probabilistic case selection (e.g., Bayesian or belief networks, clustering algorithms).

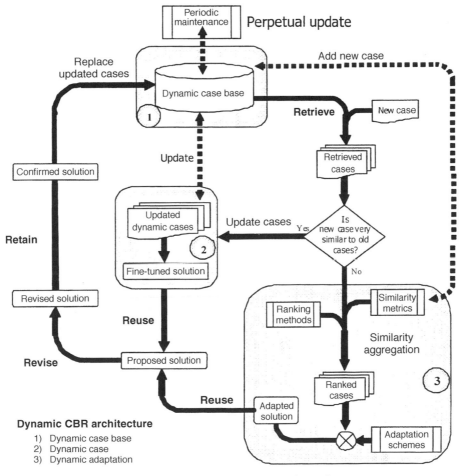

Figure 5.71 DCBR engine schematic.

Dynamic Case-Based Reasoning

- Detect an abnormal behavior that might indicate the presence of failure mode(s).
  - Use classic fault diagnostic methods for detection + DCBR *statistic vector* for fault identification.
- Compare with past cases (*statistic vector*) to narrow down the possible failure modes. (Retrieval)
- Perform specific diagnostics (use classic diagnostic methods) for suspected failure to improve confidence limits on detection.
- Categorize the problem case into a *new fault case* or *old fault case* similar to some old case(s).
- Use adaptation mechanisms to aggregate an appropriate solution scheme from the solutions of retrieved cases.
- Once the solution has been applied to the system, perform a feedback operation. This feedback can be used to
  - Modify the diagnosis and hence the proposed solution and reapply the new solution. (Feedback control)
  - Register the success/failure of the solution. (Bookkeeping for future reference)

Figure 5.72   High-level scheme for a DCBR.

- Several similarity metrics can be used to improve confidence in case rankings using an aggregate ranking mechanism (e.g., *k*-nearest-neighbor algorithm, weighted feature distance, etc.).
- An appropriate adaptation scheme can be selected from multiple adaptation schemes based on the information coded in the retrieved cases (e.g., fuzzy max-min composition, Dempster-Shafer-based aggregation, belief networks).

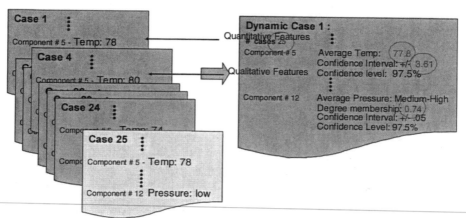

Combine several cases to form a dynamic case using the statistics based on all the cases,

e.g., 24 cases reduced to one dynamic case

Now a new "similar" case arrives whose features belong to the confidence interval of the dynamic case. Update the statistics of the dynamic case based on the features from the new case.

Figure 5.73   A dynamic case.

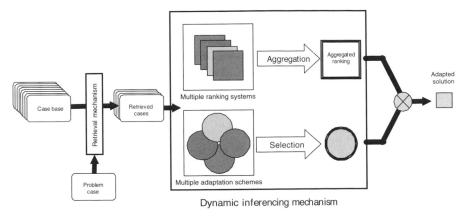

Figure 5.74 The inferencing mechanism.

- Depending on the problem case, the reasoning can be made on qualitative information only, quantitative information only, or both, which is dynamically determined.

DCBR for Fault Diagnosis. DCBR offers promise as a diagnostic and decision-support system. Of particular interest is a DCBR environment that addresses fault diagnosis at the system rather than the component level. Moreover, it can form the basis for fleet-wide or system-wide case bases that may be used for off-board prognosis, CBM, and other logistics functions. Figure 5.75 shows a two-step diagnosis using DCBR.

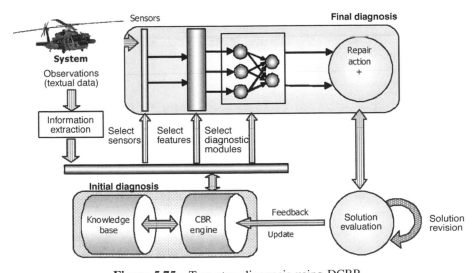

Figure 5.75 Two-step diagnosis using DCBR.

The diagnostic task is divided into two phases: In the first phase, a preliminary diagnosis is carried out, followed by a more accurate and informative phase. The case contents are selected on the basis of features extracted from data that most effectively identify and classify the faults. The initial diagnosis may be based also on a textual description of observations made by operators and aimed at defining a set of possible hypotheses. The latter carry weights that are computed based on past occurrences. In the second phase, the hypotheses are tested using relevant sensor data, features, and diagnostic algorithms until a successful solution is obtained. Qualitative information poses a particular challenge. Textual descriptions can be represented in the form of a *minimized language*. Such languages assume a set structure and follow certain rules unlike those adopted by full natural languages. A typical rule set is

- Passive-voice sentences are not allowed.
- Compound sentences joined by conjunctions are not allowed.
- A fixed vocabulary is used.
- Ambiguity in the meaning is resolved on the basis of a fixed domain (e.g., mechanical systems).

A semantic network is created from the textual fault symptoms. A set of three relations (IN, AT, and IS) is defined, and rules are employed to extract these relationships from given sentences. Such sentences are decomposed into a combination of basis units called *triads*. Figure 5.76 depicts a simple semantic network consisting of triads.

Natural-language processing begins by tagging the words in the sentences for corresponding parts of speech. A tagger program, called Tree Tagger, outputs in three columns the original word, the tag abbreviation, and the stem of the word. Figure 5.77 shows a typical program output.

Next, a set of rules is used to extract relationships between different words in the sentence. Table 5.5 lists three relations that capture the most relevant aspects of the hypothesized scenarios.

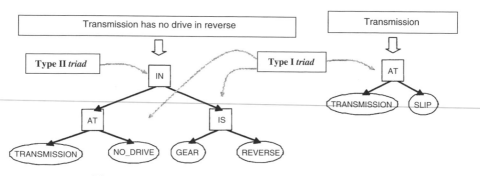

Figure 5.76   A semantic networks consists of triads.

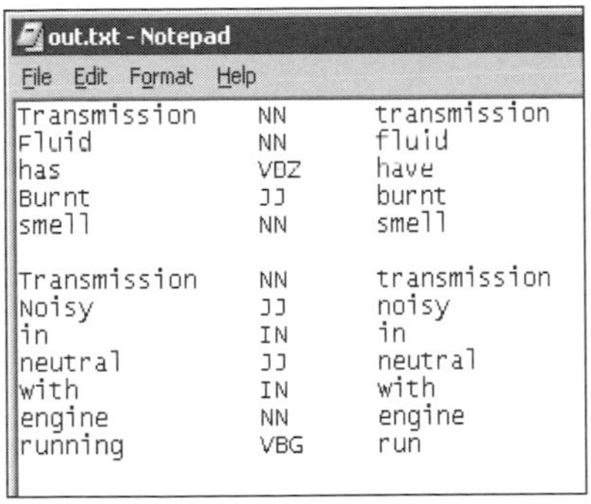

```
out.txt - Notepad
File  Edit  Format  Help
Transmission      NN      transmission
Fluid             NN      fluid
has               VDZ     have
Burnt             JJ      burnt
smell             NN      smell

Transmission      NN      transmission
Noisy             JJ      noisy
in                IN      in
neutral           JJ      neutral
with              IN      with
engine            NN      engine
running           VBG     run
```

Figure 5.77 Tree Tagger output.

The similarity assessment is based on matching the triads in the semantic network. First, the smallest level of triads is matched, e.g., the input semantic net. (Transmission—IS—Noisy)—IN—(Gear—IS—Neutral) consists of three triads: T1: (Transmission—IS—Noisy); T2: (Gear—IS—Neutral); T3: (T1—IN—T2). First, all semantic nets that contain similar type I triads (e.g., T1 or T2 in this case) are retrieved. Similarity of triads is computed based on how closely the three constituents match. With the semantic net (Transmission—AT—No_Drive)—IN—(Gear—IS—Neutral), the query net will be matched in the manner shown in Table 5.6.

For triads with link AT or IS, the weights associated with P1, P2, and L1 are 0.5, 0.3, and 0.2, respectively, acknowledging the fact that component

TABLE 5.5 Three Relations Capture Most of the Scenarios

A—IN—B → A in_condition B *or* A when B
Here, A is typically a noun or a *triad* and B is mostly a *triad*, e.g.,

```
Transmission fluid has burnt smell translates into
(Transmission_fluid)-IN-(smell-IS-burn)
```

A—IS—B → A has_property B or A is_type B
Here A is typically a noun or a triad and B is an adjective, e.g.,

```
Transmission fluid has burnt smell translates into
(Transmission_fluid)-IN-(smell-IS-burn)
```

A—AT—B → A in_state B or A exhibits_state B
Here, A is typically a noun or a triad and B is a verb, e.g.,

```
Transmission slips translates into
(Transmission)-AT-(slip)
```

TABLE 5.6 Similarity Assessment

Input	Output	SimVal
(Transmission—IS— Noisy)	(Transmission—AT— No_Drive)	$(1*0.5+0*0.2+0*0.3) = 0.5$
(Gear—IS—Neutral)	(Gear—IS—Neutral)	$(1*0.5+1*0.2+1*0.3) = 1$
(Transmission—IS— Noisy)—IN— (Gear—IS— Neutral)	(Transmission—AT— No_Drive)—IN— (Gear—IS— Neutral)	$(0.5*0.4+1*0.2+1*0.4) = 0.8$

(P1) matching is more important than its exact condition as far as localizing the fault is concerned. Fault identification may require more accurate condition matching; the emphasis here is on fault localization, whereas identification is addressed by other diagnostic algorithms. For triads containing link IN, the weights associated with P1, P2, and L1 are 0.4, 0.4, and 0.2, respectively, because both P1 and P2 convey equally important information. A detailed example of a DCBR implementation for fault diagnosis follows. Figure 5.78 is a schematic of the DCBR architecture.

The Case Structure. The following tables (5.7, 5.8) show the case structure followed by an example.

A list of different sensors, features, and diagnoses generally is available from domain experts. For this example, a list has been compiled based on the data available through the Web, textbooks, and previous experience. A representative list of these data appears in Table 5.9, 5.10, and Fig. 5.79.

A number of successful CBR systems have been fielded over the recent past for fault diagnosis and other CBM/PHM-related application domains. New advances in CBR are required to take full advantage of the capabilities this technology is offering. Its primary objective, when employed in CBM/PHM systems, is not only to store old fault cases and seek the best match of a new problem with its stored counterparts but also to provide a confidence level for the suggested solutions. The latter typically is based on the number of cases matching and the similarity of matching cases to each other. Confidence levels are derived through an aggregate of several similarity metrics. The advantages of CBR systems outweigh their disadvantages. Among the first, we recognize the reduced knowledge-acquisition task; flexibility in knowledge modeling; reasoning in domains that have not been fully understood, defined, or modeled; learning over time; providing an explanation module; and making use of both qualitative (textual descriptions) and quantitative (measurements) information. Disadvantages include large processing time to find similar cases, large storage space, and the best or optimal solution is not guaranteed, among others.

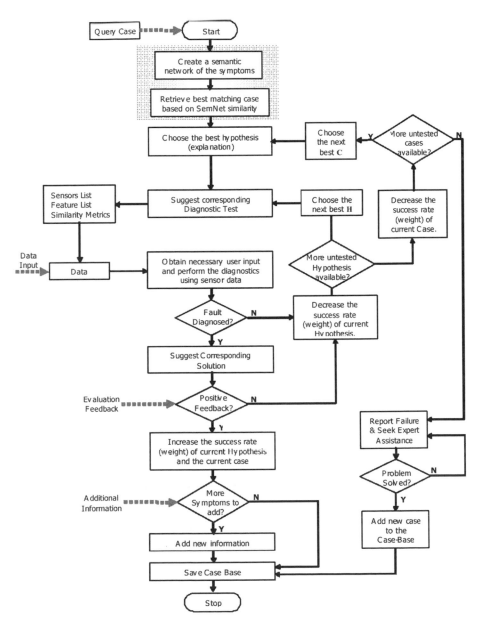

Figure 5.78   Schematic of the DCBR architecture.

TABLE 5.7 Case Structure

Case Structure						.
.ID	`Case_ID`					
.Component	`Component_Name`					
.Location	`Component_Location`					
.Symptom	`S_ID`	`Symptom`	`Weight`	`Sementic_net`	`Hypotheses`	
	1	s1	W_{s1}	SemNet1	h_1, h_2	
	2	s2	W_{s2}	SemNet2	h_1	
	3	s3	W_{s3}	SemNet3	h_2	
.Hypothesis	`H_ID`	`Hypothesis`	`Weight`	`Diagnosis`		
	1	h_1	W_{h1}	d_1		
	2	h_2	W_{h2}	d_2, d_3		
.Diagnosis	`D_ID`	`(Sensor,Feature) pairs`		`Weight`	`Solution`	
	1	$d_1: \{(S_1,F_1),(S_1,F_2),(S_3,F_1),(S_4,F_6)\}$		W_{d1}	r_1	
	2	$d_2: \{(S_1,F_2),(S_1,F_3)\}$		W_{d2}	r_2	
	3	$d_3: \{(S_5,F_1)\}$		W_{d3}	r_3	
.Repair	`R_ID`	`Repair`	`Weight`			
	1	r_1	W_{r1}			
	2	r_2	W_{r2}			
	3	r_3	W_{r3}			
.Version	`Last_Update`	`Case_Quality (weight)`	`Success`	`Failure`	`Conditions`	
	mm:dd:yy	W	nS	nF	C1, C2, C3	

5.9 OTHER METHODS FOR FAULT DIAGNOSIS

In this section we will summarize other fault diagnostic methods suggested over the years. Each one addresses a particular diagnostic system requirement, but all of them, as well as others that space does not allow us to cover in this book, present a unique technological advantage that is of interest to CBM/PHM designers.

5.9.1 Statistical Change Detection

B. Eugene Parker and colleagues of Barron Associates, Inc. (Parker et al., 2000), introduced the statistical change detection (SCD) technique used for early detection of subtle system changes. The algorithm seeks to detect situations in which a given model that describes the initial behavior of a time series eventually fails to describe that time series accurately. Given a sequence of random observations, the SCD technique aims at both online and offline

TABLE 5.8 Case Example

Case Example					
.ID	10				
.Component	Bearing				
.Location	Main_Transmission				

.Symptom	S_ID	Symptom	Weight	Sementic_net	Hypotheses
	1	Noisy in neutral with engine running	0.57	SemNet1	h_1, h_2, h_3
	2	Vibration	0.43	SemNet2	h_1, h_2, h_4

.Hypothesis	H_ID	Hypothesis	Weight	Diagnosis
	1	Primary Gear worn	0.65	d_1
	2	Primary Bearing worn	0.15	d_2, d_4
	3	Clutch Release Bearing worn	0.10	d_3, d_4
	4	Lack of oil	0.10	d_5

.Diagnosis	D_ID	(Sensor,Feature) pairs	Weight	Solution
	1	$d_1:\{(S_3,F_1),(S_3,F_4)\}$	0.75	r_1, r_3
	2	$d_2:\{(S_1,F_{15}),(S_1,F_8)\}$	0.80	r_2, r_3
	3	$d_3:\{(S_2,F_{15}),(S_2,F_8)\}$	0.80	r_3
	4	$d_3:\{(S_1,F_{13}),(S_4,F_{13})\}$	0.6	r_3
	5	$d_3:\{(S_{13},F_{17})\}$	0.6	r_3

.Repair	R_ID	Repair	Weight
	1	Change Primary Gear	0.25
	2	Change Primary Bearing	0.25
	3	Replenish oil	0.25
	4	Change Clutch Release Bearing	0.25

.Version	Last_Update	Case_Quality (wt.)	Success	Failure	Conditions
	03:16:05	0.8	8	2	Full Load Windy

fault detection. Until some (unknown) time t_p, the measurements are generated by a statistical model M_p. Thereafter, they are generated by another model, M_i. The goal of online SCD is to detect changes from M_p to M_i as quickly as possible after they occur (i.e., minimize detection delay for a given mean time between false alarms). The goal of offline SCD is to detect a change from M_p to M_i with minimum sensitivity (i.e., maximize the power of the test subject to the constraint of a fixed false-alarm probability).

For diagnostic applications, the emphasis of the SCD approach is on "quickest detection." Early fault detection benefits, in some cases, prognosis of the remaining useful life of a failing component because it allows sufficient time for prediction and follow-up corrective action when the failing compo-

TABLE 5.9 Sensors Look-Up Table

Sensor ID	Name	Description
S_1	PortRing_ACC1	Accelerometer (100 MHz)
S_2	StbdRing_ACC1	Accelerometer (100 MHz)
S_3	Input1_ACC1	Accelerometer (100 MHz)
S_4	PortRing_ACC2	Accelerometer (100 MHz)
S_5	StbdRing_ACC2	Accelerometer (100 MHz)
S_6	Input1_ACC2	Accelerometer (100 MHz)
S_7	PortRing_AEx	Acoustic emission horizontal (x)
S_8	PortRing_AEy	Acoustic emission verticle (y)
S_9	PortRing_AEz	Acoustic emission axial (z)
S_{11}	TC1	Temperature (thermocouple)
S_{12}	TC2	Temperature (thermocouple)
S_{13}	P1	Pressure (oil) in gearbox 1
S_{14}	P2	Pressure (oil) in gearbox 2
. . .	. . .	. . .

nent is endangering the operational integrity of the system. SCD takes advantage of the bispectral domain, thus avoiding the challenges posed by its spectral counterpart, such as low signal-to-noise ratio (SNR) at characteristic defect frequencies. Higher-order spectra are sensitive to non-Gaussian vibration signatures only. It is well known that third- and higher-order cumulants vanish for Gaussian processes, thus enhancing SNR for fault signatures. In the SCD scheme, failure modes involve sets of phase-coupled harmonics and sidebands because cumulants are sensitive to phase coupling among different frequencies. Barron Associates used a time-local bispectrum and the magnitude-squared bicoherence function as higher-order statistics for fault detection.

The SCD approach was applied to accelerometer data obtained from a CH-47D helicopter combiner transmission spiral bevel input pinion gear that exhibited a classic tooth-bending fatigue failure as a result of test-cell overloading. SCD detected the fault clearly at the same time as with special-purpose Steward Hughes, Ltd., Mechanical Systems Diagnostic Analysis. Details of the SCD method can be found in Parker and colleagues (2000).

5.9.2 State-Based Feature Recognition

George D. Hadden and colleagues at Honeywell International introduced a state-based feature-recognition software technique that used temporally correlated features from multiple data streams to perform anomaly detection and diagnosis (Hadden et al., 1994). Features are detected using software structures called *finite state machines* or just *state machines*. Unique advantages of this approach are

TABLE 5.10 Features Look-Up Table

F_ID	Feature Name	Description	Detection	Rating
F_1	Half_ShaftOrder	Amplitude measurement of signal average $1/2 \times$ fundamental shaft frequency	Tooth profile error	Excellent gearmesh anomaly indicator
F_2	First_ShaftOrder	Amplitude measurement of signal average fundamental shaft frequency	Shaft Imbalance	Excellent shaft imbalance indicator
F_3	Second_ShaftOrder	Amplitude measurement of signal average $2 \times$ fundamental shaft frequency	Shaft misalignment and eccentricity	Excellent misalignment, Good backup to F_2, Fair gear fault indicator
F_4	Residual_RMS	Root mean square of residual signal average	Sensor specific gear faults	Good late responding indicator
F_5	Residual_Kurtosis	Measures localized anomalies of the residual signal average	Sensor specific gear faults	Excellent early responding indicator
F_6	EneregyOperator_Kurtosis	Measures localized anomalies in the distribution of adjacent points in a signal	Energy operator analysis	Good late responding indicator
F_7	Sideband_Modulation	Measures the combined energy of the sideband frequencies	Gear eccentricity, misalignment and looseness	Good late responding indicator
F_8 F_9 F_{11} F_{12}	InnerRace_Energy OuterRace_Energy Roller_Energy Cage_Energy	The measure of the total energy present in the enveloped vibration signal associated with the inner-race, outer-race, roller and cage of a specific bearing	Bearing faults	Excellent bearing fault indicator
F_{13}	RollerCage_InteractionEnergy	The measure of the total energy present in the enveloped vibration signal associated with the interaction of roller elements and cage of a specific bearing	Bearing roller/cage faults	Good bearing fault indicator
F_{14}	Tone_Energy	The energy level associated with the predominant tones of a signal	Sensor specific bearing faults	Fair gear fault indicator, good bearing fault indicator
F_{15}	Base_Energy	The energy level associated a signal after removing the predominant tones	Bearing wear	Fair gear fault indicator, good bearing fault indicator
. . .	. . .			. . .

E. Fault Symptoms and Corresponding Diagnosis

Module: AUTOMATIC TRANSMISSION

Fluid leakage

◆ Automatic transmission fluid is usually deep red in color.
Fluid leaks should not be confused with engine oil, which
can easily be blown onto the transmission by airflow.

◆ To determine the source of a leak, first remove all built-
up dirt and grime from the transmission housing and
surrounding areas using a degreasing agent or by steam
cleaning.

Transmission fluid brown or has burned smell

◆ Transmission fluid level low or fluid in need of renewal.

General gear selection problems

Check and adjust the selector cable on automatic
transmissions.

Common problems that may be caused by a poorly adjusted
cable:

a) Engine starting in gears other than park or neutral.
b) Indicator on gear selector lever pointing to a gear
   other than the one actually being used.
c) Vehicle moves when in park or neutral.
d) Poor gear shift quality or erratic gear changes.

**Transmission will not downshift with accelerator pedal
fully depressed**

◆ Low transmission fluid level.

◆ Incorrect selector cable adjustment.

Engine will not start in any gear other than park or neutral

◆ Incorrect starter/inhibitor switch adjustment, where
applicable.

◆ Incorrect selector cable adjustment.

Transmission slips or

Transmission shifts roughly or

Transmission is noisy or

Transmission has no drive in forward or reverse gears

◆ Check the fluid level and condition of the fluid. Correct
the fluid level as necessary, or change the fluid and filter if
needed.

Figure 5.79 Fault symptoms and corresponding diagnosis.

- *Hierarchical* pattern recognition can be built on recognition of simpler patterns.
- *Small,* 100 to 300 bytes per individual state machine.
- *Dynamically downloadable* sensors can take a "closer look."
- *Parallelizable* state machines operate independently yet can use output from each other.

Failure modes are modeled as state machines whose structure simulates a state register, local and global variables, and time stamps, whereas the transition list entails state-to-state transitions, conditions, and actions. Figure 5.80 shows a general state machine structure. Honeywell applied state-based feature-recognition concepts to engine and chiller fault diagnosis, among others (Hadden and Nelson, 1991, 1994).

5.9.3 Bayesian Networks

Bayesian networks, or belief networks, knowledge maps, and probabilistic causal networks, as they are sometimes known, are helpful tools to model multifault, multisymptom dependency relations (Jensen, 2001). Probabilistic reasoning enters as a significant tool in situations in which causality seems to play a role but our understanding of actual cause-effect is incomplete. The Bayesian philosophy's foundation is Bayes' theorem, where the posterior conditional is derived as the ratio of the likelihood times the prior divided by the evidence. Applying Bayes' theorem to multiple variables requires chain rules, for example,

$$P(A, B, C, D) = P(A|B, C, D)P(B|C, D)P(C|D)P(D) \qquad (1.7\text{-}1)$$

As the number of variables increases, distributions required for the joint distribution of *all* variables increase exponentially! For example, for n binary random variables we need $2^n - 1$ joint distributions.

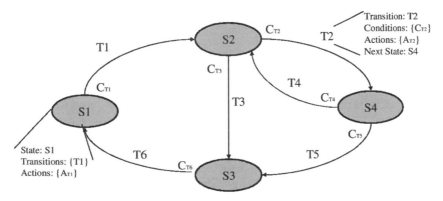

Figure 5.80 A general state machine.

Bayesian networks offer a way to compactly describe the distributions. Formally, they are directed acyclic graphs whose nodes are random variables such as causes, effects, and pieces of evidence. The network's arcs specify the dependence assumption between random variables. In designing a Bayesian network, one typically specifies the dependence of variables, the prior probabilities of "root" nodes, and the conditional probabilities of nonroot nodes given only their direct predecessors. The simple schematic in Fig. 5.81 illustrates the computational advantage gained in this approach. The network shows that A causes B and C, and C and D cause E. The quantities that are required are the probabilities $P(A)$ and $P(D)$ and the conditional probabilities $P(B|A)$, $P(C|A)$, and $P(E|D, C)$. Compared with the direct representation $P(A, B, C, D, E)$, if A, B, C, D, and E are binary (true/false type), direct representation needs $2^5 - 1 = 31$ distributions, whereas the Bayesian network needs only 5.

Advantageous properties of such network configurations include the following: (1) If the specified distributions are consistent, all probabilities calculated with be consistent. (2) The network always uniquely defines a distribution; that is, the joint distribution for the Bayesian network is uniquely defined by each random variable. Evaluating a Bayesian network is an arduous task because it is generally an NP-hard problem. Exact solutions are available only for singly connected networks. We are forced, therefore, either to convert the network from multiply connected to singly connected by clustering or to use approximate methods. Among the latter, we may randomly position values for some nodes and then use them to pick values for other nodes. Or we may keep statistics on values that the nodes take and arrive at answers from those statistics.

The superiority of Bayesian reasoning for fault diagnosis stems from the likelihood principle; that is, our inferences depend on probabilities assigned to the data that have been received, not on data that might have occurred but did not. This is in sharp contrast to classic statistical inference. In our diagnosis problems, diagnosis is based on reasoning about cause-effect or fault-symptom relations. Probabilistic reasoning remains the only viable alternative in situations where one symptom may be caused by several faults and many fault symptoms are not deterministic or well understood.

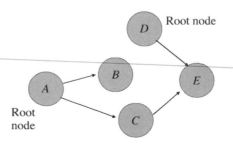

Figure 5.81   A simple Bayesian network.

5.9.4 Mechanical Face Seal and Drive Shaft Fault Diagnosis

Itzhak Green and colleagues introduced a methodology to diagnose faults for mechanical face seals and drive shafts using eddy-current proximity sensors and *xy* orbit plots (Zou and Green, 1997; Zou et al., 2000). A brief discussion of eddy-current proximity sensors is in Chap. 3, Sec. 3.24. The eddy-current sensors allow one to observe directly relative shaft motion, for example, inside clearance of fluid-film bearings. They are employed commonly in high-speed turbomachinery and often are supplied as standard equipment on these machine types. In shaft monitoring, deflections are relatively large (fractions of millimeters), whereas clearances in seal monitoring are of the order of sub-millimeters. In addition to the usual fault features in the time and frequency domains, angular orbit plots are used to detect seal absolute and/or relative misalignment. For a seal noncontacting or in normal operation, the time domain, frequency domain, and orbit plots, derived from three eddy-current proximity probes, are shown in Fig. 5.82. For a contacting seal, though, the corresponding plots are depicted in Fig. 5.83.

It is observed from these plots that noncontacting behavior and contacting behavior are distinctly different and can clearly be identified. Features can be derived in all three domains, including the tilt orbit plot to suggest contacting seal behavior. Furthermore, cracked-shaft dynamic behavior is distinctly different (having a dominant second-order harmonic component) compared with seal failure in contact (having higher harmonic oscillations). Hence multifaults (in this case, a dual fault) can be detected using a single monitoring system. This maximizes the return on the investment in such a system. Theoretical models for seal and shaft dynamic behavior and experimental results are shown in Zou and Green (1997) and Zou et al. (2000) to agree well.

5.9.5 Hidden Markov Models (HMMs)

HMMs have been studied extensively since their introduction in the late 1960s and early 1970s. An excellent tutorial on this subject can be found in Rabiner

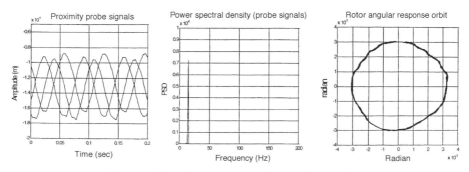

Figure 5.82 Noncontacting face seal operation.

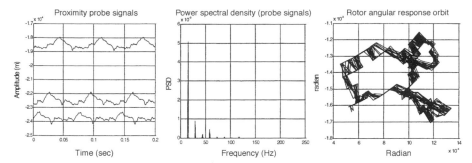

Figure 5.83 Contacting seal operation.

(1989). Because of their rich mathematical structure, they have become increasingly popular in the last several years and have found a wide range of applications. A problem of fundamental interest to fault diagnosis is characterizing signals in terms of signal models. A system may be described at any time as being in one of several distinct states. A typical Markov chain with five states and selected state transitions is shown in Fig. 5.84.

At regularly spaced, discrete times, the system undergoes a change of state (possibly from normal to a faulty condition) according to a set of probabilities associated with the state (and estimated from actual measurements). HMMs are well suited to modeling of quasi-stationary signals, such as those derived from vibration monitoring, and can be used to perform fault detection and estimation (Bunks et al., 2000). Similar to the notional description earlier,

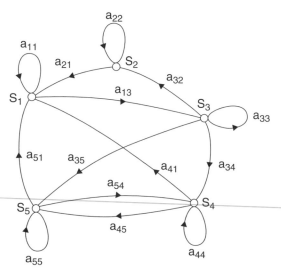

Figure 5.84 Markov chain with five states and selected state transitions.

HMMs are composed here again of two main components: the state-transition matrix which probabilistically describes how the model moves from one stationary type of behavior to another, and the probability distribution associated with each state, which describes the statistics of the particular stationary model. Details of application of the HMM approach to a Westland helicopter gearbox data for gearbox fault detection and isolation can be found in Bunks et al. (2000). If sufficient data are available to estimate the parameters of an appropriate probability density for each state (i.e., load level, fault mode, etc.) and a priori evidence is given regarding the frequency of occurrence of each fault mode and the average time spent at each of the load levels, then the HMM is a flexible and powerful tool, although technical challenges still remain in terms of suitable data sets, training of the HMM, the quasi-stationarity assumption, etc.

5.10   A DIAGNOSTIC FRAMEWORK FOR ELECTRICAL/ELECTRONIC SYSTEMS

There is a growing need to accurately detect, isolate, and predict the time to failure for components of modern electrical and electronic systems, such as avionics, weapon systems, automotive systems, computing platforms, and other systems or processes that entail electrical and electronic components. Technology is rapidly developing robust and high-confidence capabilities to diagnose incipient failure and predict the useful remaining life of electromechanical components, structures, and processes that contain mechanical components. However, achieving the same capability for predominantly electronic components lags behind their electromechanical counterparts. Built-in testing (BIT) techniques have been employed widely on board aircraft and other critical assets but are notorious for their large numbers of false alarms.

Electronic components such as power amplifiers, power converters, electronic control units, and avionics present substantial challenges to PHM designers and CBM developers. Normally, system designers assume constant failure rates for electronics or simply use laboratory test data to define mean time between failures (MTBF). However, electronics have been shown to exhibit typical failure behaviors known as the *bathtub curve*. Classic engineering perspectives on electronics describe failure rates as shown in Fig. 5.85. The conventional assumption is that electronics essentially have constant (and low) failure rates for most of their installed life after an initial period of infant mortality. The components then suffer an increasing, linear growth of failures in an "end of life" region of total life span. Convention asserts that such components' useful life ends at the point that the constant failure rate begins this increasing growth.

This assumption, while closer to reality than a constant failure rate, is also too simplistic. Electronic components are not "memoryless." The elements of a typical electronic component, such as solder joints, wires, and printed-

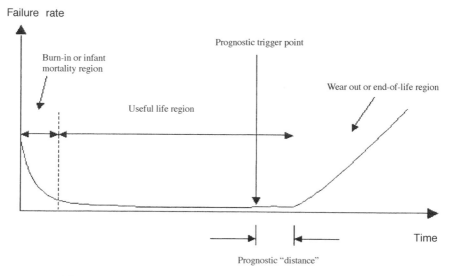

Failure rate

Figure 5.85 Bathtub curve. (Courtesy Ridgetop Group, 2004.)

circuit boards (PCBs), accumulate damage at rates depending on their environment and operational conditions. Therefore, their useful life is a function of environmental history, design tolerances, material properties, and operational conditions. Parameters that accelerate failure most significantly in electronics include

- Temperature and humidity
- Temperature variation
- Vibrations
- Shock
- Power quality
- Extreme operating conditions

The absence of globally observable parameters for electronic components is a barrier to establishing effective diagnosis and prognosis. Electronic systems must develop capabilities for sensing these parameters similar to the current state of practice for vibration monitoring in electromechanical systems. For example, thermography is being used to monitor excessive heat in electrical systems (Fig. 5.86). Some approach similar to this should be used in electronics, but practical application of thermography depends on the individual design of the electronics systems to be monitored. Because system design for electronics in a system such as radar or avionics frequently distributes the components throughout the system, thermography may have to

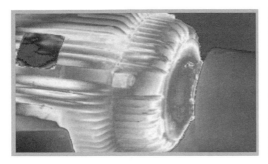

Figure 5.86 Examples of thermography.

be replaced by some other techniques such as thermally sensitive fiberoptic sensors. Thermography permits simultaneous inspection of several components but is hard to automate for board-level diagnosis and prognosis on highly integrated systems. Design tradeoffs are necessary to manage the cost and complexity of thermal or other environmental sensors in such a distributed system.

As an example, consider the radar system (with the antenna shown in Fig. 5.87) and its main modules (shown in Fig. 5.88). The radar's electrical subsystem is further decomposed into its subcomponents, shown in the lower tier of the figure. Electrical system health monitoring, as practiced today, typically relies on such industry standard methods as thermography, impedance measurement (contact and breakdown), and visual inspection.

Many systems include some form of built-in test, either inherently placed in the system or as an external device that performs tests of the system to ensure proper operation. There are few, if any, systems that use distributed sensors to continuously monitor the sources of failure, such as vibration, heat, or power quality, and from those sensors detect the onset of incipient failures.

Electronic components have other attributes that influence the requirements for diagnosis, prognosis, and health management. Aside from the failure modes of the individual elements and their basis in the underlying structure

Figure 5.87 Typical radar sytem antenna.

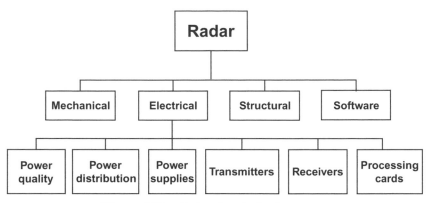

Figure 5.88 Radar electrical subsystems.

of the constituent material, each failure has an effect on adjoining components that can generate additional failure modes and cause increases in data or power loads on individual elements. The path of propagation can be variable, causing the estimation of useful life to be more complex and probabilistic.

Electronic systems are also technology-driven to a much larger extent than mechanical systems. Since the life cycle of some computer chips and other components typically is only 3 to 5 years, it is possible that some failure modes may occur only as a small fraction of the total, making the creation of diagnostic and prognostic processes for those instances ineffective from a cost perspective. A typical electronic system frequently is upgraded in capability or performance, making it difficult to manage the FMECA for a constantly changing configuration and potentially adding cost and complexity to the development of diagnostic and prognostic processes for these coexisting, multiply-variant configurations. Design analysis at the beginning of system design should examine these factors to ensure that the optimal balance of technology refresh/modernization and prognosis is achieved.

A possible framework to model system behavior and detect faults/predict useful life in electronics may include the following tools:

- *Embedded life modes.* Failure models derived from reliability studies, OEM data, etc. Such models express in a probabilistic sense the remaining life estimate of electronic components.
- *Environmental information.* Sensors can be distributed to measure temperature, humidity, and vibration. From test data, models can be developed that estimate the life impact to exposures of various levels of environmental stressors and estimate the useful life remaining.
- *Operational data.* The number of power-on cycles and run time can be an important factor in wiring and PCB failures. Tracking the operational data, as well as the power quality feeding the system during operation, can be used to measure life.

If no direct or surrogate measurements of life impact are available, the traditional approach, which employs BIT to confirm system operation, can be used to identify system faults as they occur. This approach, however has limited value for prediction of remaining life and results in a large number of false alarms. Figure 5.89 depicts a general life conception model for electronic components.

5.11 CASE STUDY: VIBRATION-BASED FAULT DETECTION AND DIAGNOSIS FOR ENGINE BEARINGS

An integrated diagnostic capability can be achieved for rolling-element bearings through the fusion of health state awareness with vibration feature-based classifiers. This type of diagnostic technique is based on event-driven integration of the vibration features extracted from the raw measurements. Since bearings have many failure modes and there are many influencing factors, a modular approach is taken in the design.

As just stated, there are many potential failure modes for rolling-element bearings, and to completely discuss all modes is impractical for this case-study description. Therefore, we will specifically focus on the rolling-element contact fatigue failure mode referred to as *spalling*. Also, other failure modes that may cause conditions that result in spalling are addressed. For instance, scratches or manufacturing defects produce localized stress concentrations that prematurely cause spalling.

The integration of diagnostic methods within a comprehensive health management system can entail the three functional steps shown in Fig. 5.90: sensed data, current bearing health, and future bearing health. The health prediction process begins with the sensed data module. Signals indicative of bearing health (i.e., vibration, oil debris, temperature, etc.) are monitored to determine the current bearing condition. Diagnostic features extracted from these signals then are passed on to a current bearing health module. These

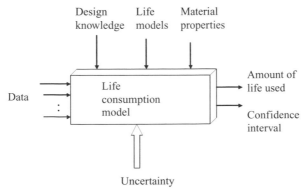

Figure 5.89 Simplified life estimation model.

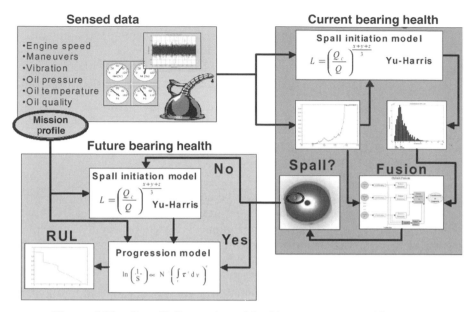

Figure 5.90 Overall diagnostic and health management architecture.

diagnostic features are low-level signal extraction type features, such as root mean square (RMS), kurtosis, and high-frequency enveloped features. In addition, engine speed and maneuver-induced loading are outputted for use as inputs to bearing health models. Also extracted are characteristic features that can be used to identify failure of a particular bearing component (i.e., ball, cage, and inner or outer raceway).

Central to the current bearing health step is a bearing component fatigue model. This model uses information from the sensed data module to calculate the cumulative damage sustained by the bearing since it was first installed. Life-limiting parameters used by the model, such as load and lubricant film thickness are derived from the sensed data using physics-based and empirical models. Using a knowledge-fusion process, this probability is combined with the extracted features that are indicative of spalling. Combining the model output with the features improves the robustness and accuracy of the prediction.

Whether or not a spall exists determines the next step of estimating future bearing health. If a spall does not currently exist, the spall-initiation module is used to forecast the time to spall initiation. This forecast is based on the same model that is used to assess the current probability of spall initiation, but instead, the model uses projected future operating conditions (loads, speeds, etc.) rather than the current conditions. Then the initiation results are passed to the progression model, which also uses the mission profile to allow an accurate prediction on the time from spall initiation to failure. If a spall currently exists, the initiation prognostic module is bypassed, and the process just described is performed directly.

5.11.1 Vibration-Based Diagnostic Indicators

Development of advanced vibration features that can detect early spalling characteristics is a critical step in the design of the integrated system just mentioned. To this end, a series of tests was conducted to provide data for diagnostic algorithm development and testing. Vibration and oil-debris data acquired from a ball bearing test rig that included a damaged bearing were used to compare the effectiveness of various diagnostic features. Vibration data acquired from a gas turbine engine running in a test cell provide more realistic information that includes multiple excitation sources and background noise. Distinguishing faulty bearing signatures from the background noise of an operating engine presents a significant technical challenge. To test the ability of the vibration analysis algorithms to observe weak bearing signatures, the bearings in the engine under test did not contain any known faults.

For the faulted bearing tests, data were collected from a miniaturized lubrication system simulator called the minisimulator located at the Air Force Research Laboratory (ARFL) at Wright Patterson Air Force Base. These data were collected with support from the University of Dayton Research Institute (UDRI) and the AFRL. The minisimulator consists of a test head, as shown in Fig. 5.91, and a lubrication sump. A pair of angular contact bearings located in the test head supports a rotating shaft. The bearings are identical to the number 2 main shaft bearing of an Allison T63 gas turbine engine.

Although designed primarily for lubrication tests, the minisimulator was used to generate accelerated bearing failures. During one of the run-to-failure

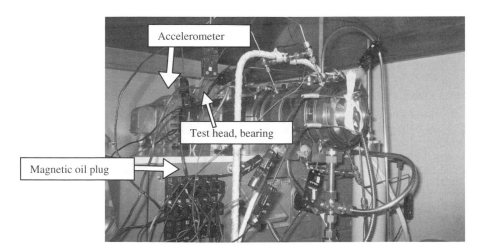

Figure 5.91 Minisimulator setup.

seeded fault tests, the bearing failure was accelerated by seeding a fault into the inner raceway of one of the bearings by means of a small hardness indentation (Brinnell mark). The bearing then was loaded to approximately 14,234 N (3200 lbf) and run at a constant speed of 12,000 rpm (200 Hz). Vibration data were collected from a cyanoacrylate-mounted (commonly called Super Glue) accelerometer, which was sampled at over 200 kHz. Also, the quantity of debris in the oil draining from the test head was measured using a magnetic chip collector (manufactured by Eateon Tedeco). The oil data were used in determining the initiation of the spall.

During a series of separate full engine tests, data were collected from a T63 engine located in a test cell, also at AFRL. The T63 tests were performed at two gas generator speed levels, a "cruise speed" of 50,000 rpm (833 Hz) and an "idle speed" of 32,000 rpm (533 Hz). Dimensions of the bearings of interest are given in Table 5.11.

5.11.2   Fault Frequencies for T63 bearing

Extraction of the vibration amplitude at the fault frequencies from a fast Fourier transform (FFT) often enables isolation of the fault to a specific bearing in an engine. High amplitude of vibration at any of these frequencies indicates a fault in the associated component.

Table 5.12 shows the calculated minisimulator bearing fault frequencies. Tables 5.13 and 5.14 summarize the two bearings of interest for the T63 engine.

Note the listings are defined by

BSF: ball spin frequency
BPFI: inner raceway frequency
BPFO: outer raceway frequency
Cage: cage frequency

During engine testing, periodic forces associated with meshing of gear teeth also excite vibrations at specific frequencies. These gear mesh frequencies (GMFs) are calculated for several of the gears in the T63 engine and are summarized in Table 5.15. The calculated GMFs for the T63 gas producer and power turbine gear trains are for an idle speed of 32,000 rpm. The speed

TABLE 5.11   Bearing Dimensions (mm)

Bearing	Ball Diameter (B_d)	Number of balls (z)	Pitch Diameter (P_d)	Contact Angle (β)
1	4.7625	8	42.0624	18.5°
2	7.9375	13	42.0624	18.5°

TABLE 5.12 Minisimulator Bearing Fault Frequencies (Hz)

Shaft Speed	BSF	BPFI	BPFO	Cage
200	512	1530	1065	82

TABLE 5.13 T63 No. 1 Bearing Fault Frequencies (Hz)

Shaft Speed	BSF	BPFI	BPFO	Cage
533	2327	2361	1931	238
833	3636	3690	2974	372

TABLE 5.14 T63 No. 2 Bearing Fault Frequencies (Hz)

Shaft Speed	BSF	BPFI	BPFO	Cage
533	1367	4085	2845	219
833	2136	6384	4456	342

TABLE 5.15 T63 Gear Mesh Frequencies (Hz)

Gear	Speed Ratio (to GP Shaft Speed)	GMF at 533 Hz
A	1.000	9061
B	0.202	3775
C	0.202	9061
D	0.08	3775
E	0.151	3777
F	0.236	3775
G	0.252	1125
H	0.073	1124
I	0.117	3128
J	0.117	2315
K	0.099	2315
L	0.082	2315
M	0.196	8758

ratio—the ratio of the gear speed to the gas producer (GP) input gear speed—may be used to calculate the gear mesh frequencies at other gas producer speeds.

5.11.3 Feature Extraction Using High-Frequency Enveloping

Although bearing characteristic frequencies are calculated easily, they are not always detected easily by conventional frequency-domain analysis techniques. Vibration amplitudes at these frequencies owing to incipient faults (and sometimes more developed faults) often are indistinguishable from background noise or are obscured by much-higher-amplitude vibration from other sources, including engine rotors, blade passing, and gear mesh, in a running engine. However, bearing faults produce impulsive forces that excite vibration at frequencies well above the background noise in an engine.

Impact Energy is a multiband, enveloping-based vibration feature-extraction technique shown in Fig. 5.92. The enveloping process consists of, first, band-pass filtering of the raw vibration signal. Second, the band-pass filtered signal is full-wave rectified to extract the envelope. Third, the rectified signal is passed through a low-pass filter to remove the high-frequency carrier signal. Finally, the signal has any dc content removed.

The Impact Energy algorithm was applied to the seeded fault-test data collected using the minisimulator. To provide the clearest identification of the fault frequencies, several band-pass and low-pass filters were used to analyze various regions of the vibration spectrum. Using multiple filters allowed in-

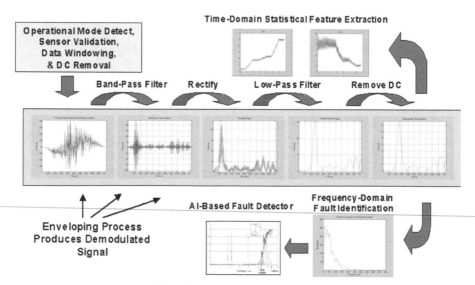

Figure 5.92 High-frequency demodulation process.

vestigation of many possible resonances of the bearing test rig and its components. A sample Impact Energy spectrum from early in the minisimulator test is shown in Fig. 5.93. Note that these data were collected prior to spall initiation (based on oil-debris data), and the feature response is due to the small "seeded" indentation on the race. For comparison, a conventional FFT with 10 frequency domain averages of vibration data also was calculated and is shown in Fig. 5.94. In this conventional frequency-domain plot, there is no peak at the inner race ball pass frequency (1530 Hz). However, the Impact Energy plot of Fig. 5.93 shows clearly defined peaks at this frequency and the second through forth harmonics of the inner race ball pass frequency. These peaks were defined using a detection window of ±5 Hz about the theoretical frequency of interest to account for bearing slippage. From the onset of the test, there is an indication of a fault.

Identification of vibration features in a running engine is very difficult owing to the high ambient noise level, which often obscures diagnostic features. Despite the fact that the bearings and gears in the engine are healthy, minor imperfections within manufacturing tolerances cause slightly elevated vibration levels at the characteristic frequencies. Raceway and ball waviness, surface-finish imperfections, and dimensional tolerances can lead to "damage-like" vibration signatures in a healthy engine. Vibration analysis techniques that are able to distinguish these signatures from a healthy bearing in a running engine are considered the most likely to detect the signature of a defective bearing because they are more sensitive than other techniques.

The most prominent features in the Impact Energy spectrum in Fig. 5.95 are harmonics of the gas generator shaft speed ($N \times$ rpm). Although less prominent, a gear mesh frequency (GMF) and the outer race characteristic

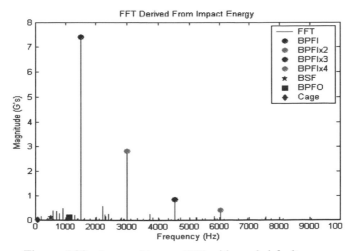

Figure 5.93 Impact Energy FFT with seeded fault.

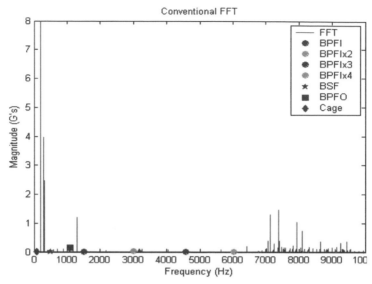

Figure 5.94 Conventional FFT of minisimulator-seeded fault.

frequency of the number 2 bearing (BPFO 2) are also observable. Low-amplitude vibration at a gear mesh frequency does not necessarily indicate that a problem exists. The low-amplitude vibration at a characteristic bearing frequency may be attributable to a benign bearing condition as well. However, observation of these features illustrates the ability of Impact Energy to detect incipient faults.

Although most of the peaks have been identified, several still are unidentified. In a turbine engine, there are many gears and even more bearings that

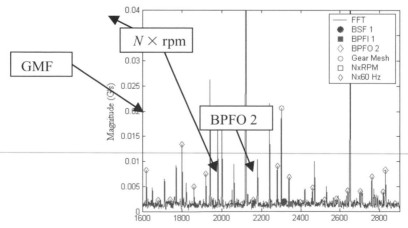

Figure 5.95 T63 Impact Energy FFT showing frequencies at calculated frequencies using measured magnitudes.

have distinct frequencies. In addition, there are frequencies associated with the compressor and turbine blades.

In addition to the preceding features, conventional vibration features are also used. Although traditional statistical-based features such as RMS, kurtosis, crest value, and peak value are valuable to the prognosis, they are not represented. These other features can be used in combination with the more sensitive Impact Energy features to avoid false alarms.

Hence, in order to achieve a comprehensive diagnostic capability throughout the life of critical engine components such as bearings, sensor-based (or feature-based) diagnostic methods complement the model-based prediction by updating the model to reflect the fact that fault conditions may have occurred. These measurement-based approaches can provide direct correlation of the component condition that can be used to update the modeling assumptions and reduce the uncertainty in the remaining useful life (RUL) predictions.

Real-time algorithms for predicting and detecting bearing failures are currently being developed in parallel with emerging flight-capable sensor technologies, including inline oil-debris/condition monitors and vibration-analysis MEMS. These advanced diagnostic algorithms use intelligent data-fusion architectures to optimally combine sensor data with probabilistic component models to achieve the best decisions on the overall health of oil-wetted components. By using a combination of health monitoring data and model-based techniques, a comprehensive component prognostic capability can be achieved throughout a component's life, employing model-based estimates when no diagnostic indicators are present and monitored features such as oil debris and vibration at later stages when failure indications are detectable.

5.12 REFERENCES

A. Aamodt and E. Plaza, "Case-Based Reasoning: Foundational Issues, Methodological Variations, and System Approaches," *AI-Communications* **7**(i):39–59, 1994.

M. S. Arulampalam, S. Maskell, N. Gordon, and T. Clapp, "A Tutorial on Particle Filters for Online Nonlinear/Non-Gaussian Bayesian Tracking," *IEEE Transactions on Signal Processing* **50**(3):174–188, 2002.

C. Bunks, D. McCarthy, and T. Ani, "Condition-Based Maintenance of Machines Using Hidden Markov Models," *Mechanical Systems and Signal Processing,* Oakland, CA, **14**(4):597–612, 2000.

R. H. Chen, H. K. Ng, J. L. Speyer, and D. L. Mingori, "Fault Detection, Identification and Reconstruction for Ground Vehicles," *ITSC 2001, IEEE Intelligent Transportation Systems, Proceedings* (cat. No. 01TH8585), Piscataway, NJ, pp. 924–929, 2001.

H. Chestnut, *Systems Engineering Tools.* New York: Wiley, 1965.

R. W. Clough, "The Finite Element Method in Plane Stress Analysis," *Proceedings of ASCE Conference on Electronic Computation,* Pittsburg, PA, IEEE, 1960.

Cornell Fracture Group, *FRANC3D Concepts and Users Guide,* Version 2.6. Ithaca, NY: Cornell Fracture Group, Cornell University, 2003.

R. Courant, "Variational Methods for the Solution of Problems of Equilibrium and Vibrations," *Bulletin of the American Mathematical Society* **49**:1–23, 1943.

R. Davis, "Diagnostic Reasoning Based on Structure and Behavior," *Artificial Intelligence* **24**:347–410, 1984.

J. De Kleer and B. C. Williams, "Diagnosing Multiple Faults," *Artificial Intelligence* **32**:97–130, 1987.

G. J. DeSalvo and R. W. Gorman, *ANSYS Engineering Analysis System User's Manual,* For ANSYS Revision 4.3. Houston, PA: Swanson Analysis Systems, 1987.

D. Doucet, N. de Freitas, and N. Gordon, *Sequential Monte Carlo Methods in Practice.* New York: Springer-Verlag, 2001.

R. C. Eisenmann, Sr., and R. C. Eisenmann, Jr., *Machinery Malfunction Diagnosis and Correction.* Englewood Cliffs, NJ: Prentice-Hall, 1998.

S. J. Engel, B. J. Gilmartin, K. Bongort, and A. Hess, "Prognostics, the Real Issues Involved with Prediting Life Remaining," *Aerospace Conference Proceedings,* Vol. 6, Big Sky, Montana, March 2000, pp. 457–469.

P. Fouche and B. J. Kuipers, "Reasoning about Energy in Qualitative Simulation," *IEEE Transactions on Systems, Man and Cybernetics* **22**(1):47–63, 1992.

R. Friedman and H. L. Eidenoft, "A Two-stage Fatigue Life Prediction Method." Grumman Advanced Development Report, 1990.

F. Filippetti, G. Franceschini, C. Tassoni, and P. Vas, "Recent Developments of Induction Motor Drives Fault Diagnosis Using AI Techniques," *IEEE Transaction on Industrial Electronics,* vol. 47, 5, pp. 994–1004, 2000.

M. Gandy and K. Line, "Joint Strike Fighter: Prognostics and Health Management (PHM)," Lockheed Martin Aeronautics Company, Public Release AER200307014, 2000.

P. R. Gill, W. Murray, and M. H. Wright, "The Levenberg-Marquardt Method," in *Practical Optimization.* London: Academic Press, 1981, pp. 136–137.

M. Goto and H. Nisitani, "Crack Initiation and Propagation Behavior of a Heat Treated Carbon Steel in Corrosion Fatigue," *Fatigue Fracture Engineering Material Structure* **15**(4):353–363, 1992.

G. Hadden and K. Nelson, "A State-Based Approach to Trend Recognition and Failure Prediction for the Space Station Freedom," *Proceedings of the Space Operations, Applications, and Research Symposium (SOAR-91),* 1991.

G. Hadden, K. Nelson, and T. Edwards, "State-Based Feature Recognition," 17th Annual AAS Guidance and Navigation Conference, Scottsdale, AZ, 1994.

G. Hadden, P. Bergstrom, B. Bennett, G. Vachtsevanos, and J. Van Dyke, "Shipboard Machinery Diagnostics and Prognostics/Condition Based Maintenance: A Progress Report," *MARCON 99, Maintenance and Reliability Conference,* Gatlinburg, TN, May 9–12, 1999a, pp. 73.01–73.16.

G. Hadden, G. Vachtsevanos, B. Bennett, and J. Van Dyke, "Machinery Diagnostics and Prognostics/Condition Based Maintenance: A Progress Report, Failure Analysis: A Foundation for Diagnostics and Prognostics Development," *Proceedings of the 53rd Meeting of the Society for Machinery Failure Prevention Technology,* Virginia Beach, VA, April 1999b.

W. Hardman, A. Hess, and J. Sheaffer, "SH-60 Helicopter Integrated Diagnostic System (HIDS) Program: Diagnostic and Prognostic Development Experiences," *Proceedings of Aerospace Conference,* Big Sky, MT, Vol. 1, 1999, pp. 473–491.

S. J. Henkind and M. C. Harrison, "An Analysis of Four Uncertainty Calculi," *IEEE Transactions on Systems, Man, and Cybernetics* **18**(5):700–714, 1988.

S. Haykin, *Neural Networks*. New York: IEEE Press, 1994.

R. Isermann, "Process Fault Detection Based on Modeling and Estimation Methods: A Survey," *Automatica* **20**(4):387–404, 1984.

A. G. Ivakhnenko and S. A. Petukhova, "Objective Choice of Optimal Clustering of Data Sampling under Nonrobust Random Disturbances Compensation," *Soviet Journal of Automation and Information Sciences* **26**(4):58–65, 1993.

F. V. Jensen, *Bayesian Networks and Decision Graphs*. New York: Springer-Verlag, 2001.

H. L. Jones, "Failure Detection in Linear Systems," Ph.D. thesis, Department of Aeronautics and Astronautics, Massachusetts Institute of Technology, Cambridge, MA, 1973.

P. Kallappa, C. Byington, P. Kalgren, M. Dechristopher, and S. Amin, "High-Frequency Incipient Fault Detection for Engine Bearing Components," *Proceedings of GT2005 ASME Turbo Expo 2005: Power or Land, Sea and Air,* Reno–Tahoe, NV, June 6–9, 2005.

J. Keller and P. Grabill, "Vibration Monitoring of a UH-60A Main Transmission Planetary Carrier Fault," American Helicopter Society 59th Annual Forum, Phoenix, Arizona, May 6–8, 2003.

J. L. Kolodner, *Case-Based Reasoning*. San Mateo, CA: Morgan Kaufmann, 1973.

Y. Kondo, "Prediction of Fatigue Crack Initiation Life Based on Pit Growth," *Corrosion Science* **45**(1):7–11, 1989.

H. Konrad and R. Isermann, "Diagnosis of Different Faults in Milling Using Drive Signals and Process Models," *Proceedings of the 13th World Congress, International Federation of Automatic Control,* Vol. B., *Manufacturing,* San Francisco, CA, 30 June–5 July 1996, pp. 91–96.

B. Kosko, *Fuzzy Engineering*. Englwood Cliffs, NJ: Prentice-Hall, 1997.

F. L. Lewis, *Optimal Estimation: With an Introduction to Stochastic Control Theory.* New York: Wiley, 1986.

F. L. Lewis, *Applied Optimal Control and Estimation,* Englewood Cliffs, NJ: Prentice-Hall, 1992.

F. L. Lewis, S. Jagannathan, and A. Yesildirek, *Neural Network Control of Robot Manipulators and Nonlinear Systems*. London: Taylor & Francis, 1999.

R. P. Lippmann, "An Introduction to Computing with Neural Networks," *IEEE ASSP Magazine,* April 1987, pp. 4–22.

L. Ljung, System Identification: Theory for the User, Prentice-Hall, New Jersey, 2d ed., 1999.

K. A. Marko, J. V. James, T. M. Feldkamp, C. V. Puskorius, J. A. Feldkamp, and D. Roller, "Applications of Neural Networks to the Construction of "Virtual" Sensors and Model-Based Diagnostics," *Proceedings ISATA 29th International Symposium on Automotive Technology and Automation,* Florence, Italy, 1996, pp. 133–138.

MATLAB Reference Guide. Cambridge, MA: MathWorks, Inc., 1994.

MATLAB Statistics Toolbox, Version 4. Cambridge, MA: MathWorks, Inc., 2002.

MATLAB Neural Network Toolbox. Cambridge, MA: MathWorks, Inc., 1994.

MATLAB Fuzzy Logic Toolbox, Version 2, Cambridge, MA: MathWorks, Inc., 2002.

P. D. McFadden and J. D. Smith, "An Explanation for the Asymmetry of the Modulation Sidebands about the Tooth Meshing Frequency in Epicyclic Gear Vibration," *Proceedings of the Institution of Mechanical Engineers, Part C: Mechanical Engineering Science,* **199**(1):65–70, 1985.

S. Moaveni, *Finite Element Analysis: Theory and Application with ANSYS,* 2d ed. Upper Saddle River, NJ, Prentice-Hall, 2003.

M. Mufti and G. Vachtsevanos, "Automated Fault Detection and Identification Using a Fuzzy Wavelet Analysis Technique," *Proceedings of AUTOTESTCON '95 Conference,* Atlanta, GA, August 1995, pp. 169–175.

D. Mylaraswamy and V. Venkatasubramanian, "A Hybrid Framework for Large Scale Process Fault Diagnosis," *Computers & Chemical Engineering* 21:S935–940, 1997.

J. C. Newman, *FASTRAN-II: A Fatigue Crack Growth Structural Analysis Program,* revised copy. Langley, VA: NASA Technical Memorandum 104159, February 1992.

A. V. Oppenheim and R. W. Schafer, *Digital Signal Processing,* Englewood Cliffs, NJ: Prentice-Hall, 1975.

B. E. Parker, Jr., H. A. Ware, D. P. Wipf, W. R. Tompkins, B. R. Clark, E. C. Larson, and H. V. Poor, "Fault Diagnostics Using Statistical Change Detection in the Bispectral Domain," *Mechanical Systems & Signal Processing* **14**(4):561–570, 2000.

N. Propes and G. Vachtsevanos, "Fuzzy Petri Net Based Mode Identification Algorithm for Fault Diagnosis of Complex Systems," *System Diagnosis and Prognosis: Security and Condition Monitoring Issues III, SPIE's 17th Annual AeroSense Symposium,* Orlando, FL, April 21, 2003, pp. 44–53.

N. Propes, S. Lee, G. Zhang, I. Barlas, Y. Zhao, G. Vachtsevanos, A. Thakker, and T. Galie, "A Real-Time Architecture for Prognostic Enhancements to Diagnostic Systems," MARCON, Knoxville, TN, May 6–8, 2002.

L. R. Rabiner, "A Tutorial on Hidden Markov Models and Selected Applications in Speech Recognition," *Proceedings of the IEEE* **77**(2):257–286, 1989.

S. S. Rao, *Mechanical Vibrations,* 2nd ed. Reading, MA: Addison-Wesley, 1990.

C. K. Reisbec and R. C. Schank, *Inside Case-Based Reasoning.* Hillsdale, NJ: Lawrence Earlbaum Associates, 1989.

Ridgetop Group, Inc., Tucson, Arizona, "Electronic Prognostics for dc to dc Converters," Electronic Prognostic Workshop, Integrated Diagnostics Committee, National Defense Industrial Association, Denver, CO, June 15–16, 2004.

B. J. Rosenberg, "The Navy IDSS Program: Adaptive Diagnostics and Feedback Analysis: Precursors to a Fault Prognostics Capability," *Proceedings of the IEEE 1989 National Aerospace and Electronics Conference NAECON* 1989, Dayton, OH, vol. 3, May 1989, pp. 1334–1338.

T. J. Ross, *Fuzzy Logic with Engineering Applications.* New York: McGraw-Hill, 1995.

H. Sehitoglu, "Thermal and Thermal-Mechanical Fatigue of Structural Alloys," *ASM Handbook,* Vol. 19: *Fatigue and Fracture.* New York: ASM International, 1997, pp. 527–556.

G. Shafer, *A Mathematical Theory of Evidence.* Princeton, NJ: Princeton University Press, 1976.

Q. Shen and R. Leitch, "Diagnosing Continuous Dynamic Systems using Qualitative Simulation," *International Conference on Control,* Vol. 2, March 1991, pp. 1000–1006.

Q. Shen and R. Leitch, "Fuzzy Qualitative Simulation," *IEEE Transactions on Systems, Man and Cybernetics,* Part A **35**(2):298–305, 2005.

V. A. Skormin, J. Apone, and J. J. Dunphy, "On-Line Diagnostics of a Self-Contained Flight Actuator," *IEEE Transactions Aerospace and Electronic Systems* **30**(1):186–196, 1994.

G. Strang and G. J. Fix, *An Analysis of the Finite Element Method.* Englewood Cliffs, NJ: Prentice-Hall, 1973.

D. W. Tong, C. H. Jolly, and K. C. Zalondek, "Multibranched Diagnostic Trees," *Conference Proceedings, IEEE International Conference on Systems, Man and Cybernetics,* Cambridge, MA: IEEE, Vol 1, Nov. 1989, pp. 92–98.

M. J. Turner, R. W. Clough, H. C. Martin, and L. J. Topp, "Stiffness and Deflection Analysis of Complex Structures," *J. Aero. Sci.* **23**:805–823, 1956.

G. Vachtsevanos and P. Wang, "A Wavelet Network Framework for Diagnostics of Complex Engineered Systems," *Proceedings of MARCON 98 Maintenance and Reliability Conference,* Knoxville, TN, May 12–14, 1998.

J. Wagner and R. Shoureshi, "Failure Detection Diagnostics for Thermofluid Systems," *Journal of Dynamic Systems, Measurements and Control* **114**(4):699–706, 1992.

I. Watson and F. Marir, "Case-Based Reasoning: A Review," The Knowledge Engineering Review, 9–4, 327–354, 1994.

A. S. Willsky, "A Survey of Design Methods for Failure Detection in Dynamic Systems," *Automatica* **12**(6):601–611, 1976.

B. Wu, A. Saxena, T. S. Khawaja, R. Patrick, G. Vachtsevanos, and P. Sparis, "Data Analysis and Feature Selection for Fault Diagnosis of Helicopter Planetary Gears," IEEE Autotestcon 2004, San Antonio, TX.

Q. Zhang and A. Benveniste, "Wavelet Networks," *IEEE Transactions on Neural Networks* **3**(6):889–898, 1992.

G. Zhang, S. Lee, N. Propes, Y. Zhao, G. Vachtsevanos, A. Thakker, and T. Galie, "A Novel Architecture for an Integrated Fault Diagnostic/Prognostic System," AAAI Symposium, Stanford, CA, March 25–27, 2002.

L. A. Zadeh, "Outline of a New Approach to the Analysis of Complex Systems and Decision Processes," *IEEE Transactions on Systems, Man, and Cybernetics* **SMC-3**:28–44, 1973.

M. Zou and I. Green, "Real-Time Condition Monitoring of Mechanical Face Seal," *Proceedings of the 24th Leeds-Lyon Symposium on Tribology,* London, Imperial College, September 4–6, 1997, pp. 423–430.

M. Zou, J. Dayan, and I. Green, "Dynamic Simulation and Monitoring of a Noncontacting Flexibly Mounted Rotor Mechanical Face Seal," *IMechE, Proceedings Instn. Mech. Engineers* **214**(C9):1195–1206, 2000.

CHAPTER 6

FAULT PROGNOSIS

6.1 INTRODUCTION

Prognosis is the ability to predict accurately and precisely the remaining useful life of a failing component or subsystem. It is the Achilles' heel of the condition-based maintenance/prognostic health management (CBM/PHM) system and presents major challenges to the CBM/PHM system designer primarily because it entails large-grain uncertainty. Long-term prediction of the fault evolution to the point that may result in a failure requires means to represent and manage the inherent uncertainty. *Uncertainty representation* implies the ability to model various forms of uncertainty stemming from a variety of sources, whereas *uncertainty management* concerns itself with the methodologies and tools needed to continuously "shrink" the uncertainty bounds as more data become available. We will elaborate on these issues in this chapter. Moreover, accurate and precise prognosis demands good probabilistic models of the fault growth and statistically sufficient samples of failure data to assist in training, validating, and fine-tuning prognostic algorithms. Prognosis performance metrics, robust algorithms, and test platforms that may provide needed data have been the target of CBM/PHM researchers in the recent past. Many accomplishments have been reported, but major challenges still remain to be addressed. This chapter reviews the key challenges in prognosis and introduces the theoretical and notional foundations. It concludes with selected application examples.

6.1.1 Fault Prognosis: The Challenges

Prognosis is a composite word consisting of the Greek words *pro* and *gnosis,* literally translating into the ability to acquire knowledge (*gnosis*) about events before (*pro*) they actually occur. The Delphi Oracle is characteristic of such a foretelling activity, with the inherent ambiguity captured in the famous statement *"Ibis redibis non morieris in bello."*

The term *prognosis* has been used widely in medical practice to imply the foretelling of the probable course of a disease (Taylor, 1953). In the industrial and manufacturing arenas, prognosis is interpreted to answer the question, "What is the remaining useful lifetime of a machine or a component once an impending failure condition is detected, isolated, and identified?"

To illustrate some of the issues surrounding fault propagation and failure prediction, examine Fig. 6.1, which shows how a fault might propagate. If one catches the fault at 4 percent severity, one need replace only the component. If the fault is not caught until 10 percent severity, the subsystem must be replaced, and at failure, the entire system must be replaced. Clearly, predictions about fault severity and impending failures are essential.

Foretelling the future in a wide area of disciplines from business and finance to weather forecasting, among many others, has attracted the interest of researchers and practitioners over the past decades, with results that vary from disappointing to promising. Consider the following situation: A typical system amenable to CBM/PHM technologies is a gearbox found in many transportation and other critical systems. Suppose that a fault—a crack in the races of a bearing, a missing gear tooth, or a shaft imbalance—has been detected and isolated. The gearbox typically is instrumented with accelerometers located on its frame for monitoring vibration levels. The acceler-

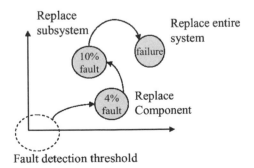

Figure 6.1   Fault propagation.

ometers respond to vibrations that may originate at a number of possible sources inside the gearbox. Suppose further that the detected single fault has been isolated to originate from a specific *line-replaceable unit* (LRU) of the gear assembly. For purposes of illustration, suppose that the fault is a crack on the gear plate of the main transmission gearbox of a helicopter, as shown in Fig. 6.2.

The task of the prognostic module is to monitor and track the time evolution (growth) of the fault. The obvious solution would be the placement of a "crack length meter" on the affected component that could transmit crack length data continuously to a remote processor. Ideally, the latter would be endowed with the capability to project the fault dimension (crack length) over time, determine the time to failure of the component, and prescribe confidence bounds, that is, some sense as to the earliest and latest time that maintenance or repair must be performed or the mission (task) be terminated to avoid a

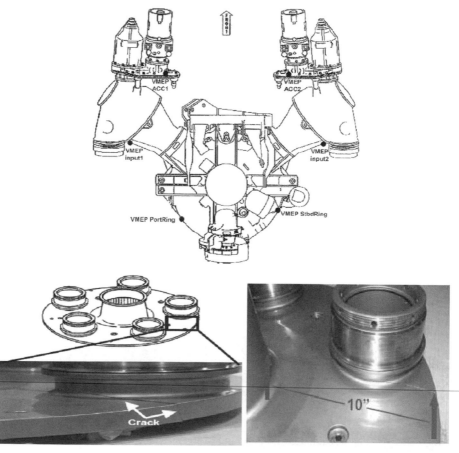

Figure 6.2 Crack in the gear plate of the main transmission gearbox of a helicopter.

catastrophic event. Unfortunately, there is no such sensing instrument as a "crack length meter."

> *Challenge 1:* In the absence of a "crack length meter," how can we infer the actual crack dimension over time? Is it possible to estimate the crack length from measurable quantities such as accelerometer readings, temperature, speed, etc.? This setting suggests some form of a virtual or soft sensor that maps known measurements from available sensors to quantifiers that may be difficult or impossible to obtain via direct sensor readings.
>
> *Challenge 2:* How do we predict accurately and precisely the temporal progression of the fault? Here, a suitable algorithm is sought that takes as inputs the available sensor measurements and system models and then outputs in a probabilistic sense the fault evolution to the point of actual failure.
>
> *Challenge 3:* How do we prescribe the uncertainty bounds or confidence limits associated with the prediction? In this case, we are seeking probabilistic or other means that may describe the uncertainty envelope or distribution as time progresses. This provides a feeling as to the accuracy of and confidence to be placed in the fault evolution prediction.
>
> *Challenge 4:* Once we have predicted the time evolution of the fault and prescribed the initial uncertainty bounds, how do we improve on such performance metrics as prediction accuracy, confidence, and precision? A learning strategy may be suggested in this case that takes advantage of new data as they stream into the processing unit and then provides corrections based on the predicted and actual fault state progression. Reliable prognosis algorithms also may require accounting for the impact of system usage patterns, operating modes, and manufacturing variability, thus complicating further the prediction task.

This chapter will address these challenges and provide a framework for fault prognosis.

6.1.2 The Prediction Problem: A Notional Framework

In the engineering disciplines, fault prognosis has been approached via a variety of techniques ranging from Bayesian estimation and other probabilistic/statistical methods to artificial intelligence tools and methodologies based on notions from the computational intelligence arena. Specific enabling technologies include multistep adaptive Kalman filtering (Lewis, 1986), autoregressive moving-average models (Lewis, 1992), stochastic autoregressive integrated-moving-average models (Jardim-Goncalves and Martins-Barata, 1996), Weibull models (Groer, 2000), forecasting by pattern and cluster search (Frelicot, 1996), and parameter estimation methods (Ljung, 1999). From the artificial intelligence domain, case-based reasoning (Aha, 1997), intelligent decision-based models, and min-max graphs have been considered as potential candidates for prognostic algorithms. Other methodologies, such as Petri nets, neural networks, fuzzy systems, and neuro-fuzzy systems (Studer and Masulli, 1996), have found ample utility as prognostic tools. Physics-based fatigue models (Ray and Tangirala, 1996) have been employed extensively to represent the initiation and propagation of structural anomalies.

Some issues involved in failure prediction are shown in Fig. 6.3. This figure shows the *evolution of a fault feature* of interest and introduces the notion of prescribed *alarm bounds* or *fault tolerance limits* outside of which a failure is declared. Also emphasized is the importance of *confidence limits* on the feature predicted values. When the confidence limits reach the tolerance limits, an alarm is declared.

We will detail in this chapter a representative sample of the multitude of enabling technologies. To organize this topical area of such great importance to CBM/PHM, we will classify prognosis schemes into three categories: model-based, probability-based, and data-driven. First, we will discuss an important and common situation in prognosis, where one has available certain sensor readings or measurements that may not yield exactly the information that one requires for the prediction of relevant quantities.

6.1.3 Prognosis Algorithm Approaches

The remaining sections of this chapter will provide an overview of a representative sample of the multitude of enabling technologies used in prognosis. Figure 6.4 summarizes the range of possible prognosis approaches as a function of the applicability to various systems and their relative implementation cost. Prognosis technologies typically use measured or inferred features, as

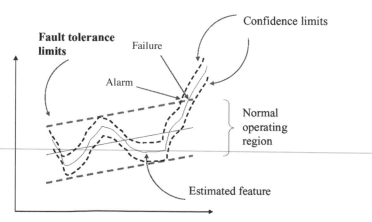

Figure 6.3 Fault trend and failure prediction.

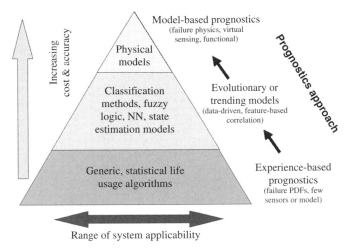

Figure 6.4 Prognosis technical approaches.

well as data-driven and/or physics-based models, to predict the condition of the system at some future time. Inherently probabilistic or uncertain in nature, prognosis can be applied to failure modes governed by material condition or by functional loss. Prognosis algorithms can be generic in design but specific in terms of application. Prognosis system developers have implemented various approaches and associated algorithmic libraries for customizing applications that range in fidelity from simple historical/usage models to approaches that use advanced feature analysis or physics-of-failure models.

Depending on the criticality of the LRU or subsystem being monitored, various levels of data, models, and historical information will be needed to develop and implement the desired prognostic approach. Table 6.1 provides an overview of the recommended models and information necessary for implementing specific approaches. Of course, the resolution of this table only illustrates three levels of algorithms, from the simplest experienced-based (reliability) methods to the most advanced physics-of-failure approaches that are calibrated by sensor data.

6.2 MODEL-BASED PROGNOSIS TECHNIQUES

Model-based prognostic schemes include those that employ a dynamic model of the process being predicted. These can include physics-based models, autoregressive moving-average (ARMA) techniques, Kalman/particle filtering, and empirical-based methods.

Model-based methods provide a technically comprehensive approach that has been used traditionally to understand component failure mode progression. Implementing physics-based models can provide a means to calculate

TABLE 6.1 Prognostic Accuracy

| | Prognostic accuracy ⟶ | | |
	Experienced-Based	Evolutionary	Physics-Based
Engineering model	Not required	Beneficial	Required
Failure history	Required	Not required	Beneficial
Past operating conditions	Beneficial	Not required	Required
Current conditions	Beneficial	Required	Required
Identified fault patterns	Not required	Required	Required
Maintenance history	Beneficial	Not required	Beneficial
In general	No sensors/no model	Sensors/no model	Sensors and model

the damage to critical components as a function of operating conditions and assess the cumulative effects in terms of component life usage. By integrating physical and stochastic modeling techniques, the model can be used to evaluate the distribution of remaining useful component life as a function of uncertainties in component strength/stress properties, loading, or lubrication conditions for a particular fault. Statistical representations of historical operational profiles serve as the basis for calculating future damage accumulation. The results from such a model then can be used for real-time failure prognostic predictions with specified confidence bounds. A block diagram of this prognostic modeling approach is given in Fig. 6.5. As illustrated at the core of this figure, the physics-based model uses the critical, life-dependent uncertainties so that current health assessment and future remaining useful life (RUL) projections can be examined with respect to a risk level.

Model-based approaches to prognosis differ from feature-based approaches in that they can make RUL estimates in the absence of any measurable events, but when related diagnostic information is present (such as the feature described previously), the model often can be calibrated based on this new information. Therefore, a combination or fusion of the feature-based and model-based approaches provides full prognostic ability over the entire life of the component, thus providing valuable information for planning which components to inspect during specific overhaul periods. While failure modes may be unique from component to component, this combined model-based and feature-based methodology can remain consistent across different types of critical components or LRUs.

To perform a prognosis with a physics-based model, an operational profile prediction must be developed using the steady-state and transient loads, tem-

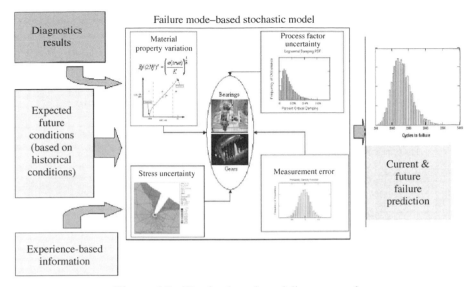

Figure 6.5 Physics-based modeling approach.

peratures, or other online measurements. With this capability, probabilistic critical-component models then can be "run into the future" by creating statistical simulations of future operating profiles from the statistics of past operational profiles or expected future operating profiles.

Based on the authors' past experience correlating operational profile statistics and component or LRU life usage, the nonlinear nature associated with many damage mechanisms depends on both the inherent characteristics of the profiles and the operational mix types. Significant component damage resulting from the large variability in the operating environment and severity of the missions directly affects the system component lifetimes. Very often, component lives driven by fatigue failure modes are dominated by unique operational usage profiles or a few rare, severe, randomly occurring events, including abnormal operating conditions, random damage occurrences, etc. For this reason, we recommend a statistical characterization of loads, speeds, and conditions for the projected future usage in the prognostic models, as shown in Fig. 6.6.

6.2.1 Physics-Based Fatigue Models

In such critical systems as aircraft and industrial and manufacturing processes, defect (e.g., cracks and anomalies) initiation and propagation must be estimated for effective fault prognosis. We will repeat here for easy reference basic notions in fatigue modeling introduced in Chapter 5 Section 5.6.1 since they form the foundation for physics-based prognosis. Dynamic models of fatigue and fault progression can be derived by using principles of physics.

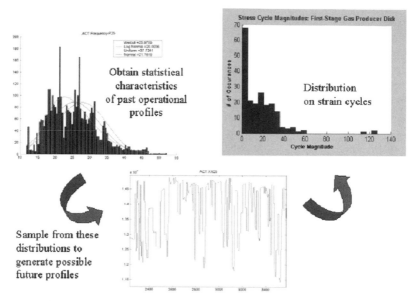

Figure 6.6   Operation profile and loading model for prognosis.

Recent studies in materials research have focused on microstructural characterization of materials and modeling frameworks that describe robustly and reliably the anomaly initiation and propagation as a function of time (Friedman and Eidenoff, 1990). Such models addressing the constitutive material properties down to the single particle level generally are not suitable for real-time treatment owing to their immense computational complexity. Reduced-order models, called *metamodels,* usually are derived from the microstructural characterizations so that they may be used to estimate in almost real-time the crack initiation and propagation with conventional computing platforms (Newman, 1992). Crack initiation models must account for variations in temperature, cycle frequency, stress ratio, sustained load hold time, and interactions of damage mechanisms such as fatigue, environment, and creep. A thorough description of the crack initiation constitutive equations can be found in Chestnut (1965).

Crack propagation models, clearly of interest from the prognosis viewpoint, are separated into two major categories: deterministic and stochastic models. Fatigue crack growth in such typical machinery components as bearings, gears, shafts, and aircraft wings is affected by a variety of factors, including stress states, material properties, temperature, lubrication, and other environmental effects. Variations of available empirical and deterministic fatigue crack propagation models are based on Paris' formula:

$$\frac{d\alpha}{dN} = C_0(\Delta K)^n$$

where α = instantaneous length of dominant crack

$\quad N$ = running cycles

$\quad C_0, n$ = material dependent constants

$\quad \Delta K$ = range of stress-intensity factor over one loading cycle

Model parameters typically are estimated via a nonlinear recursive least-squares scheme with a forgetting factor (Ljung, 1999).

Stochastic crack growth models consider all parameters as random quantities. Thus the resulting growth equation is a stochastic differential equation. Here, means to estimate parameter distributions include Monte Carlo simulations, probabilistic neural networks, and others.

6.2.2 ARMA, ARMAX, and ARIMA Methods

In many cases involving complex systems, it is very difficult or impossible to derive dynamic models based on all the physical processes involved. In such cases, it is possible to assume a certain form for the dynamic model and then use observed inputs and outputs of the system to determine the model parameters needed so that the model indeed serves as an accurate surrogate for the system. This is known as *model identification*. To introduce this method, we will begin by considering a classic identification/prediction problem: Given input-output data of a dynamic system, how do we develop a model to predict the system's future behavior? The intent here is to establish some fundamental notions in prognosis that will set the stage for more advanced methods and algorithms to be detailed in the rest of the chapter.

Model Identification Given Observed Data. Consider the abstract representation of a system shown in Fig. 6.7. The input to the system is u_k, whereas the output is designated by y_k, and v_k is an unmeasured disturbance. Let us assume that a series of experiments is designed so that a sufficient data set can be obtained to fit a model of the system. A classic system identification loop is depicted in Fig. 6.8.

Prior knowledge guides the design of experiments, and data are derived through a series of test runs. A system model is hypothesized together with a suitable criterion of fit. Test data, the model set, and the criterion of fit are

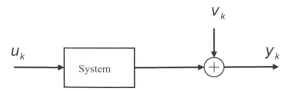

Figure 6.7 A generalized system representation.

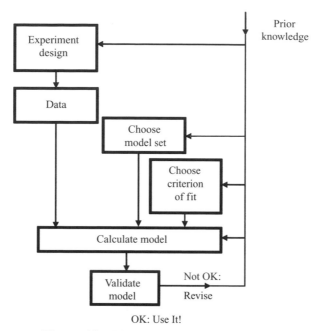

Figure 6.8   The system identification loop.

used next to calculate the model parameters. Once this step is completed, the model is validated using previously unseen data. The identification task is terminated if the validation step is successful; otherwise, we loop back to the initial step until we arrive at a satisfactory solution.

System Dynamic Model. Let the disturbance v_k be a stationary process with spectrum

$$\Phi_v(\omega) = \lambda |R(e^{j\omega})|^2$$

It is assumed that $R(z)$ has no poles and no zeros on or outside the unit circle (for stability); then the input-output relation is given by

$$v_k = R(q)e_k$$

where e_k is white noise with variance λ. The transfer function $R(q)$ is the ratio of two polynomials in q, that is, $R(q) = C(q)/A(q)$ with

$$A(q) = 1 + a_1 q^{-1} + \cdots + a_{n_a} q^{-n_a}$$

$$C(q) = 1 + c_1 q^{-1} + \cdots + c_{n_c} q^{-n_c}$$

where q^{-1} is a unit delay. Then one has the discrete-time input-output recursion

$$v_k + a_1 v_{k-1} + \cdots + a_{n_a} v_{k-n_a} = e_k + c_1 e_{k-1} + \cdots + c_{n_c} e_{k-n_c}$$

Such a dynamic system representation relating the function v_k at time k to its past history, as well as the time history of the noise e_k up to time k, is known as an *autoregressive moving-average (ARMA) model.* It can be viewed as projecting (predicting) the value of v at time k from the past histories of v_k and the noise e_k. This equation constitutes a good modeling framework for data, disturbances, etc. that is derived from the spectrum only.

The input-output relationship

$$y_k + a_1 y_{k-1} + \cdots + a_{n_a} y_{k-n_a} = b_1 u_{k-1} + \cdots + b_{n_b} u_{k-n_b} + e_k$$

is known as the *equation error model* (Ljung, 1999). The coefficients of such a model are unknown parameters that must be adjusted so that the input-output behavior of the model corresponds to the observed input-output data sets. The adjustable parameters can be collected into a vector of unknown parameters $\boldsymbol{\theta} = [a_1, a_2, \ldots a_{na}, b_1 \ldots b_{nb}]^T$ so that one may write

$$y_k = [-y_{k-1} - y_{k-2} \cdots -y_{k-n_a} \, u_{k-1} \cdots u_{t-n_b}] \, \boldsymbol{\theta} + e_k \equiv \boldsymbol{h}_k^T \boldsymbol{\theta} + e_k$$

where $\boldsymbol{h}$ is a *regression vector* that contains all the observed data, including past values of the input u_k and the output y_k.

The equation error model lacks the freedom of fully describing the properties of the disturbance term, as required by the ARMA model where both current and previous values of the noise are needed. A variation of the ARMA model called *ARMAX* describes the equation error as a moving average of white noise according to

$$y_k + a_1 y_{k-1} + \cdots + a_{n_a} y_{k-n_a} = b_1 u_{k-1} + \ldots + b_{n_b} u_{k\ n_b} + e_k$$
$$+ c_1 e_{k-1} + \cdots + c_{nc} e_{k-n_c}$$

where now the adjustable unknown parameters are

$$\boldsymbol{\theta} = [\alpha_1 \ldots \alpha_{na} \, b_1 \ldots b_{nb} \, c_1 \ldots c_{nc}]^T$$

The regression vector now contains previous values of e_k as well. Finally, a version with an enforced integration (ARIMA) is useful in describing systems with slow disturbances.

Parameter Identification and System Prediction. The model-identification problem in ARMA-type representation boils down to determining the model parameter vector $\boldsymbol{\theta}$ given the observed input-output data sequences. Given a sufficient set of test measurements, a *least-squares method* may be called on to determine $\boldsymbol{\theta}$ (Ljung, 1999). System-identification techniques such as least-squares and others have been developed extensively and are by now available, for example, in MATLAB toolboxes.

ARMA models and their derivatives have found extensive utility in various application domains as *predictors,* that is, employing the model to predict the expected system states from observed historical data. This, of course, makes them very useful in prognosis, where one desires to predict the states to estimate fault progression. Prediction is accomplished by using the observed data up to time t to estimate the parameter vector $\boldsymbol{\theta}$, using, for example, a least-squares technique. Then, once the model parameters are known, the dynamic ARMAX model can be used to predict the future outputs.

6.2.3 The Particle Filter Framework for Failure Prognosis

Bayesian estimation techniques are finding application domains in machinery fault diagnosis and prognosis of the remaining useful life of a failing component/subsystem. This section introduces a methodology for accurate and precise prediction of a failing component based on particle filtering and learning strategies developed at the Georgia Institute of Technology (Orchard et al., 2005). This novel approach employs a state dynamic model and a measurement model to predict the posterior probability density function of the state, that is, to predict the time evolution of a fault or fatigue damage. It avoids the linearity and Gaussian noise assumption of Kalman filtering and provides a robust framework for long-term prognosis while accounting effectively for uncertainties. Correction terms are estimated in a learning paradigm to improve the accuracy and precision of the algorithm for long-term prediction. The approach is applied to a crack fault, and the results support its robustness and superiority. The particle filter is used for diagnosis in Chap. 5. A brief description of particle-filtering approach is presented in Chap. 5 as well.

Prediction of the evolution of a fault or fault indicator entails large-grain uncertainty. Accurate and precise prognosis of the time to failure of a failing component/subsystem must consider critical-state variables such as crack length, corrosion pitting, etc. as random variables with associated probability distribution vectors. Once the probability distribution of the failure is estimated, other important prognosis attributes—such as confidence intervals—can be computed. These facts suggest a possible solution to the prognosis problem based on recursive Bayesian estimation techniques that combine both the information from fault-growth models and online data obtained from sensors monitoring key fault parameters (observations).

Prognosis or long-term prediction for the failure evolution is based on both an accurate estimation of the current state and a model describing the fault progression. If the incipient failure is detected and isolated at the early stages of the fault initiation, it is reasonable to assume that sensor data will be available for a certain time window allowing for corrective measures to be taken, that is, improvements in model parameter estimates so that prognosis will provide accurate and precise prediction of the time to failure. At the end of the observation window, the prediction outcome is passed on to the user (operator, maintainer), and additional adjustments are not feasible because corrective action must be taken to avoid a catastrophic event.

Figure 6.9 depicts a conceptual schematic of a particle-filtering framework aimed at addressing the fault prognosis problem. Available and recommended CBM/PHM sensors and the feature-extraction module provide the sequential observation (or measurement) data of the fault growth process z_k at time instant k. We assume that the fault progression can be explained through the state-evolution model and the measurement model:

$$x_k = f_k(x_{k-1}, \omega_k) \quad \leftrightarrow \quad p(x_k|x_{k-1})$$

$$z_k = h_k(x_k, \nu_k) \quad \leftrightarrow \quad p(z_k|x_k)$$

where x_k is the state of the fault dimension (such as the crack size), the changing environment parameters that affect fault growth, ω_k and ν_k, are the non-Gaussian noises, and f_k and g_k are nonlinear functions.

The first part of the approach is state estimation, that is, estimating the current fault dimension as well as other important changing parameters in the environment. This part is similar to what is used for diagnostic purposes, as described in Chap. 5. The second part of the approach is the long-term prediction based on the current estimate of the fault dimension and the fault-growth model with parameters refined in the posteriori state estimation. A novel recursive integration process based on both importance sampling and PDF approximation through Kernel functions is then applied to generate state predictions from $(k + 1)$ to $(k + p)$:

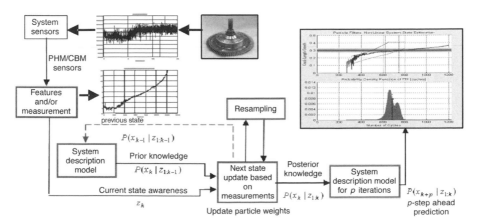

Figure 6.9 Fault prognosis based on particle filtering.

$$p(x_{k+p}|z_{1:k}) = \int p(x_k|z_{0:k}) \prod_{j=k+1}^{k+p} p(x_j|x_{j-1}) \, dx_{k:k+p-1}$$

$$= \sum_{i=1}^{N} \tilde{w}_k^{(i)} \int \cdots \int p(x_{k+1}|x_k^{(i)}) \prod_{j=k+2}^{k+p} p(x_j|x_{j-1}) \, dx_{k+1:k+p-1}$$

Long-term predictions can be used to estimate the probability of failure in a process, given a hazard zone that is defined by its lower and upper bounds (H_{lb} and H_{up}, respectively). The prognosis confidence interval as well as the expected time to failure (TTF) can be deduced from the TTF PDF:

$$p_{\mathrm{TTF}}(\mathrm{TTF}) = \sum_{i=1}^{N} \Pr\{H_{\mathrm{lb}} \le x_{\mathrm{TTF}}^{(i)} \le H_{\mathrm{up}}\} \, \tilde{w}_{\mathrm{TTF}}^{(i)}$$

The uncertainty usually increases as the prediction farther into the future is made. To reduce the uncertainty in the particle-filter-based failure prognosis, an additional learning paradigm for correction of the estimated TTF expectation is used. This learning paradigm computes a correction term C_n consisting of the difference between the current TTF estimation and the previous one at every iteration (Fig. 6.10). Once p correction terms are obtained, a linear regression model is build to establish a relationship between all past correction terms. The linear model then is used to generate an estimation of all future corrections that would be applied if the process were to be wide sense stationary (WSS), thus improving the accuracy of long-term predictions.

$$T_k^{(c)} = T_k - \sum_{l=k+1}^{k+p} C_l$$

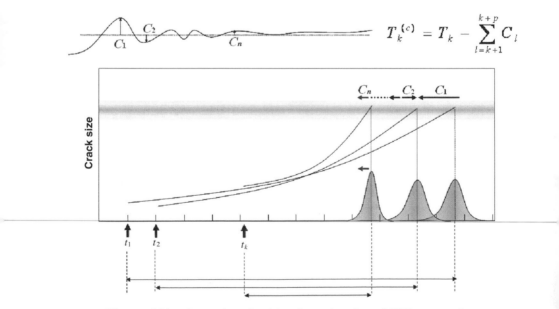

Figure 6.10 Correction algorithm for estimation of TTF expectation.

Example 6.1 Analysis of Crack Length Growth in a Turbine Engine.
The proposed methodology just mentioned has been used for the analysis of
a crack in blades of a turbine engine, and the results obtained are shown next.
The fault-growth model is assumed to be

$$N = P \cdot L^6 + f(L^5)$$

$$\frac{dL}{dN} = \frac{1}{6P \cdot L^5 + df/dL} + \omega(N)$$

where L is the length of the crack (in inches), N is the number of stress cycles
applied to the material, P is a random variable with known expectation, f is
a fifth order polynomial function derived from a FRANC-3D model (low-
cycle-fatigue assumption), and $\omega(N)$ is independent and identically distributed
(I.I.D.) white noise. The measurement model is considered to be corrupted
with Gaussian noise $v(N)$, $y(N) = L(N) + v(N)$. Figure 6.11 shows the simulation results of the prognosis. The TTF estimate, corrected by the learning
paradigm, is shown in the second plot as the lighter shaded line. The prognosis result converges to the true values as time increases.

*Example 6.2 Analysis of Crack Length Growth in a Planetary Gear
Plate.* In the case of the analysis of the crack growth in a plate of a planetary
gearbox, ANSYS was used to determine the crack tip stress intensity factors

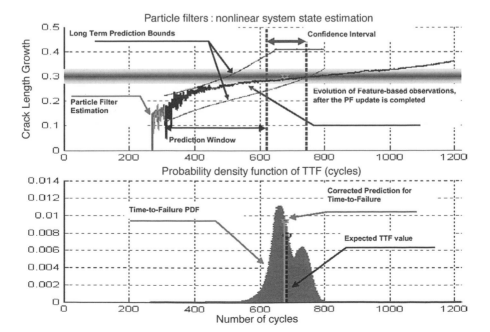

Figure 6.11 Prognosis results for crack growth analysis.

for different crack sizes. These values, if correct, can be used to determine the crack growth rate with an empirical equation based on Paris's crack growth model with an unknown parameter, that is,

$$\frac{dL}{dN} = C \cdot \alpha \cdot [(\Delta K_{\text{inboard}})^m + (\Delta K_{\text{outboard}})^m]$$

where n is a material constant obtainable from tables.

The term ΔK is also known as the crack tip stress and is often denoted by ΔK_{eff}. Once the stress profile is generated, the value of ΔK at different torque values is computed by considering variations in stress that result from the combined effect of the ground-air-ground (GAG) cycle shape and the first-harmonic component (whose magnitude was validated by using test baseline data).

This offline procedure allows generation of long-term predictions based only on the dimensions and material properties of the plate, given an initial condition for the crack. The uncertainty associated with the unknown parameter, though, will be importantly reflected in those predictions. One solution for this problem is to use features obtained from vibration data into a particle-filter framework for failure prognosis, as Fig. 6.12 shows.

Under this approach, the use of a feature—crack length nonlinear mapping for the generation of noisy estimates of crack length based on the vibration data helps to identify the trend of the failure evolution. This information, together with knowledge of both the load profile and the initial crack length, is fed into the particle-filter module to update the estimation of the unknown

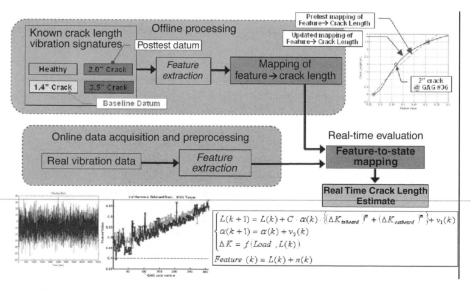

Figure 6.12   Feature-based particle-filter approach for failure prognosis.

parameter in the model, thus improving the current estimation of the crack length and the long-term prediction generated from the current particle population.

$$\begin{cases} L(k+1) = L(k) + C \cdot \alpha(k) \cdot \{(\Delta K_{\text{inboard}})^m + (\Delta K_{\text{outboard}})^m\} + \upsilon_1(k) \\ \alpha(k+1) = \alpha(k) + \upsilon_2(k) \\ \Delta K = f(\text{Load}, L(k)) \end{cases}$$

$$\text{Feature}(k) = L(k) + n(k)$$

As a final result, it is possible to obtain a very accurate prediction for the growth of the crack after the inclusion of 100 data points (GAG cycles) for the vibration-based feature. The evolution of the particle population allows generating TTF PDFs for a given hazard zone (Fig. 6.13), as discussed previously, or for several given thresholds of interest (Fig. 6.14). In the latter case, the resulting PDFs will reflect directly the weight distribution among the particles, thus also giving information about the most probable paths that the system is going to follow as time passes.

6.3 PROBABILITY-BASED PROGNOSIS TECHNIQUES

Determining a complete dynamic model in terms of differential equations that relate the inputs and outputs of the system being considered may be unnecessary or impractical in some instances. Often, historical data from previous

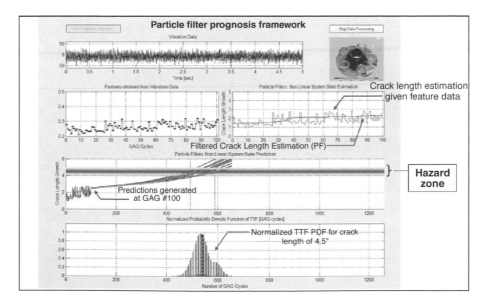

Figure 6.13 Particle-filter approach for crack in plate prognosis.

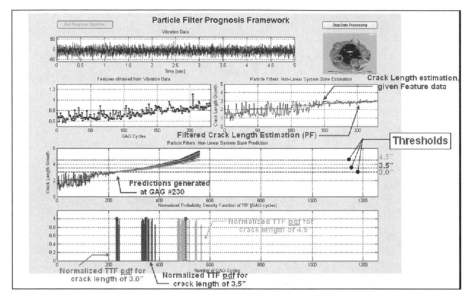

Figure 6.14 Particle-filter threshold approach for crack in plate prognosis.

failures for a given class of machinery takes the statistical form shown in Fig. 6.15, where each point represents one previous failure instance. Probabilistic methods for prognosis are very effective in such situations. These methods require less detailed information than model-based techniques because the information needed for prognosis resides in various probability density functions (PDFs), not in dynamic differential equations. Advantages are that the required PDFs can be found from observed statistical data and that the PDFs are sufficient to predict the quantities of interest in prognosis. Moreover, these

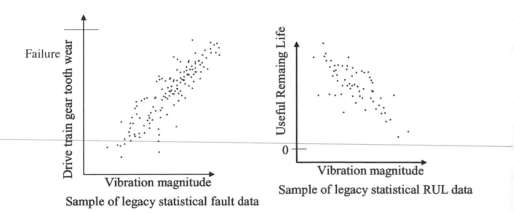

Figure 6.15 Historical data from previous failures.

methods also generally give *confidence limits* about the results, which are important in giving a feeling for the accuracy and precision of the predictions.

A typical probability of failure curve is known as the *hazard function* (Chestnut, 1965), depicted in Fig. 6.16 as the *bathtub curve*. Very early during the life of a machine, the rate of failure is relatively high (so-called infant-mortality failures); after all components settle, the failure rate is relatively constant and low. Then, after some time of operation, the failure rate again begins to increase (so-called wearout failures), until all components or devices will have failed. Predicting failure based on system characteristics is further complicated by the fact that one must take into account *manufacturing variability, mission history variations,* and *life degradation,* as shown in Fig. 6.17. All these are probabilistic effects. Finally, in predicting failures, one must take into account and reduce as far as possible the *false-alarm probability*.

6.3.1 Bayesian Probability Theory

Various methods of confronting uncertainty include Bayesian methods, the Dempster-Shafer theory, and fuzzy logic. Probability methods generally are based on Bayes' theorem (Lewis, 1986), which says that

$$f_{X,Y}(x, y) = f_{X/Y}(x/y)f_Y(y) = f_{Y/X}(y/x)f_X(x)$$

where $f_{X,Y}(x, y)$ is the joint probability density function (PDF), the marginal PDFs are $f_X(x)$ and $f_Y(y)$, and the conditional PDFs are $f_{X/Y}(x/y)$ and $f_{Y/X}(y/x)$. One computes the marginal PDFs from the joint PDF by using

$$f_X(x) = \int f_{X,Y}(x, y)\, dy \qquad f_Y(y) = \int f_{X,Y}(x, y)\, dx$$

where the integrals are over the entire region of definition of the PDF. Therefore, one can write

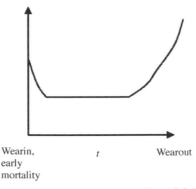

Wearin, early mortality t Wearout

Figure 6.16   The hazard function—probability of failure versus time.

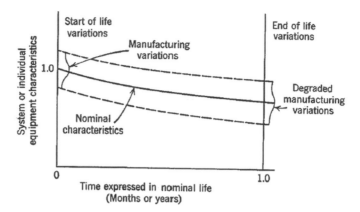

Figure 6.17 Manufacturing variability and mission history effects. (Chestnut, 1965.)

$$f_{Y/X}(y/x) = \frac{f_{X,Y}(x, y)}{\int f_{X,Y}(x, y) \, dy}$$

and similarly for $f_{X/Y}(x/y)$.

The mean or expected value of a random variable is given in terms of the PDF by

$$\bar{x} = E(x) = \int x f_X(x) \, dx$$

and the conditional mean is given by

$$E(Y/X) = \int y f_{Y/X}(y/x) \, dy = \frac{\int y f_{X,Y}(x, y) \, dy}{\int f_{X,Y}(x, y) \, dy}$$

Clearly, all these quantities may be computed from the joint PDF. Therefore, in applications, the computation of the joint PDF given the available data is of high priority. Figures 6.18 and 6.19 show the geometric meaning of the manipulations involved in obtaining the marginal and conditional PDF from the joint PDF.

In estimation applications, one may interpret the unknown quantity of interest to be x and the measured output to be y, as in Fig. 6.20. Then many decision-making techniques rely on $f_{Y/X}(y/x)$, which is known as the *likelihood function*.

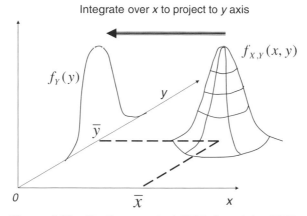

Figure 6.18 Finding marginal PDFs from joint PDF.

6.3.2 The Weibull Model: Analysis of Time to Failure

Reliability analysis of a wide variety of engineered products and systems suggests that time-to-failure data tend to fit well a Weibull distribution (Groer, 2000; Schömig and Rose, 2003). Thus Weibull models have found considerable appeal in the fault-prognosis community, and many researchers have employed such models to predict the remaining useful life (RUL) of failing components. A general form of a three-parameter Weibull distribution is given by (Weibull, 1951)

$$h(t;\, \beta,\, \eta,\, \gamma) = \frac{\beta}{\eta} \left(\frac{t - \gamma}{\eta} \right)^{\beta - 1} \qquad \text{for hazard rate}$$

$$f(t;\, \beta,\, \eta,\, \gamma) = \frac{\beta}{\eta} \left(\frac{t - \gamma}{\eta} \right)^{\beta - 1} \exp\left[-\left(\frac{t - \gamma}{\eta} \right)^{\beta} \right] \qquad \text{for failure density}$$

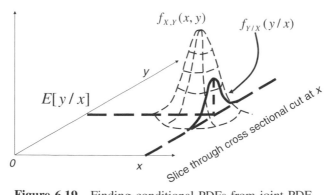

Figure 6.19 Finding conditional PDFs from joint PDF.

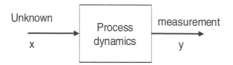

Figure 6.20 Estimation problem formulation.

for the range $\gamma \leq t < \infty$, where η is the characteristic life, β is the shape factor, and γ is the location parameter. If γ is set to zero, the two-parameter distribution is obtained. Note that the hazard rate or the instantaneous failure rate is the failure rate normalized to the number of equipment surviving, whereas the failure density or failure distribution function is the failure rate normalized to the original number of equipment. The significance of the shape factor is

$$\beta < 1 \qquad \text{Hazard rate decreasing}$$
$$\beta = 1 \qquad \text{Hazard rate constant}$$
$$\beta > 1 \qquad \text{Hazard rate increasing}$$

The bathtub curve depicted in Fig. 6.16 can be described by the Weibull distribution for hazard rate with $\beta < 1$ at the wearin, $\beta = 1$ at the steady, and $\beta > 1$ at the wearout stages.

There are several other versions of the two-parameter Weibull distribution, one of which is given by (Groer, 2000)

$$h(t|\alpha,\lambda) = \alpha\lambda^\alpha t^{\alpha-1}$$
$$f(t|\alpha, \lambda) = \alpha\lambda^\alpha t^{\alpha-1} \exp(-\lambda^\alpha t^\alpha)$$

and is related to the first form given by $\lambda = 1/\eta$ and $\alpha = \beta$.

Failure data in the form of times to failure or hazard rates may be fitted to the Weibull distribution by parameter estimation methods (Lees, 1996) or using Bayesian analysis regarding the model parameters as random variables as well. For the former, three principal methods can be used (1) least squares, (2) moments, and (3) maximum likelihood. In practice, parameter estimation is a challenge when considering large-scale dynamic systems such as gearboxes and gas turbines because it requires the availability of a sufficient sample of actual failure data. That is, one requires time-to-failure recordings originating from either extensive testing or from actual historical operating records. We will illustrate the utility of the Weibull model as a prognostic tool via an example from Groer (2000).

Example 6.3 Weibull Distribution for Estimating Remaining Useful Life or Time to Failure. Suppose that a set of failure data D contains the following failure times:

$$t_f = \{1617., 1761., 2014., 2093., 2284., 2423., 2780.,$$

$$2912., 2947., 2948., 2985., 3185.\}$$

We consider the two-parameter Weibull distribution for failure density given by the preceding equation. The likelihood of failure data D conditioned to the model parameters α and λ is given by

$$L(D|\alpha, \lambda) = \prod_{k=1}^{12} f(t_k|\alpha, \lambda) = \alpha^{12}\lambda^{12\alpha} \prod_{k=1}^{12} t_k^{\alpha-1} e^{-\lambda^\alpha \Sigma_k^{12} t_k^\alpha}$$

where the product runs over all failure times.

To use Bayesian analysis, it is required to identify the model parameters α and λ as random variables with the prior joint distribution $p(\alpha, \lambda)$. The posterior probability of α and λ based on the field data D can be calculated by

$$p(\alpha, \lambda|D) = L(D|\alpha, \lambda)p(\alpha, \lambda)$$

Integration of this density over α gives the marginal density for λ, and vise versa.

Figure 6.21 depicts a plot of the joint posterior density of $\alpha(1 \le \alpha \le 15)$ and $\lambda(0.0003 \le \lambda \le 0.0004)$, whereas Fig. 6.22 shows the marginal density for the shape parameters α and λ, respectively.

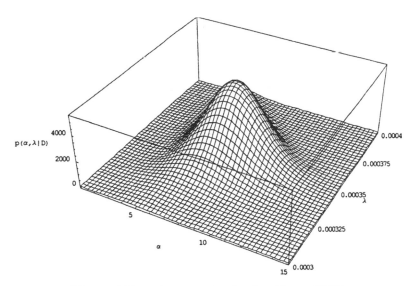

Figure 6.21 Joint posterior density. (Groer, 2000.)

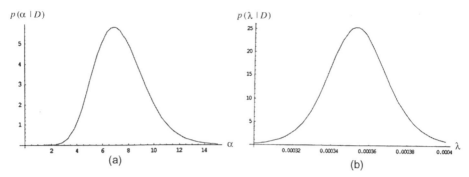

Figure 6.22   The marginal density for the shape (*a*) and scale (*b*) parameters. (Groer, 2000.)

Integrating over both parameters, we arrive at the predictive density conditioned to the failure data *D*:

$$p(t_f|D) = \int\int p(t_f|\alpha, \lambda)p(\alpha, \lambda|D) \, d\alpha \, d\lambda$$

Parameter uncertainty is incorporated into the prediction because the integral considers all possible values of α and λ properly weighted by the probability density $p(\alpha, \lambda|D)$. A plot of the predictive density for the future time t_f is shown in Fig. 6.23. Integration of this density over t_f provides predicted failure probabilities for the time interval corresponding to the limits of integration. For example, the probability that a future failure will occur by $t_f = 2500$ is 0.34, and the expectation of t_f is 2650.

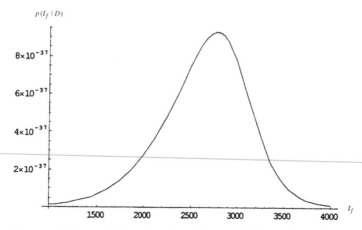

Figure 6.23   Predictive density of time to failure. (Groer, 2000.)

6.3.3 Remaining Useful Life Probability Density Function

A complete viewpoint on probabilistic techniques in prognosis as related to predicting the remaining useful life (RUL) is given by Engel and colleagues (2000). The seminal notions presented in this paper serve to clarify our thinking about remaining useful life prediction. A key concept in this framework is the *remaining useful life failure probability density function*. A component must be removed from service prior to attaining a high probability of failure. This concept is depicted in Fig. 6.24 in terms of the RUL PDF, where a *just-in-time point* is defined for removal from service that corresponds to a 95 percent probability that the component has not yet failed.

A key issue, unfortunately, is that the RUL PDF is actually a *conditional PDF* that changes as time advances. In fact, one must recompute the RUL PDF at each time *t* based on the new information that the *component has not yet failed at that time*. This concept is shown in Fig. 6.25. One starts with an a priori PDF similar to the hazard function. Then, as time passes, one must recompute the a posteriori RUL PDF based on the fact that the failure has not yet occurred. This involves renormalizing the PDF at each time so that its area is equal to 1. As time passes, the variance of the RUL PDF decreases; that is, the PDF becomes narrower. This corresponds to the fact that as time passes and one approaches the failure point, one becomes more and more certain about the time of failure, and its predicted value becomes more accurate. The final figure that illustrates these notions and fixes them in one's mind extremely clearly is Fig. 6.26.

Example 6.4 Remaining Useful Life Prediction for SH-60 Helicopter Gearbox. In Engel and colleagues (2000), these techniques were applied to the RUL analysis of a failing SH-60 helicopter gearbox. The evolution of automated diagnosis for helicopter mechanical systems has been assisted by a Navy program known as Helicopter Integrated Diagnostic Systems (HIDS). The Navy conducted a series of tests on an iron bird test stand (SH-60) and seeded faults by implanting an electronic discharge machine (EDM) notch in

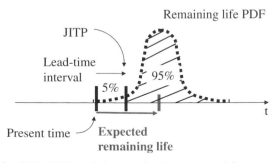

Figure 6.24 RUL PDF and the just-in-time removal-from-service point.

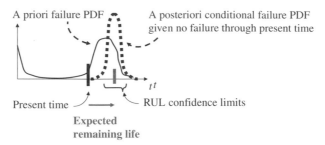

Figure 6.25 The RUL PDF as a time-varying conditional PDF.

the gear tooth root of the helicopter's gearbox. This action creates a localized stress concentration at the tooth root that eventually leads to the formation and growth of a crack. A test then was run at 100 percent tail power for a total of 2 million cycles. Testing was terminated prior to gearbox failure when a gross change in the raw FFT spectra was observed on an HP36650 spectrum analyzer. During this test, data were recorded at 100 kHz (using 16-bit integer precision) from two nearby accelerometers. A total of 36 recordings were made during the run to failure, each lasting for 30 seconds and spaced 15 minutes apart. This data set provided a rare and nearly continuous view of gear damage progression from healthy to condemnable conditions. During postmortem analysis, a crack was found that extended from the tooth root through the gear web and stopped at the bearing support diameter.

A set of four prognostic features that traditionally have been used for gear fault detection were chosen to illustrate the approach. Initially, accelerometer

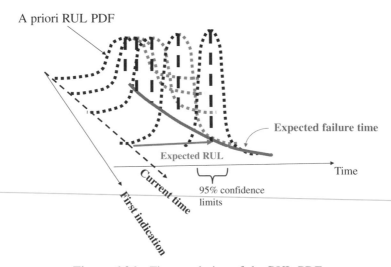

Figure 6.26 Time evolution of the RUL PDF.

data were time-synchronously averaged to reduce noise and incoherent vibration from unrelated sources. Normally occurring frequencies (i.e., primary gear mesh frequencies and their harmonics) then were removed to create a residue. The signal resulting from this step is referred to as the *time synchronous average residue* (TSAR). The four chosen features—normalized kurtosis, RMS value of the TSAR signal, variance, and proportional energy—are simply different ways of characterizing the residue signal. These features were analyzed from both repeatability and temporal viewpoints. For repeatability analysis, within each 30-second recording interval, 21 nonoverlapping windows, each of 5 seconds' duration, were used to compute the features. If the values are not close to each other, the feature may be too brittle to extrapolate without further processing. The degree of spread, combined with the uniqueness of the range of values, provides a measure of the robustness of each feature for this data set.

Figure 6.27 shows the TSAR proportional energy values. The feature is reasonably well behaved, largely monotonic, and has a favorable first derivative throughout the progression of damage. The error bars at each time instance on the plots represent the spread of these 21 values for each recording. Using bootstrap statistical methods, RUL estimates were computed at each recording time, from which errors and distributions were computed.

Next, for temporal analysis, the sequence of values from healthy to failure for each feature was modeled as a polynomial. A subsequence of four consecutive values then were fit to a polynomial model (in a least-squares sense) to derive an expected remaining life. Figure 6.28 shows the just-in-time Point (using the maximum LTI) for the feature of TSAR proportional energy as a function of time, the actual remaining life. The expected remaining life and 95 percent confidence intervals were computed as a function of time by considering the PDF distributions attributable to predictions generated by the given feature. The figure suggests that as data accumulate and indications become easier to interpret (over time), the expected remaining life and lead-time estimates approach the actual remaining life.

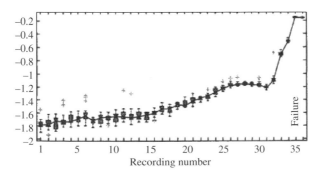

Figure 6.27 Proportional energy (healthy to failure conditions). (Engel et al., 2000.)

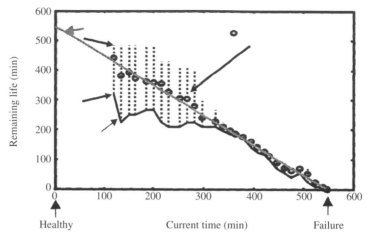

Figure 6.28 Expected RUL and just-in-time line for TSAR proportional energy. (Engel et al., 2000.)

6.4 DATA-DRIVEN PREDICTION TECHNIQUES

In many instances, one has historical fault/failure data in terms of time plots of various signals leading up to failure, or statistical data sets. In such cases, it is very difficult to determine any sort of model for prediction purposes. In such situations, one may use *nonlinear network approximators* that can be tuned using well-established formal algorithms to provide desired outputs directly in terms of the data. Nonlinear networks include the *neural network*, which is based on signal-processing techniques in biologic nervous systems, and *fuzzy-logic systems*, which are based on the linguistic and reasoning abilities of humans. These are similar in that they provide structured nonlinear function mappings with very desirable properties between the available data and the desired outputs.

Figure 6.29 represents the issues involved in prognosis. What are required are both the fault evolution *prediction* and the *confidence limits* of that prediction. Also needed are prescribed *hazard zones* beyond which an alarm is activated. Clearly, the issue is to use the available data most efficiently to obtain the fault prediction as well as its confidence limits. Both can be accomplished directly from the available data by using neural networks.

In prediction, artificial neural networks (ANNs), fuzzy systems, and other computational intelligence methods have provided an alternative tool for both forecasting researchers and practitioners (Sharda, 1994). Werbos (1988) reported that ANNs trained with the backpropagation algorithm outperform traditional statistical methods such as regression and Box-Jenkins approaches. In a recent forecasting competition organized by Weigend and Gerhenfeld (1993) through the Santa Fe Institute, all winners of each set of data used

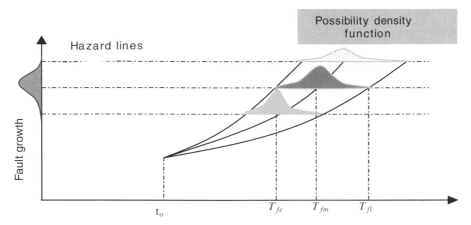

Figure 6.29 Interaction between alarm boundaries and trend prediction.

ANNs. Unlike the traditional model-based methods, ANNs are data-driven and self-adaptive, and they make very few assumptions about the models for problems under study. ANNs learn from examples and attempt to capture the subtle functional relationship among the data. Thus ANNs are well suited for practical problems, where it is easier to have data than knowledge governing the underlying system being studied. Generally, they can be viewed as one of many multivariate nonlinear and nonparametric statistical methods (Cheng and Titerington, 1994). The main problem of ANNs is that their decisions are not always evident. Nevertheless, they provide a feasible tool for practical prediction problems.

Unfortunately, standard statistical methods for *confidence estimation* are not directly applicable to neural networks. Although research efforts on ANNs for forecasting are vast and growing, few publications describe the development of novel methods for confidence estimation that can be applied to neural networks. Chryssolouris and colleagues (1996) used an assumption of normally distributed error associated with a neural network to compute a confidence interval. Veaux and colleagues (1998) used nonlinear regression to estimate prediction intervals by using weight decay to fit the network. Leonard and colleagues (1992) introduced the "validity index network," which is an extension of radial basis function networks (RBFNs) that calculates the confidence of its output. Following the structure of RBFN, Spetch (1990, 1991) formulated his probabilistic neural network and, subsequently, the generalized regression neural network; both of them have the built-in confidence-estimation capability. More recently, the confidence prediction neural network (CPNN) has been developed, which has a unique confidence estimator node that takes the confidence level from each pattern sample and constructs a possibility distribution using a method similar to Parzen's distribution estimator. A splitting algorithm is employed to split a multimodal distribution into branches. Iterated time-series prediction of these branches yields a prediction tree whose branches represent possible trends of prediction. The uncertainty represented by these branches can be managed via an additional learning scheme that prunes the prediction tree by eliminating unnecessary or inactive branches (Khiripet, 2001).

6.4.1 Neural Network

Section 5.3 in Chap. 5 gives an introduction of neural networks for classification purposes. We will describe in this section a more detailed mathematical model of a static feedforward NN, as shown in Fig. 6.30, and its learning methods. This NN has two layers of adjustable weights and is known as a *two-layer NN*. The values x_k are the NN inputs, and y_i are its outputs. Function $\sigma(\cdot)$ is a nonlinear *activation function* contained in the *hidden layer* (the middle layer of nodes) of the NN. The hidden-layer weights are v_{jk}, and the output-layer weights are w_{ij}. The hidden-layer thresholds are θ_{vj}, and the output-layer thresholds are θ_{wi}. The number of hidden-layer neurons is L.

A mathematical formula describing the NN is given by

$$y_i = \sum_{j=1}^{L} w_{ij}\sigma\left(\sum_{k=1}^{n} v_{jk}x_k + \theta_{vj}\right) + \theta_{wi} \qquad i = 1, 2, \ldots, m$$

We can streamline this by defining the weight matrices:

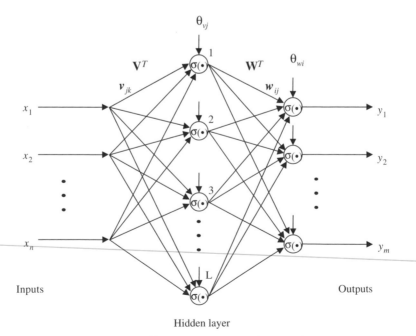

Figure 6.30 Two-layer neural network.

$$\mathbf{V}^T = \begin{bmatrix} \theta_{v1} & v_{11} & \cdots & v_{1n} \\ \theta_{v2} & v_{21} & \cdots & v_{2n} \\ \vdots & \vdots & & \vdots \\ \theta_{vL} & v_{L1} & \cdots & v_{Ln} \end{bmatrix} \qquad \mathbf{W}^T = \begin{bmatrix} \theta_{w1} & w_{11} & \cdots & w_{1L} \\ \theta_{w2} & w_{21} & \cdots & w_{2L} \\ \vdots & \vdots & & \vdots \\ \theta_{wm} & w_{m1} & \cdots & w_{mL} \end{bmatrix}$$

which contain the thresholds as the first columns. Then the NN can be written as

$$\mathbf{y} = \mathbf{W}^T \boldsymbol{\sigma}(\mathbf{V}^T \mathbf{x})$$

where the output vector is $\mathbf{y} = [y_1 \quad y_2 \quad \cdots \quad y_m]^T$.

Note that owing to the fact that the thresholds appear as the first columns of the weight matrices, one must define the input vector augmented by a 1 as

$$\mathbf{x} = [1 \quad x_1 \quad x_2 \quad \cdots \quad x_n]^T$$

for then one has the jth row of $\mathbf{V}^T \mathbf{x}$ given by

$$[\theta_{vj} \quad v_{j1} \quad v_{j2} \quad \cdots \quad v_{jn}] \begin{bmatrix} 1 \\ x_1 \\ x_2 \\ \vdots \\ x_n \end{bmatrix} = \theta_{vj} + \sum_{k=1}^{n} v_{jk} x_k$$

as required. Likewise, the $\boldsymbol{\sigma}(\cdot)$ used in the equation describing the NN is the augmented hidden-layer function vector, defined for a vector $\mathbf{w} = [w_1 \quad w_2 \quad \cdots \quad w_L]^T$ as

$$\boldsymbol{\sigma}(\mathbf{w}) = [1 \quad \sigma(w_1) \quad \cdots \quad \sigma(w_L)]^T$$

The computing power of the NN comes from the fact that the activation functions $\boldsymbol{\sigma}(\cdot)$ are nonlinear and the fact that the weights $\mathbf{W}$ and $\mathbf{V}$ can be modified or tuned through some learning procedure. Typical choices for the activation function appear in Fig. 6.31. It is important to note that the thresholds θ_{vj} shift the activation function at node j, whereas the hidden-layer weights v_{jk} scale the function. More information on NNs appears in Lewis, Jagannathan, and Yesildirek (1999).

An important feature of ANNs is that they can *learn* or be *trained* to capture desired knowledge, including modeling and prediction for unknown systems. Without the learning attribute, the neural network cannot generalize useful information at its output from observed data. The ANN learns by modifying the weights through tuning in such a fashion that the ANN better accomplishes the prescribed goals. The NN weights may be tuned by one of many techniques. A common weight-tuning algorithm is the *gradient algorithm* based on the *backpropagated error* (Werbos, 1988), which has been an

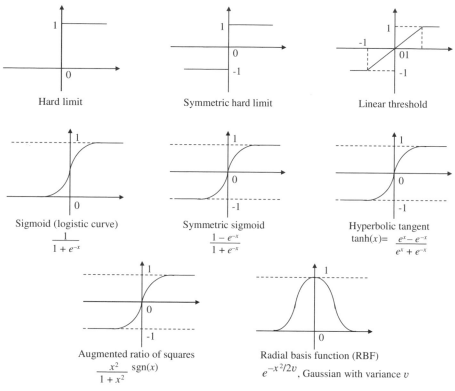

Figure 6.31   Some common choices for the activation function.

integral part of multilayer perceptron networks since their early inception. This algorithm is given by

$$\mathbf{W}_{t+1} = \mathbf{W}_t + \mathbf{F}\boldsymbol{\sigma}(\mathbf{V}_t^T \mathbf{x}_d)\mathbf{E}_t^T$$

$$\mathbf{V}_{t+1} = \mathbf{V}_t + \mathbf{G}\mathbf{x}_d\,(\boldsymbol{\sigma}_t'^T \mathbf{W}_t\mathbf{E}_t)^T$$

This is a discrete-time tuning algorithm where the time index is t. The desired NN output in response to the reference input $\mathbf{x}_d \in R^n$ is prescribed as $\mathbf{y}_d \in R^m$. The NN output error at time t is $\mathbf{E}_t = \mathbf{y}_d - \mathbf{y}_t$, where $\mathbf{y}_t \in R^m$ is the actual NN output at time t. $\mathbf{F}$ and $\mathbf{G}$ are weighting matrices selected by the designer that determine the speed of convergence of the algorithm.

The term $\boldsymbol{\sigma}^d(\cdot)$ is the derivative of the activation function $\boldsymbol{\sigma}(\cdot)$, which may be computed easily. For the sigmoid activation function, for instance, the hidden-layer output gradient is

$$\boldsymbol{\sigma}^d \equiv \mathrm{diag}\{\boldsymbol{\sigma}(\mathbf{V}^T \mathbf{x}_d)\}[\mathbf{I} - \mathrm{diag}\{\boldsymbol{\sigma}(\mathbf{V}^T \mathbf{x}_d)\}]$$

where $\mathrm{diag}\{\cdot\}$ is a diagonal matrix with the indicated elements on the diagonal.

The continuous-time version of backpropagation tuning is given as

$$\dot{\mathbf{W}} = \mathbf{F}\boldsymbol{\sigma}(\mathbf{V}^T x_d)\mathbf{E}^T$$

$$\dot{\mathbf{V}} = \mathbf{G}x_d(\boldsymbol{\sigma}^{dT}\,\mathbf{WE})^T$$

A simplified weight tuning technique is the *Hebbian algorithm,* which is given in continuous-time as

$$\dot{W} = \mathbf{F}\boldsymbol{\sigma}(\mathbf{V}^T x)\mathbf{E}^T$$

$$\dot{\mathbf{V}} = \mathbf{G}x[\sigma(\mathbf{V}^T x)]^T$$

In Hebbian tuning it is not necessary to compute the derivative σ^d. Instead, the weights in each layer are tuned based on the outer product of the input and output signals of that layer. It is difficult to guarantee NN learning performance using Hebbian tuning.

By now, learning and tuning-in ANNs are well-understood and highly efficient numerical algorithms that are available from standard sources, including MATLAB. These can be applied directly in prognosis applications. The Levenberg-Marquart algorithm has been used with feedforward networks for nonlinear estimation (Hagan and Menhaj, 1994). Classification of learning strategies can be based on the so-called lazy/eager distinction (Aha, 1997). The learning methods mentioned above are eager methods, which implies that the observed data are compiled into an intermediate form (such as synaptic weights), and then the data are discarded. The description is used to respond to information requests. In contrast, data are never discarded in the lazy learning methods, and no derived description of the data is defined (Atkeson et al., 1997). The data are used as the description of the model. It defers data processing until an explicit request for information is received. Lazy learning is based on nearest-neighbor techniques and nonparametric statistics. It is also known as *memory-based learning, exemplar-based learning, case-based learning* (Aha, 1997), or *just-in-time learning* (Cybenko, 1996).

One of the most popular classes of learning methods is called *reinforcement learning.* The term *reinforcement learning* appears as early as 1961 (Minsky, 1961; Waltz and Fu, 1965) and refers to learning of mappings from situations to actions in order to maximize a scalar reward or reinforcement criterion. The learner is not told which action to take but instead must discover which actions yield the highest reward by trying them (Sutton, 1992). This learning method has been combined recently with a field of mathematics called *dynamic programming* (Bertsekas, 1995). The interesting property of the combination is that a learner can learn to take actions that may affect not only the immediate reward but also the next situation and through that all subsequent rewards (Sutton, 1992). Several other learning methods based on this approach are Q-learning (Watins and Dayan, 1992), temporal differences (TD) (Sutton, 1988), and advantage learning (Harmon et al., 1995). In practice, the applicability of reinforcement learning to many important domains is limited by the enormous size of the state and action spaces. To overcome this bottleneck, neural networks are used to approximate the reinforcement "cost to go" function. This approach is called neurodynamic programming (Bertsekas and Tsitsiklis, 1996).

6.4.2 Dynamic Wavelet Neural Networks

The wavelet neural network (WNN) described in Chap. 5 is a static model in the sense that it establishes a static relation between its inputs and outputs. All signals flow in a forward direction with this configuration. Dynamic or recurrent neural networks, on the other hand, are required to model the time evolution of dynamic systems as required in prognosis applications. Signals in such a network configuration can flow not only in the forward direction but also backwards, in a feedback sense, from the output to the input nodes. Dynamic wavelet neural networks (DWNNs) have been proposed recently to address the prediction/classification issues. A multiresolution dynamic predictor was designed and employed for multistep prediction of the intracranial pressure signal (Tsui et al., 1995). The predictor uses the discrete wavelet transform and recurrent neural networks.

The basic structure of a DWNN is shown in Fig. 6.32. Delayed versions of the input and output now augment the input feature vector, and the resulting construct can be formulated as

$$y_{k+1} = \text{WNN}(y_k, \ldots, y_{k-M}, u_k, \ldots, u_{k-N})$$

where u_k is the external input, y_k is the output, M is the number of past output values retained in memory, N is the number of past external inputs retained

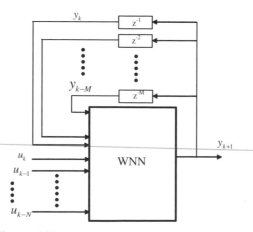

Figure 6.32 Dynamic wavelet neural network.

in memory, and function WNN stands for the static WNN just described. When measuring the angular velocity of a servo motor, for example, y_k could be the velocity to be measured, and u_k could be the regulating voltage that controls the motor's rotation at time k. The preceding equation forms a recursive evolving or prediction model that dynamically maps the historical and current data into the future. Compared with traditional prediction techniques such as ARIMA, DWNNs offer, in a systematic manner, more flexibility in terms of nonlinear mapping, parallel processing, heuristics-based learning, and efficient hardware implementation.

The DWNN can be trained in a time-dependent way using either a gradient-descent technique such as the Levenberg-Marquardt algorithm or an evolutionary one such as the genetic algorithm (Wang, 2001).

A major advantage of the DWNN is that the input vector can be composed of signals of different sorts and even of mixed time signals and frequency spectra as functions of time. This means that the DWNN is also a sort of *information fusion device* that combines the available information, which is of great value in prognosis. Figure 6.33 illustrates a fault predictor based on the DWNN that takes time-dependent features extracted from system sensor data and other relevant information such as fault dimensions, failure rates, trending information, temperature, and component ID as the inputs and generates the fault growth as the output.

Example 6.5   Bearing Crack Size Prediction and RUL for an Industrial Chiller. Industrial chillers are typical processes found in many critical applications. These devices support electronics, communications, etc. on a navy ship, computing and communications in commercial enterprises, and refrigeration and other functions in food processing. Of special interest is the fact that their design incorporates a diverse assemblage of common components such as pumps, motors, compressors, etc. A large number of failure modes are observed on such equipment ranging from vibration-induced faults to electrical failures and a multitude of process-related failure events. Most chillers are well instrumented, monitoring vibrations, temperature, fluid pressure, and flow. Figure 6.34 shows a U.S. Navy centrifugal chiller and its possible fault modes.

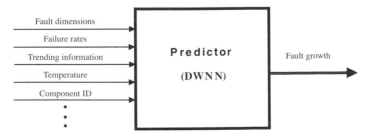

Figure 6.33   A schematic representation of the DWNN as the predictor.

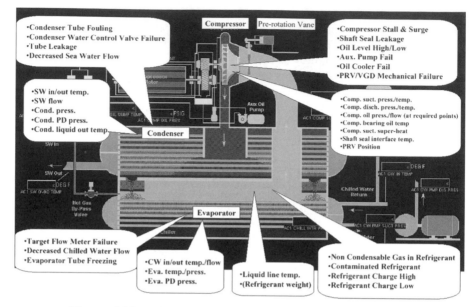

Figure 6.34 U.S. Navy centrifugal chiller and its fault modes.

Many mechanical faults exhibit symptoms that are sensed via vibration measurements. For example, a water pump will vibrate if its motor bearing is defective, if its shaft is misaligned, or if its mounting is somewhat loose. A rolling-element bearing fault in an industrial chiller is used as an example to demonstrate the effectiveness of prognostic algorithms based on DWNN.

Triaxial vibration signals caused by a bearing with a crack in its inner race were collected via accelerometers by Shiroishi and colleagues (1997). The initial crack was seeded, and the vibration data were collected for a period of time while the motor was running. The motor then was stopped, and the crack size was increased, which was followed by a second run. This procedure was repeated until the bearing failed. The crack sizes were organized in ascending order, whereas time information was assumed to be uniformly distributed among the crack sizes. Time segments of vibration signals from a good bearing and a defective one are shown in Fig. 6.35. Their corresponding power spectral densities (PSDs) are shown in Fig. 6.36.

The raw vibration data were windowed, with each window containing 1000 time points. The maximum values of the vibration signals in each window were recorded as time goes on, which are shown in Fig. 6.37. The peak values of PSD were calculated as another feature of the vibration data, as shown in Fig. 6.38, as a function of time. Figure 6.39 shows the corresponding crack sizes during the run-to-failure experiment. Crack size information at intermediate points was generated via interpolation to reduce the number of test runs.

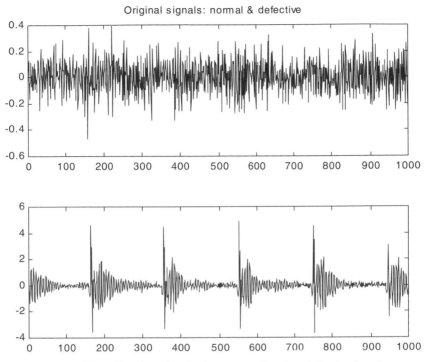

Figure 6.35   Vibration signals from a good and a defective bearing.

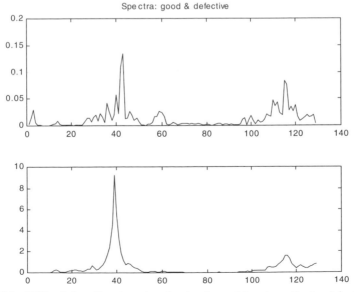

Figure 6.36   PSD of the vibration signals: (*top*) good bearing; (*bottom*) bad bearing.

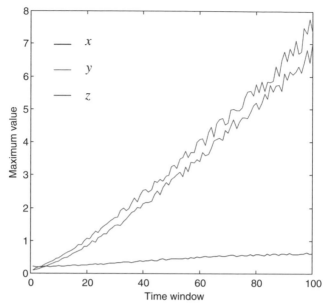

Figure 6.37 The peak values of the original signals.

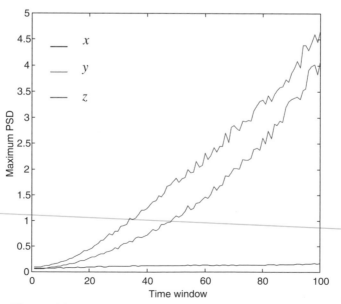

Figure 6.38 The maximum PSDs of the original signals.

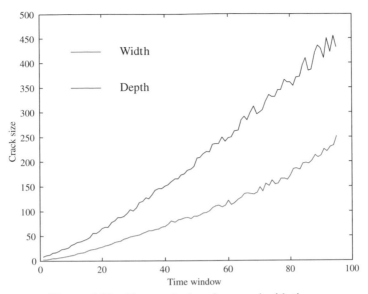

Figure 6.39 The crack sizes increased with time.

The features chosen for the prognosis purpose were the maximum signal values and the maximum signal PSDs for all three axes, that is, (MaxSx MaxSy MaxSz) and (MaxPSDx MaxPSDy MaxPSDz). Mexican hats were used as mother wavelets in the WNN and the DWNN. The virtual sensor, implemented as the WNN with seven hidden nodes or neurons, uses the maximum signal amplitudes and the maximum signal PSDs as inputs and measures virtually the crack sizes, as shown in Fig. 6.40. The measured width and length of the crack match the true values very well.

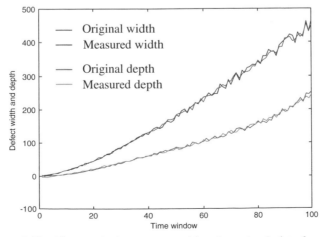

Figure 6.40 The crack sizes measured by the trained virtual sensor.

The DWNN is first trained using fault data up to the one-hundredth time window, after which it predicts the feature evolution until the final bearing failure. The WNN is used to map the feature evolution to the crack growth. The results are shown in Figs. 6.41 and 6.42. The predicted values match the true values well. In Fig. 6.42, a failure hazard threshold was established on the basis of empirical evidence corresponding to Crack_Width = 2000 μm or Crack_Depth = 1000 μm. The crack reaches this hazard condition at the one-hundred and seventy-fourth time window, and the Crack_Depth criterion is reached first.

These results are intended to illustrate the data-driven prognosis methodology. In practice, a substantially larger database is required for data storage, feature extraction, training, and validation of the DWNN. Such a database will permit a series of sensitivity studies that may lead to more conclusive results as to the capabilities and the effectiveness of the DWNN as a prediction tool.

6.4.3   The Confidence Prediction Neural Network (CPNN)

As we have seen, confidence limits are essential for prognosis applications. Standard ANNs can act as predictors after training, but they do not provide confidence limits of the predictions. Several techniques have been developed to address this issue and provide confidence limits. The general regression neural network (GRNN) takes advantage of the concept that given a known joint continuous probability density function $f(\mathbf{X}, y)$ of a vector input $\mathbf{X}$ and a scalar output y, the expected value of y given $\mathbf{X}$ can be computed as

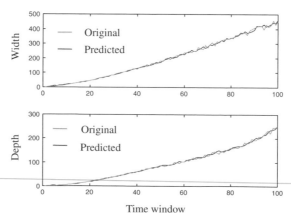

Figure 6.41   The crack growth predicted by the trained predictor within one-hundredth time window.

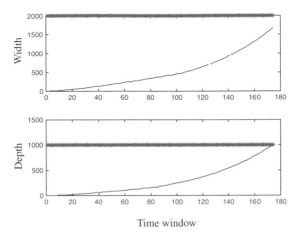

Time window

Figure 6.42 The crack growth predicted by the trained predictor beyond one-hundredth time window.

$$E(y/\mathbf{X}) = \frac{\int_{-\infty}^{\infty} y f(\mathbf{X}, y)\, dy}{\int_{-\infty}^{\infty} f(y, \mathbf{X})\, dy}$$

Given a statistical data set of points, the GRNN estimates the joint probability density using the Parzen estimator. The Parzen estimator approximates probability density functions using Parzen windows centered at the points in the training set. A collection of n samples of a population will have the estimated density function

$$g(\mathbf{X}) = \frac{1}{n\sigma} \sum_{i=1}^{n} W\left(\frac{\mathbf{X} - \mathbf{X}_i}{\sigma}\right)$$

where X_i is a sample in the population, σ is the scaling parameter, and $W(d)$ is a kernel function, which has its largest value at $d = 0$. The kernel functions that are employed most often are the Gaussian function and the Cauchy function. For example, if the Gaussian function is chosen as the kernel, the estimate $\hat{f}(\mathbf{X}, y)$ of the joint PDF is computed as

$$\hat{f}(\mathbf{X}, y) = \frac{1}{2\pi^{(P+1)/2}\sigma^{p+1}} \frac{1}{n} \sum_{i=1}^{n} e^{-\left[\frac{(\mathbf{x} - \mathbf{x}_i)^T (\mathbf{x} - \mathbf{x}_i)}{2\sigma^2}\right]} e^{-\left[\frac{(y - y_i)^2}{2\sigma^2}\right]}$$

where P is the dimension of the vector $\mathbf{X}$. Using the estimated $\hat{f}(\mathbf{X}, y)$, one may compute the desired expected value $\hat{y}(\mathbf{X}) = E[y/\mathbf{X}]$. Evaluating the required integrals, the expected value of y is computed as

$$\hat{y}(\mathbf{X}) = \frac{\sum_{i=1}^{n} y_i\, e^{-D_i^2/2\sigma^2}}{\sum_{i=1}^{n} e^{-D_i^2/2\sigma^2}}$$

where the scalar distance function is defined as $D_i^2 = (\mathbf{X} - \mathbf{X}_i)^T (\mathbf{X} - \mathbf{X}_i)$. The estimate of y can be visualized as a weighted average of all the observed values y_i.

Example 6.6   Parzen Estimator for Joint PDF. Suppose that one has statistical historical data for past faults, as shown in Fig. 6.43, where a point at $(x, y) = (13, 5)$, for instance, might mean that if vibration readings reach the level of 13 units, then time to failure is 5 hours. The data show that as vibration increases, time to failure decreases. Let us take Gaussian kernels to estimate the joint PDF of the data. The Gaussian kernels centered at the given points are shown in Fig. 6.44. Adding these together results in the joint PDF estimate shown in Fig. 6.45. The contour plot of the estimated PDF is given in Fig. 6.46. Clearly, the PDF estimate is highly sensitive to the choice of the variances used for the Gaussian basis functions. Note that although the basis functions are Gaussian, the estimated joint PDF is not necessarily Gaussian. From the joint PDF estimate, one may compute the conditional expected value of time to failure given current vibration measurements and other quantities of interest, including marginal PDFs and the RUL PDF.

In the GRNN, the confidence interval can be obtained from the training set by using the concept of conditional variance. The variance should be low

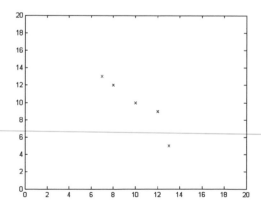

Figure 6.43   Statistical test data available.

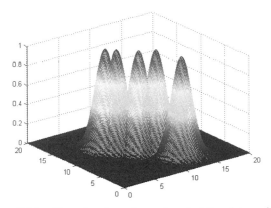

Figure 6.44 Gaussian kernels centered at the data points.

in regions where the training-set points have similar values to the input values and should increase in areas where the training-set points vary greatly from the input values. Since the estimate is based on values from the training set, the confidence in the accuracy of the estimate will vary according to the variances of the points used to create the estimate. The conditional variance, where y is the output and x is the input of the predictor, is computed as

$$
\text{var}[y/\mathbf{X}] = E[y^2/\mathbf{X}] - (E[y/\mathbf{X}])^2 = \frac{\sum\limits_{i=1}^{n} f_{X_i}(x)E_{y_i}[y^2]}{\sum\limits_{i=1}^{n} f_{X_i}(x)} - \left[\frac{\sum\limits_{i=1}^{n} f_{X_i}(x)E_{y_i}[y]}{\sum\limits_{i=1}^{n} f_{X_i}(x)}\right]^2
$$

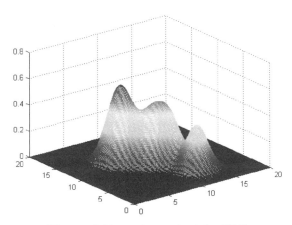

Figure 6.45 Estimate for joint PDF.

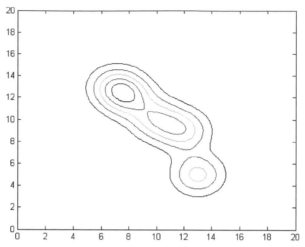

Figure 6.46   Contour plot of joint PDF estimate.

The final single number output from the GRNN and its confidence interval can be extended to represent the confidence distribution in the confidence prediction neural network (CPNN), as described below.

For one-step-ahead prediction, a buffer of a finite number P of recent values X forms an input vector to the network. The input vector is compared with patterns extracted from historical data. Each comparison receives a degree of matching $c(\mathbf{X}, y_i)$ and a candidate output y_i. This is exactly the first step of the GRNN, in arriving at an estimated probability density function $\hat{f}(\mathbf{X}, y)$, which approximates the output of an input vector. Similarly, in CPNN, after the comparison process is completed, each candidate output is weighted by its degree of matching to give a final contribution of confidence as a scalar output.

However, as stated previously, the purpose of the new methodology is not only to obtain a single output but also to estimate the confidence distribution of that output as well. For this purpose, the confidence distribution function $CD(\mathbf{X}, y)$ is defined, using an idea from the Parzen estimator, by

$$CD(\mathbf{X}, Y) = \frac{1}{2\pi\sigma_{CD}} \frac{1}{n} \sum_{i=1}^{n} C(\mathbf{X}, y_i) e^{-((y-y_i)^2/2\sigma_{CD}^2)}$$

This distribution can be multimodal. To simplify the solution, the peaks of the distribution are kept, if possible, one-step-ahead of prediction outputs. Iteratively applying the procedure yields a multistep prediction scheme that shows multiple trends, as shown in Fig. 6.47.

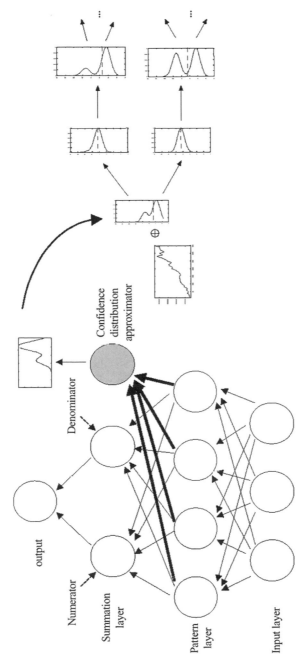

Figure 6.47 CPNN structure and single-step to multistep prediction.

6.4.4 Managing Uncertainty via a Neurodynamic Q-Learning

The uncertainty of the output of the CPNN will grow larger (wider confidence distribution) in later prediction steps. A major performance metric for prognostic algorithms is precision, that is, how precise or how narrow the prediction distribution is. Thus it is desirable to manage this uncertainty, especially when newly arrived actual data can be used to validate the output of the most recent prediction step. The tasks of managing uncertainty by incorporating the new data into the predictor's pattern and further enhancing the predictor capability are accomplished by an extra learning module. This learning module will prune the prediction tree by eliminating those unnecessary or inactive branches. Reduction of the number of branches provides a less complex tree structure, which yields a better prediction precision and makes longer-step prediction possible.

The learning scheme uses an assumption that the prediction tree output of the CPNN is similar to a game tree in a game situation. In each particular game state, a player has a set of actions to choose from. Best actions will result in a victory for the player. A modified game-learning algorithm called *Q-Learning* is employed that addresses the continuous state and continuous space problems by a neurodynamic approach. Q-Learning is a reinforcement learning scheme that will reward or punish a player when an action is taken and the result is observed. A similar situation arises in the prediction task. When new actual data arrives, some correct branches must be emphasized for future use, whereas incorrect ones need to be deemphasized or eliminated. Q-learning works with the CPNN by having the CPNN as a "cost to go" function of the Q-learning. At the same time, Q-learning will update the CPNN to produce better results by giving a reward to the correct output branches. To guarantee that the prediction of the CPNN is improved over time, the reward that Q-learning provides to the CPNN is constructed from various performance measures. They are an accuracy measure, a precision measure, a confidence measure, and a complexity measure. The accuracy measure is the measure of how close the predicted value is to the actual one. Precision refers to how narrow the prediction distribution is. A narrower distribution implies less uncertainty in the prediction. Confidence is the confidence level, ranging from 0 to 1, at the actual value, and complexity is a measure of how dense (how many branches) the prediction tree is. An overall performance measure is computed as the weighted sum of the preceding performance measures and is used both for improving the prediction of the CPNN by the Q-learning and for measuring the effectiveness of the CPNN at the end of the prediction horizon.

Reinforcement Learning. Q-learning is among a series of reinforcement learning (RL) algorithms with a common objective of successfully solving problems that require complex decision processing. Reinforcement learning is a combination of fields such as dynamic programming and supervised learning that lead to powerful intelligent machine learning systems. One of the biggest advantages of these techniques is the model-free aspect. The applicability of reinforcement learning algorithms is independent of the structure of the system taken into consideration.

Q-learning has been integrated into many computational environments to reinforce decision processes and address the evolvable aspect of a particular automated paradigm. Suppose that a maintainer is presented with a problem or symptom that leads to the choice of multiple solutions that may not necessarily lead to an effective maintenance action. No matter which solution the maintainer decides to perform, the reinforcement learning system records these problem-solution relationships and their outcome. During the course of time the RL system ultimately will lead the maintainer to choose the most-effective solution to the given problem.

The following relationships formally introduce the reinforcement learning model.

S: A discrete set of environment states
A: A discrete set of agent actions
R: A set of scalar reinforcement signals

An agent is connected to its environment through actions. For every interaction, an agent receives an indication of the current state s of the environment. Once the agent knows the current probable state, it chooses among a set of actions. The selected action a consequently will change the state s of the environment. During the transition from the current state to the new state, a scalar reinforcement signal r is used to reward or penalize the action based on its outcome. After numerous interactions and transitions, the agent will gain useful experience that will lead to choosing the optimal action a for a given current state s. Once the system has learned over time, a policy then can be used to choose which actions should be performed when a state is encountered. The value of the state is defined as the sum of the reinforcements over a discrete number of interactions. The optimal policy can be calculated by indexing the mapping between state and actions that maximizes the sum of the reinforcements.

The advantage of Q-learning is that initially, when the optimal state to action relationships are not entirely known, the agent may have the option of randomly selecting an action. Q-learning allows arbitrary experimentation and simultaneously preserves the current best estimate state values. The final results in Q-learning are not contaminated by experimental actions, which only ultimately will reward or penalize the state-action pairs. In order to obtain meaningful values, it is necessary for the agent to try out each action in every state multiple times, allowing the reinforcement updates to assign penalties or rewards based on the outcome.

The Q-learning equation is

$$Q(s_t, a_t) = (1 - \alpha)Q(s_t, a_t) + \alpha[r(s_t, a_t) + \gamma \max Q(s_{t+1}, a)]$$

where Q = the sum of all reinforcements for the state-action pair (s, a) at a given time (t)

α = the learning-rate parameter $[0, 1]$

γ = the discount parameter $[0, 1]$

The learning-rate parameter α carries a significant importance in adjusting the algorithm's learning behavior. α controls the weighting on current and previous knowledge represented in the state-action pair values. As the learning-rate parameter approaches 1, new experience outweighs previous experience. As it approaches 0, previous experience outweighs new experience. For the state-action pairs, representing task effectiveness of a maintenance action relative to a fault α is set closer to 0, outweighing previous experience to new, which translates in assigning more importance to the maintenance task history performance.

The discount parameter γ controls the weighting on the new state-action pair. For the current application, the parameter is fixed to a constant and provides scaling to the state-action pair values, which does not affect the overall results of the Q-learning algorithm. The mapping between state and action pairs can be represented in an $n \times m$ matrix, where n represents the number of possible states, and m represents the number of possible actions (Fig. 6.48).

The Q-matrix may be a representation of the task effectiveness of a particular maintenance action a directly connected to a failure mode s. The outcome of the performed action leads to a state transition. The state transitions are defined in the following two cases and illustrated in Fig. 6.49.

- *Case 1.* Given a failure mode s, if any particular selected action is successful, the state will transition from the current fault state to a healthy state.
- *Case 2.* Given a failure mode s, if any particular selected action is not successful, the state will transition from the current fault state back to its current fault state or to a different fault state. The transition to a different faulty state could be an indication that the failure mode was not clearly identified until after the maintenance action was performed.

Q - Matrix			
s1	Q(s1,a1)	Q(s1,a2)	Q(s1,a3)	Q(s1,a4)
s2	Q(s2,a1)	Q(s2,a2)	Q(s2,a3)	Q(s2,a4)
s3	Q(s3,a1)	Q(s3,a2)	Q(s3,a3)	Q(s3,a4)
	a1	a2	a3	a4

Figure 6.48 Q-learning matrix.

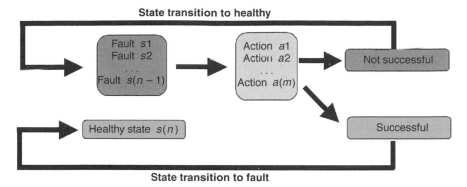

State transition to healthy

State transition to fault

Figure 6.49 Decision process flowchart.

6.5 CASE STUDIES

6.5.1 Gas Turbine Performance Prognosis

Fouling degradation of gas turbine engine compressors causes significant efficiency loss, which incurs operational losses through increased fuel usage or reduced power output. Scheduling maintenance actions based on predicted condition minimizes unnecessary washes and saves maintenance dollars. The effect of the various maintenance tasks (washing and overhaul) on gas turbine engine efficiency is shown in Fig. 6.50.

An example of a prognostic algorithm for compressor performance can consist of data preprocessing and specific algorithms for assessing current and future compressor conditions. A data preprocessor algorithm examines the engine operating data and automatically calculates key corrected performance parameters such as pressure ratios and efficiencies at specific power levels.

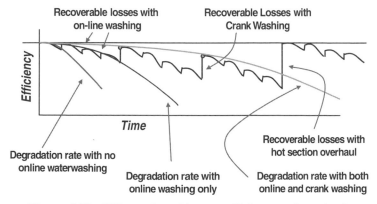

Figure 6.50 Effects of washing on efficiency and overhaul.

As fouling starts to occur in service, probabilistic classifiers match up corresponding parameter shifts to fouling severity levels attained from existing models/tests with corresponding degrees of confidence.

As is the case for many prognostic implementations, a probabilistic-based technique typically is developed that uses the known information about the degradation of measured parameters over time to assess the current severity of parameter distribution shifts and to project their future state. The parameter space for this application can be populated by two main components. These are the current condition and the expected degradation path. Both are multivariate probability density functions (PDFs) or three-dimensional statistical distributions. Figure 6.51 shows a top view of these distributions. The highest degree of overlap between the expected degradation path and the current condition is the best estimate of compressor fouling.

The top two plots illustrate the distributions of pressure ratio and fuel flow, respectively, whereas the bottom two provide the joint probability distributions. The dotted line represents the path (either current or projected) of degradation for the compressor. The compressor inlet temperature (CIT), outlet temperature (CDT), inlet total pressure (CIPT), and discharge total pressure (CDPT) typically can be used to find compressor efficiency. In the event that total pressure measurements are not available, other methods can be used to

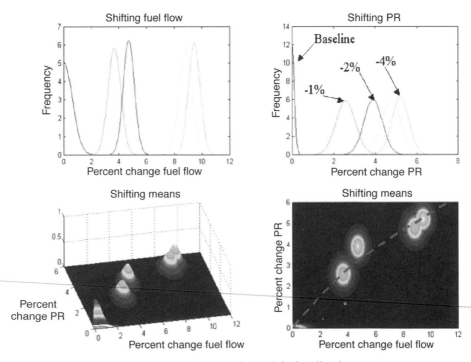

Figure 6.51 Prognostic model visualization.

approximate this efficiency calculation with other specific sets of sensors. A typical efficiency calculation is

$$\eta_{\text{adb}} = \frac{\left[\left(\dfrac{\text{CDP}_T}{\text{CIP}_T}\right)^{\gamma - 1/\gamma} - 1\right]}{\left[\left(\dfrac{\text{CDT}}{\text{CIT}}\right) - 1\right]}$$

Once the statistical performance degradation path is realized along with the capability to assess current degradation severity, we need to implement the predictive capability. The actual engine-specific fouling rate is combined with historical fouling rates with a double-exponential smoothing method. This time-series technique weights the two most recent data points over past observations. The following equations give the general formulation for the exponential smoothing prediction (Bowerman and Koehler, 1993):

$$S_T = \alpha y_T + (1 - \alpha)S_{T-1}$$

$$S^{[2]}_T = \alpha S_T + (1 - \alpha)S^{[2]}_{T-1}$$

$$\hat{y}_{T+\tau}(T) = \left(2 + \frac{\alpha\tau}{(1 - \alpha)}\right)S_T - \left(1 + \frac{\alpha\tau}{1 - \alpha}\right)S^{[2]}_T$$

Analysis of the degradation requires the simulation to predict the range of conditions that might exist, given the measurement and modeling uncertainties. This is accomplished using a Monte Carlo simulation with the mean and 2σ uncertainties. The resulting distribution is the range of time-to-wash predictions. Appropriate statistical confidence intervals can be applied to identify the mean predicted value. This estimate is updated with a weighted fusion from the predicted value and the historical degradation level derived from the fouling data. The results of this process are shown in Fig. 6.52.

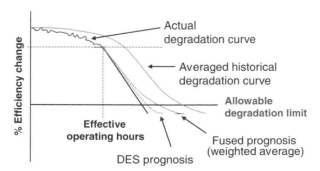

Figure 6.52 Prognostic model visualization.

6.5.2 Bearing Prognosis

An integrated approach to bearing prognosis is described next that builds on technologies such as advanced oil-debris monitoring, vibration analysis, and empirical/physics-based modeling to practically achieve its objectives. An important aspect of this approach is that it is flexible enough to accept input from many different sources of diagnostic/prognostic information in order to contribute to better fault isolation and prediction on bearing remaining useful life.

The block diagram shown in Fig. 6.53 illustrates an abstract representation of the integrated system architecture for the bearing prognosis module described herein. Within this architecture, measured parameters from health monitoring systems such as oil-condition/debris monitor outputs and vibration signatures can be accommodated within the anomaly-detection/diagnostic fusion center. Based on these outputs, specific triggering points within the prognostic module can be processed so that effective transition associated with various failure mode models (i.e., spall initiation model to a spall progression model) can be accomplished.

Also shown in Fig. 6.53, various sources of diagnostic information are combined in the model-based and feature-based prognostic integration algorithms for the specific bearing under investigation using a probabilistic update process. Knowledge on how the specific bearing is being loaded, historical failure-mode information, and inspection data feedback also can be accommodated within this generic prognostic architecture. Finally, based on the overall component health assessment and prognosis, specific information re-

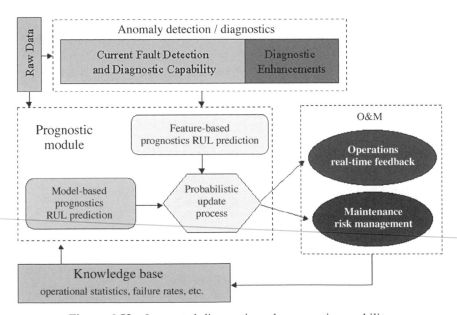

Figure 6.53 Integrated diagnostic and prognostic capability.

lated to remaining useful life and associated risk will be passed to operations and maintenance systems as an important input to that decision-making processes (covered in the last section of this chapter).

A more specific block diagram illustrating the implementation of this generic prognostic architecture as applied to a machinery bearing is given in Fig. 6.54. In this figure, features from oil-analysis monitors are combined with vibration features in the anomaly and diagnosis module at the top. Once the appropriate feature extraction and fusion is performed, an output is passed to the physical stochastic model that determines the current "state of health" of the bearing in terms of failure-mode progression. This health rating or status is used by the model to probabilistically weigh the current "life-line location" in terms of the overall remaining useful life of the bearing. The fused features are also used as an independent measure of RUL that can be incorporated in the "update" process. The output of the module is a continuously updated description of the component's current health state, as well as a projected RUL estimate based on its individual "signature" usage profile.

Early Fault Detection for Rolling-Element Bearings. Early fault detection is an essential requirement for reliable prognosis. In Chapter 5, Section 5.11, we introduced an engine bearing case study to illustrate fault detection and diagnosis tools. It is instructive and helpful to repeat here the brief discussion on the diagnostic architecture, the vibration-based diagnostic indicators and the feature extraction technique using high-frequency enveloping so that the reader can realize the continuity to the bearing fault initiation and propagation models elaborated in this chapter. Development of specialized vibration or other features that can detect the incipient formation of a spall on a bearing

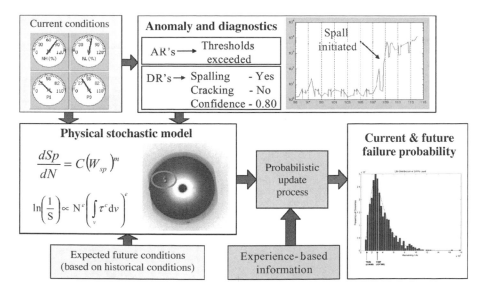

Figure 6.54 Bearing prognostic and health management module.

race or rolling element is a critical step in the design of the integrated prognosis system mentioned earlier. To this end, a series of tests was designed and conducted to provide data for algorithm development, testing, and validation. Vibration and oil-debris data were acquired from the ball-bearing test rig shown in Fig. 6.55. A damaged bearing was installed and used to compare the effectiveness of various diagnostic features extracted from the measured data. The bearings are identical to the number 2 main shaft bearing of an Allison T63 gas turbine engine and their dimensions are specified in Table 6.2.

Although originally designed for lubrication testing, the test rig was used to generate accelerated bearing failures owing to spalling. To accelerate this type of bearing failure, a fault was seeded into the inner raceway of one of the bearings by means of a small hardness indentation (Brinnell mark). The bearing then was loaded to approximately 14,234 N (3200 lbf) and run at a constant speed of 12000 rpm (200 Hz). Vibration data were collected from a cyanoacrylate–mounted (commonly called Super Glue) accelerometer, which was sampled at over 200 kHz. Also, the quantity of debris in the oil draining from the test head was measured using a magnetic chip collector (manufactured by Eateon Tedeco). The oil data were used in determining the severity of the spall progression.

The fundamental pass frequencies of the components of a bearing can be calculated easily with standard equations. Extraction of the vibration amplitude at these frequencies from a fast Fourier transform (FFT) often enables isolation of the fault to a specific bearing in an engine. High amplitude of vibration at any of these frequencies indicates a fault in the associated component. Table 6.3 shows the calculated bearing fault frequencies.

Note the listings are defined by

BSF: ball spin frequency
BPFI: inner raceway frequency

Figure 6.55   Bearing test rig setup.

TABLE 6.2   Test Rig Bearing Dimensions (mm)

Bearing	Ball Diameter (B_d)	Number of Balls (z)	Pitch Diameter (P_d)	Contact Angle (β)
1	4.7625	8	42.0624	18.5°
2	7.9375	13	42.0624	18.5°

BPFO: outer raceway frequency

Cage: cage frequency

High-Frequency Enveloping Analysis.   Although bearing characteristic frequencies are calculated easily, they are not always detected easily by conventional frequency-domain analysis techniques. Vibration amplitudes at these frequencies owing to incipient faults (and sometimes more developed faults) are often indistinguishable from background noise or obscured by much-higher-amplitude vibration from other sources, including engine rotors, blade passing, and gear mesh in a running engine. However, bearing faults produce impulsive forces that excite vibration at frequencies well above the background noise in an engine.

Impact Energy is an enveloping-based vibration feature-extraction technique that consists of, first, band-pass filtering of the raw vibration signal. Second, the band-pass filtered signal is full-waved-rectified to extract the envelope. Third, the rectified signal is passed through a low-pass filter to remove the high-frequency carrier signal. Finally, the signal's dc content is removed.

The Impact Energy process was applied to the seeded-fault test data collected using the bearing test rig. To provide the clearest identification of the fault frequencies, several band-pass and low-pass filters were used to analyze various regions of the vibration spectrum. Using multiple filters allowed investigation of many possible resonances of the bearing test rig and its components. A sample Impact Energy spectrum from early in the rig test is shown in Fig. 6.56. Note that these data were collected prior to spall initiation (based on oil-debris data), and the feature response is due to the indentation on the race.

For comparison, a conventional FFT (10 frequency-domain averages) of vibration data also was calculated and is shown in Fig. 6.57. In the conventional frequency-domain plot (Fig. 6.57), there is no peak at the inner-race ball pass frequency (1530 Hz). However, the Impact Energy plot (Fig. 6.56) shows clearly defined peaks at this frequency and the second through forth

TABLE 6.3   Test Rig Bearing Fault Frequencies (Hz)

Shaft Speed	BSF	BPFI	BPFO	Cage
200	512	1530	1065	82

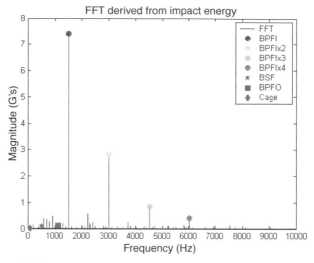

Figure 6.56   Impact Energy FFT of test rig bearing, seeded fault.

harmonics of the inner-race ball pass frequency. These peaks were defined using a detection window of ± 5 Hz about the theoretical frequency of interest to account for bearing slippage. From the onset of the test, there is an indication of a fault.

Model-Based Analysis for Prognosis. A physics-based model is a technically comprehensive modeling approach that has been used traditionally to

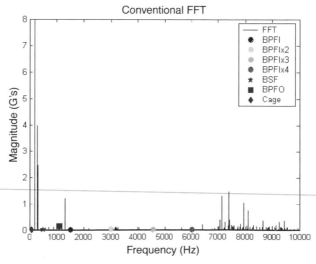

Figure 6.57   Conventional FFT of test rig bearing, seeded fault.

understand component failure-mode progression. Physics-based models provide a means to calculate the damage to critical components as a function of operating conditions and assess the cumulative effects in terms of component life usage. By integrating physical and stochastic modeling techniques, the model can be used to evaluate the distribution of remaining useful component life as a function of uncertainties in component strength/stress properties, loading, or lubrication conditions for a particular fault. Statistical representations of historical operational profiles serve as the basis for calculating future damage accumulation. The results from such a model then can be used for real-time failure prognostic predictions with specified confidence bounds. A block diagram of this prognostic modeling approach is given in Fig. 6.58. As illustrated at the core of this figure, the physics-based model uses the critical-life-dependent uncertainties so that current health assessment and future RUL projections can be examined with respect to a risk level.

Model-based approaches to prognosis differ from feature-based approaches in that they can make RUL estimates in the absence of any measurable events, but when related diagnostic information is present (such as the feature described previously), the model often can be calibrated based on this new information. Therefore, a combination or fusion of the feature-based and model-based approaches provides full prognostic ability over the entire life of the component, thus providing valuable information for planning which components to inspect during specific overhaul periods. While failure modes may be unique from component to component, this combined model-based and feature-based methodology can remain consistent across different types of oil-wetted components.

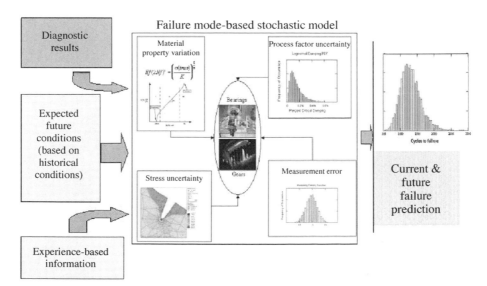

Figure 6.58 Prognostic bearing model approach.

Bearing Spall Initiation and Propagation Models. Rolling-element contact fatigue is considered the primary cause of failure in bearings that are properly loaded, well aligned, and receive an adequate supply of uncontaminated lubricant. Rolling contact fatigue causes material to flake off from the load-bearing surfaces of rolling elements and raceways, leaving a pit or spall, as shown in Fig. 6.59.

Spalling of well-lubricated bearings typically begins as a crack below the load-carrying surface that propagates to the surface. Once initiated, a spall usually grows relatively quickly, producing high vibration levels and debris in the oil. Owing to the relatively short remaining life following spall initiation, the appearance of a spall typically serves as the criterion for failure of bearings in critical applications.

Spall Initiation Model. A number of theories exist for predicting spall initiation from bearing dimensions, loads, lubricant quality, and a few empirical constants. Many modern theories are based on the Lundberg-Palmgren (L-P) model that was developed in the 1940s. A model proposed by Ioannides and Harris (I-H) (1985) improved on the L-P model by accounting for the evidence of fatigue limits for bearings. Yu and Harris (Y-H) (2001) proposed a stress-based theory in which relatively simple equations are used to determine the fatigue life purely from the induced stress. This approach depends to a lesser extent on empirical constants, and the remaining constants may be obtained from elemental testing rather than complete bearing testing, as required by L-P.

The fundamental equation of the Y-H model relates the survival rate S of the bearing to a stress weighted volume integral, as shown below. The model uses a new material property for the stress exponent c to represent the material

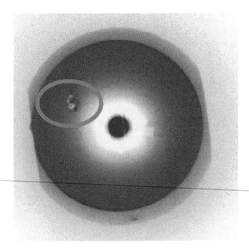

Figure 6.59 Ball bearing with spall.

fatigue strength and the conventional Weibull slope parameter to account for dispersion in the number of cycles N. The fatigue initiating stress τ may be expressed using Sines multiaxial fatigue criterion for combined alternating and mean stresses, or as a simple Hertz stress. Thus

$$\ln\left(\frac{1}{S}\right) \propto N^e\left(\int_v \tau^c \, dv\right)^e$$

For simple Hertz stress, a power law is used to express the stress-weighted volume. In the following equation, λ is the circumference of the contact surface, and a and b are the major and minor axes of the contact-surface ellipse. The exponent values were determined by Yu and Harris for $b/a \approx 0.1$ to be $x = 0.65$, $y = 0.65$, and $z = 10.61$. Yu and Harris assume that these values are independent of the bearing material. Thus

$$\int_A \tau^c \, dA \cdot \lambda \propto a^x b^y \tau^z \lambda$$

According to the Y-H model, the life L_{10} of a bearing is a function of the basic dynamic capacity Q_c and the applied load, as stated below, where the basic dynamic capacity is also given. A lubrication-effect factor may be introduced to account for variations in film thickness owing to temperature, viscosity, and pressure. Thus

$$L_{10} = \left(\frac{Q_c}{Q}\right)^{(x+y+z)/3}$$

$$Q_C = A_1 \Phi D^{(2z-x-y)/(z+x+y)}$$

$$\Phi = \left[\left(\frac{T}{T_1}\right)^z \frac{u(D\Sigma\rho)^{(2z-x-y)/3}}{(a^*)^{z-x}(b^*)^{z-y}} \frac{d}{D}\right]^{-3/(z+x+y)}$$

where A_1 = material property
T = a function of the contact-surface dimensions
T_1 = value of T when $a/b = 1$
u = number of stress cycles per revolution
D = ball diameter
ρ = curvature (inverse radii of component)
d = component (raceway) diameter
a^* = function of contact ellipse dimensions
b^* = function of contact ellipse dimensions

A stochastic bearing fatigue model based on Y-H theory has been developed to predict the probability of spall initiation in a ball/V-ring test. The stochastic model accounts for uncertainty in loading, misalignment, lubrication, and manufacturing that are not included in the Y-H model. Uncertainties in these quantities are represented using normal and log-normal probability distributions to describe the input parameters to the model. As a result, the stochastic Y-H model produces a probability distribution representing the likelihood of various lives.

Spall Progression Model. Once initiated, a spall usually grows relatively quickly, producing increased amounts of oil debris, high vibration levels, and elevated temperatures that eventually lead to bearing failure. While spall progression typically occurs more quickly than spall initiation, a study by Kotzalas and Harris showed that 3 to 20 percent of a particular bearing's useful life remains after spall initiation. The study identified two spall progression regions. Stable spall progression is characterized by gradual spall growth and exhibits low broadband vibration amplitudes. The onset of unstable spall progression coincides with increasing broadband vibration amplitudes.

Kotzalas and Harris also presented a spall progression model. The model relates the spall progression rate dSp/dN to the spall similitude W_{sp} using two constants C and m, as shown below. The spall similitude is defined in terms of the maximum stress σ_{max}, average shearing stress τ_{avg}, and the spall length Sp. Thus

$$\frac{dSp}{dN} = C(W_{sp})^m$$

$$W_{sp} = (\sigma_{max} + \tau_{avg})\sqrt{\pi Sp}$$

Spall progression data were collected during the Kotzalas-Harris study using a ball/V-ring test device. The data were collected at the Pennsylvania State University. Each of the lines shown in Fig. 6.60 represents spall progression of a ball bearing. For one ball, indicated by multiple markers, the test was suspended periodically to measure the spall-length. Only two spall-length measurements, initial and final values, were taken for the other balls. The figure shows a large dispersion of the spall initiation life, which is indicated by the lowest point on each line. The figure also shows a large dispersion of the spall progression life, which can be inferred from the slope of the lines.

The progression model also was applied to the bearing test rig data acquired. A correlation between the recorded oil-particle quantity and the spall size was determined from the initial and final quantities and spall sizes. This allowed scaling of the oil-particle quantity to approximate spall size. The model agreed well with the spall progression during the test, as seen in Fig. 6.61.

A stochastic spall progression model based on Kotzalas-Harris theory was developed to predict the probability of spall progression in the ball/V-ring

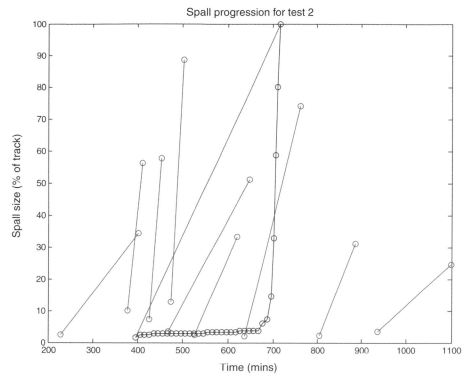

Figure 6.60   Spall progression data from phase I analysis.

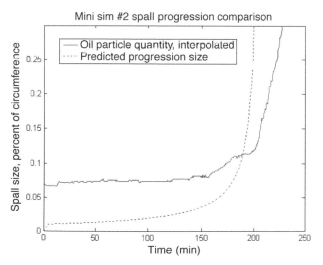

Figure 6.61   Bearing test rig spall progression.

test. The stochastic model accounts for uncertainty in loading, misalignment, lubrication, and manufacturing that is not included in the Kotzalas-Harris model. As a result, the stochastic Kotzalas-Harris model produces a probability distribution representing the likelihood of various lives.

Prognosis Model Validation. Validation of the spall initiation model requires a comparison of actual fatigue life values to predicted model values. Acquiring sufficient numbers of actual values is not a trivial task. Under normal conditions, it is not uncommon for a bearing life value to extend past 100 million cycles, prohibiting normal run-to-failure testing.

Accelerated-life testing is one method used to rapidly generate many bearing failures. By subjecting a bearing to high speed, load, and/or temperature, rapid failure can be induced. There are many test apparatus used for accelerated-life testing, including ball- and rod-type test rigs. One such test rig is operated by UES, Inc., at the Air Force Research Laboratory (AFRL) at Wright Patterson Air Force Base in Dayton, OH. A simple schematic of the device is shown in Fig. 6.62 with dimensions given in millimeters. This rig consists of three 12.7-mm-diameter balls contacting a 9.5-mm rotating central rod; see Table 6.4 for dimensions. The three radially loaded balls are pressed against the central rotating rod by two tapered bearing races that are

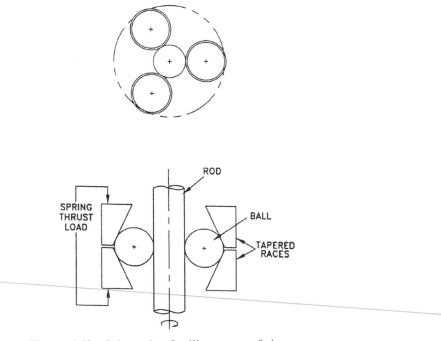

Figure 6.62 Schematic of rolling contact fatigue tester.

TABLE 6.4 Rolling Contact Fatigue Tester Dimensions (mm)

Rod diameter (D_r)	9.52
Ball diameter (D_b)	12.70
Pitch diameter (D_m)	22.23

thrust loaded by three compressive springs. A photo of the test rig is shown in Fig. 6.63. Notice the accelerometers mounted on the top of the unit. The larger accelerometer is used to automatically shut down the test when a threshold vibration level is reached; the other measures vibration data for analysis.

By design, the rod is subjected to high contact stresses. Owing to the geometry of the test device, the 222-N (50-lb) load applied by the springs translates to a 942-N (211-lb) load per ball on the center rod. Assuming Hertzian contact for balls and rod made of M50 bearing steel, the 942-N radial load results in a maximum stress of approximately 4.8 GPa (696 ksi). This extremely high stress causes rapid fatigue of the bearing components and can initiate a spall in less than 100 hours, depending on test conditions, including lubrication, temperature, etc. Since failures occur relatively quickly, it is possible to generate statistically significant numbers of events in a timely manner.

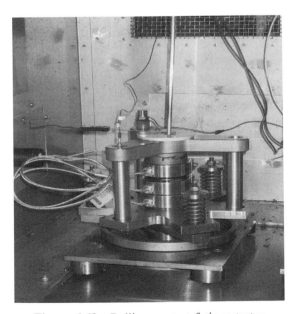

Figure 6.63 Rolling contact fatigue tester.

For validation purposes, M50 rods and balls were tested at room temperature (23°C). The results of these tests are shown in Table 6.5. A summary plot is shown in Fig. 6.64.

As stated earlier, one of the issues with empirical/physics-based models is their inherent uncertainty. Assumptions and simplifications are made in all modeling, and not all the model variables are known exactly. Often stochastic techniques are used to account for the implicit uncertainty in a model's results. Statistical methods are used to generate numerous possible values for each input.

A Monte Carlo simulation was used in the calculation of bearing life distribution. Inputs to the model were represented by normal or log-normal distributions to approximate the uncertainty of the input values. Sample input distributions to the model are shown in Fig. 6.65.

The Yu-Harris model was used to simulate the room-temperature M50 RCF tests. Figure 6.66 shows the results for a series of the room-temperature RCF tests on the M50 bearing material. This test was run at 3600 rpm at room temperature with the 7808K lubricant. The y axis is the number of central rod failures, and the x axis is the millions of cycles to failure. The predicted life from the model is also shown in Fig. 6.66, superimposed on the actual test results. This predicted distribution shown in the lighter shaded curve was calculated from the model using 1 million Monte Carlo points.

In Fig. 6.67, the median ranks of the actual lives (dots) are plotted against the cumulative distribution function (CDF) of the predicted lives (solid line). The model-predicted lives are slightly more conservative (in the sense that the predicted life is shorter than the observed life) once the cumulative probability of failure exceeds 70 percent. However, since bearings are a critical component, the main interest is in the left-most region of the distribution, where the first failures occur, and the model correlates better.

Calculation of median ranks is a standard statistical procedure for plotting failure data. During run-to-failure testing, there are often tests that either are stopped prematurely before failure or a failure occurs of a component other than the test specimen. Although the data generated during these failures are the mode of interest, they provide a lower bound on the fatigue lives owing to the failure mode of interest. One method for including these data is by median ranking.

The median rank was determined using Benard's median ranking method, which is stated in the equation below. This method accounts for tests that did not end in the failure mode of interest (suspensions). In the case of the ball and rod RCF test rig, the failure mode of interest is creation of a spall on the

TABLE 6.5   RCF Fatigue Life Results

Failures (no.)	Susp (no.)	Susp. Time (cycles)
29	1 (ball failed)	83.33

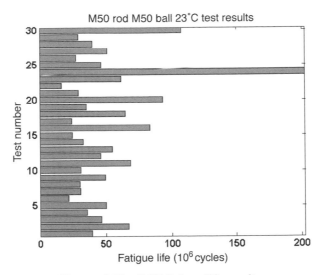

Figure 6.64   RCF fatigue life results.

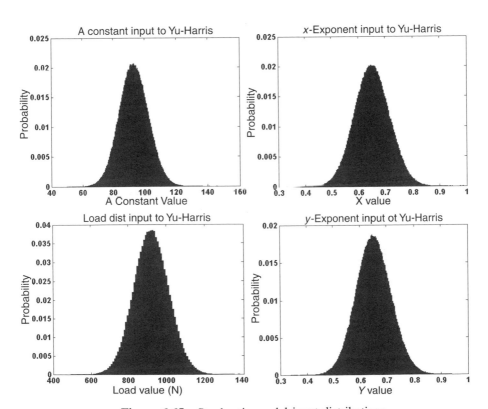

Figure 6.65   Stochastic model input distributions.

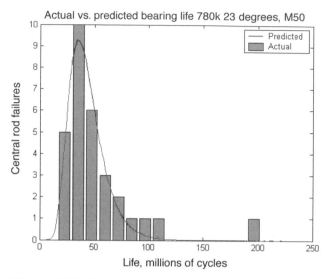

Figure 6.66 Room-temperature results versus predicted.

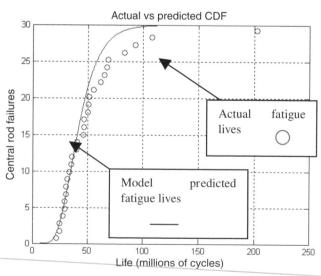

Figure 6.67 Actual life versus predicted life.

inner rod. The time to suspension provides a lower bound for the life of the test article (under the failure mode of interest), which can be used in reliability calculations. During the testing on the RCF test rig, significant portions of the tests were terminated without failure after reaching 10 times the L_{10} life. There also were several tests that ended owing to development of a spall on one of the balls rather than on the central rod.

$$\text{Benard's median rank} = \frac{(AR - 0.3)}{(N + 0.4)}$$

where AR = adjusted rank
N = number of suspensions and failures

The adjusted rank is

$$AR = \frac{(\text{reverse rank}) \times (\text{previous adjusted rank}) + (N + 1)}{\text{reverse rank} + 1}$$

Although the test does not simulate an actual bearing assembly in an engine, it does simulate similar conditions. Materials and the geometry of the bearing and the lubricants are the same for the test rig as they are in the T63 engine. The test rig results validate the model's ability to predict the fatigue life of the material under similar conditions to an operating engine.

Example Bearing Prognosis Discussion. To achieve a comprehensive prognostic capability throughout the life of a rolling-element bearing, model-based information can be used to predict the initiation of a fault before any diagnostic indicators are present. In most cases, these predictions will prompt "just in time" maintenance actions to prevent the fault from developing. However, owing to modeling uncertainties, incipient faults occasionally may develop earlier than predicted. In these situations, sensor-based diagnosis complements the model-based prediction by updating the model to reflect the fact that fault initiation has occurred. Sensor-based approaches provide direct measures of component condition that can be used to update the modeling assumptions and reduce the uncertainty in the RUL predictions. Subsequent predictions of the remaining useful component life will be based on fault progression rather than initiation models.

6.6 REFERENCES

D. W. Aha, "Special Issue on Lazy Learning," *Artificial Intelligence Review* **11**(1–5): 1–6, 1997.

M. S. Arulampalam, S. Maskell, N. Gordon, and T. Clapp, "A Tutorial on Particle Filters for Online Nonlinear/Non-Gaussian Bayesian Tracking," *IEEE Transactions on Signal Processing* **50**(3):174–188, 2002.

C. G. Atkeson, A. W. Moore, and S. Schaal, "Locally Weighted Learning," *Artificial Intelligence Review* **11**(1–5):11–73, 1997.

D. P. Bertsekas, *Dynamic Programming and Optimal Control.* Belmont, MA: Athena Scientific, 1995.

D. P. Bertsekas and J. Tsitsiklis, *Neuro-Dynamic Programming.* Belmont, MA: Athena Scientific, 1996.

B. L. Bowerman and A. B. Koehler, "Using the Power of the Computer to Choose Regression Models with Squared and Interaction Terms," *Decision Sciences Institute Proceedings,* 1993, Washington, DC: Decision Sciences Institute, 1169–1181.

J. Carpenter, P. Clifford, and P. Fearnhead, "An Improved Particle Filter for Non-Linear Problems," *IEEE Proceedings Radar, Sonar and Navigation* **146**:2–7, 1999.

J. Cheng and D. M. Titerington, "Neural Networks: A Review from a Statistical Perspective," *Statistical Science* **9**(1):2–54, 1994.

H. Chestnut, *Systems Engineering Tools.* New York: Wiley, 1965.

G. Chryssolouris, M. Lee, and A. Ramsey, "Confidence Interval Prediction for Neural Network Models," *IEEE Transactions on Neural Networks* **7**(1):229–32, 1996.

G. Cybenko. "Just-in-Time Learning and Estimation," in eds. S. Bittanti and G. Picci (eds.), *Identification, Adaptation, Learning: The Science of Learning Models from Data.* New York: Springer, 1996, pp. 423–434.

A. Doucet, N. Gordon, and V. Krishnamurthy, "Particle Filters for State Estimation of Jump Markov Linear System," Technical Report-CUED/F-INFENG/TR 359, Engineering Department, Cambridge University, Cambridge, UK, 1999.

M. Drexel and J. H. Ginsberg, "Mode Isolation: A New Algorithm for Modal Parameter Identification," *Journal of the Acoustical Society of America* **110**(3):1371–1378, 2001.

S. J. Engel, B. J. Gilmartin, K. Bongort, and A. Hess, "Prognosis, the Real Issues Involved with Predicting Life Remaining," *IEEE Aerospace Conference Proceedings,* Vol. 6. March 2000, Big Sky, MT, pp. 457–469.

C. Frelicot, "A Fuzzy-Based Prognostic Adaptive System," RAIRO-APII-JESA, *Journal European des Systemes Automatises* **30**(2–3):281–299, 1996.

R. Friedman and H. L. Eidenoff, "A Two-Stage Fatigue Life Prediction Method," Grumman Advanced Development Report, 1990, Bethpage, NY.

P. G. Groer, "Analysis of Time-to-Failure with a Weibull Model," *Proceedings of the Maintenance and Reliability Conference, MARCON* 2000, Knoxville, TN.

M. T. Hagan and M. Menhaj, "Training Feedforward Networks with the Marquardt Algorithm," *IEEE Transactions on Neural Networks* **5**(6):989–993, 1994.

D. G. Harlow and R. P. Wei, "Probabilities of Occurence and Detection of Damage in Airframe Materials," *Fatigue & Fracture Enginerring Materials & Structure* **22**(5):427–436, 1999.

M. E. Harmon, L. C. Baird, and A. H. Klopf, "Reinforcement Learning Applied to a Differential Game," *Adaptive Behavior* **4**(1):3–28, 1995.

E. Ioannides and T. A. Harris, "A New Fatigue Life Model for Rolling Bearings," *Jour. of Tribology,* 107, 3, 1985, 367–378.

R. Jardim-Goncalves, M. Martins-Barata, J. Alvaro Assis-Lopes, and A. Steiger-Garcao, "Application of Stochastic Modeling to Support Predictive Maintenance for Industrial Environments," *Proceedings of 1996 IEEE International Conference on Systems, Man and Cybernetics, Information Intelligence and Systems,* Vol. 1. IEEE Beijing, China, 1996, pp. 117–122.

N. Khiripet, "An Architecture for Intelligent Time Series Prediction with Causal Information," Ph.D thesis, Georgia Institute of Technology, May 2001.

M. Kotzalas and T. A. Harris, "Fatigue Failure Progression in Ball Bearings," *Jour. of Triboligy,* 123, 2, 2001, 238–242.

F. P. Lees, *Loss Prevention in the Process Industries,* 2d ed. Boston: Butterworth-Heinemann, 1996.

J. A. Leonard, M. A. Kramer, and L. H. Ungar, "A Neural Network Architecture That Computes Its Own Reliability," *Computers in Chemical Engineering* **16**(9):819–835, 1992.

F. L. Lewis, *Optimal Estimation: With an Introduction to Stochastic Control Theory.* New York: Wiley, 1986.

F. L. Lewis, *Applied Optimal Control and Estimation: Digital Design and Implementation.* Englewood Cliffs, NJ: Prentice-Hall, TI Series, 1992.

F. L. Lewis, S. Jagannathan, and A. Yesildirek, *Neural Network Control of Robot Manipulators and Nonlinear Systems,* Taylor and Francis, London, 1999.

J. S. Liu and R. Chen, "Sequential Monte Carlo Methods for Dynamical Systems," *Journal of the American Statistical Association* **93**:1032–1044, 1998.

L. Ljung, *System Identification: Theory for the User,* 2d ed. Englewood Cliffs, NJ: Prentice-Hall, 1999.

K. A. Marko, J. V. James, T. M. Feldkamp, C. V. Puskorius, J. A. Feldkamp, and D. Roller, "Applications of Neural Networks to the Construction of "Virtual" Sensors and Model-Based Diagnostics," *Proceedings of ISATA 29th International Symposium on Automotive Technology and Automation.* Florence, Italy, 1996, pp. 133–138.

M. L. Minsky, "Step Toward Artificial Intelligence," *Proceedings IRE* **49**:8–30, 1961.

J. C. Newman, "FASTRAN-II: A Fatigue Crack Growth Structural Analysis Program," NASA Technical Memorandum, 1992 Langley Research Center, Hampton, VA.

M. Orchard, B. Wu, and G. Vachtsevanos, "A Particle Filter Framework for Failure Prognosis," *Proceedings of WTC2005, World Tribology Congress III,* Washington, DC, 2005.

J. Ray, and S. Tangirala, "Stochastic Modeling of Fatigue Crack Dynamics for On-Line Failure Prognostics," *IEEE Transactions on Control Systems Technology* **4**(4): 443–51, 1996.

J. R. Schauz, "Wavelet Neural Networks for EEG Modeling and Classification," Ph.D. thesis, Georgia Institute of Technology, 1996.

A. Schömig and O. Rose, "On the Suitability of the Weibull Distribution for the Approximation of Machine Failures," *Proceedings of the 2003 Industrial Engineering Research Conference,* Portland, OR, May 18–20, 2003.

H. Sehitoglu, "Thermal and Thermal-Mechanical Fatigue of Structural Alloys," *ASM Handbook,* Vol. 19, *Fatigue and Fracture.* ASM International, Materials Park, OH. 1997, pp. 527–556.

R. Sharda, "Neural Network for the MS/OR Analyst: An Application Bibliography," *Interfaces* **24**(2):116–130, 1994.

J. Shiroishi, Y. Li, S. Liang, T. Kurfess, and S. Danyluk, "Bearing Condition Diagnostics via Vibration and Acoustic Emission Measurements," *Mechanical Systems and Signal Processing* **11**(5):693–705, 1997.

D. F. Specht, "Probabilistic Neural Networks," *Neural Networks* **3**:109–118, 1990.

D. F. Specht, "A General Regression Neural Network," *IEEE Transactions on Neural Networks* **2**(6):568–76, 1991.

L. Studer and F. Masulli, "On the Structure of a Neuro-Fuzzy System to Forecast Chaotic Time Series," *International Symposium on Neuro-Fuzzy Systems,* Aug. 29–31, 1996, pp. 103–110, EPFL-Lausanne, Switzerland.

R. S. Sutton, "Learning to Predict by the Methods of Temporal Differences," *Machine Learning* **3**:9–44, 1988, Advanced and Applied Technologies Institute.

R. S. Sutton, "Introduction: The Challenge of Reinforcement Learning," *Machine Learning* **8**:225–227, 1992.

N. B. Taylor, *Stedman's Medical Dictionary,* 18th ed., Baltimore: Williams & Wilkins, 1953.

F. C. Tsui, M. G. Sun, C. C. Li, and R. J. Sclabassi, "A Wavelet-Based Neural Network for Prediction of ICP Signal," *Proceedings of IEEE Engineering in Medicine and Biology,* Vol. 2. Montreal, Canada IEEE 1995, pp. 1045–1046.

D. S. J. Veaux, J. Schweinsberg, and J. Ungar, "Prediction Intervals for Neural Networks via Nonlinear Regression," *Technometrics* **40**(4):273–282, 1998.

M. D. Waltz and K. S. Fu, "A Heuristic Approach to Reinforcement Learning Control System," *IEEE Transaction on Automatic Control* **AC-10**:390–398, 1965.

P. Wang and G. Vachtsevanos, "Fault Prognostics Using Dynamic Wavelet Neural Networks," *Artificial Intelligence for Engineering Design, Analysis and Manufacturing* **15**(4):349–365, 2001.

C. J. C. H. Watins and P. Dayan, "Q-Learning," *Machine Learning* **8**(3):279–292, 1992.

W. Weibull, "A Statistical Distribution Function of Wide Applicability," *Journal of Applied Mechanics* **18**:293, 1951.

A. S. Weigend and N. A. Gershenfeld, *Time Series Prediction: Forecasting the Future and Understanding the Past.* Reading, MA: Addison-Wesley, 1993.

P. J. Werbos, "Generalization of Backpropagation with Application to Recurrent Gas Market Model," *Neural Networks* **1**:339–356, 1988.

W. K. Yu and T. A. Harris, "A New Stress-Based Fatigue Life Model for Ball Bearings," Tribology Transactions, Vol. 44, pp. 11–18, 2001.

CHAPTER 7

FAULT DIAGNOSIS AND PROGNOSIS PERFORMANCE METRICS

7.1 INTRODUCTION

This chapter addresses an important issue in condition-based maintenance/prognostics and health management (CBM/PHM) systems. It attempts to provide answers to the question: How do we assess the performance and effectiveness of CBM/PHM systems? In assessing the technical and economic feasibility of CBM/PHM systems, it is important to consider the combined impact of diagnostic and prognostic routines and the cost and performance of an "integrated system" intended to monitor key process variables. Currently, there are no accepted metrics or specifications for such assessment strategies. Researchers and CBM/PHM practitioners have explored in the recent past the development and application of appropriate technoeconomic performance measures for specific problem domains. It is generally recognized that an integrated CBM/PHM architecture entails elements of sensing and the selection/extraction of features or characteristic indicators of the presence of a fault condition, as well as the means to diagnose an impending failure and to foretell the remaining useful life or time to failure of the failing component/subsystem. Of paramount importance, therefore, is selection of the *best* features for the fault at hand; the application of diagnostic algorithms that are capable of detecting, isolating, and identifying a fault at the earliest possible stages of its initiation; and the accurate and precise prediction of the remaining useful life of the failing component. Technical performance metrics must account for means to evaluate the feature effectiveness as well as the accuracy and precision of diagnostic and prognostic algorithms. Arguably, prognosis presents unique challenges owing to the inherent uncertainty associated with long-term predictions. Performance metrics here must account for the presence of uncertainty and must entail means to assess confidence bounds relating to the prediction profile. Of major concern is also the cost associated with the realization of CBM/PHM practices.

It is generally anticipated that maximum life-cycle technoeconomic benefit from the implementation of CBM/PHM technologies will be derived from fleet-wide or system-wide practices where on-board and off-board capabilities share common equipment, data, and methods. The emerging nature of these technologies and the absence of life-cycle experiences necessitate a preliminary view of performance metrics and the reliance on "avoided cost" techniques to arrive at "estimates" of potential costs and benefits. Moreover, there are currently no available military or commercial standards to support a systematic and consistent approach to assessing the performance and effectiveness of CBM/PHM technologies. Technology developers and practitioners are embroiled in a continuous controversy in an attempt to define an agreeable set of metrics. We introduce in this chapter basic notions for the technical assessment of the performance and effectiveness of CBM/PHM technologies and suggest possible approaches for the inclusion of financial metrics toward a holistic and integrated performance assessment methodology. Our objective is to present the current status of such approaches and illustrate them with examples from specific application domains. We are also inviting a dialogue within the CBM/PHM community that eventually may lead to the establishment of standard metrics for both military and civilian applications.

7.2 CBM/PHM REQUIREMENTS DEFINITION

Although performance measures for fault diagnosis and prognosis of the remaining useful lifetime of a failing component/subsystem may vary from one application environment to the next, the CBM/PHM system must meet certain general requirements when designed and implemented on critical military or industrial processes as an "embedded" module in the process functionality. The candidate list of such requirements may include the following:

- The CBM/PHM system must ensure enhanced maintainability, safety, and reduced operational and support (O&S) costs over the life of critical system/processes.

- The CBM/PHM system must be designed as an open-system architecture that maximizes ease of subsystem and component changes, upgrades, and replacement while minimizing system/process interface change.
- The CBM/PHM system weight must be controlled closely.
- CBM/PHM system reliability, availability, maintainability, and durability (RAM-D) requirements must be met; for example, durability requirements specified as meantime between failure (MTBF) of the CBM/PHM system must be greater than x hours.
- Health monitoring requirements, that is, which faults, monitoring, display, communications, and types/protocols, must be associated with the process at hand.
- Structural and environmental requirements
- Scalability requirements
- Cost requirements
- User requirements (display, graphical user interface, etc.)
- Power requirements
- Compatibility requirements with existing sensors, components, devices, etc.

Other general requirements may be considered specific to the system at hand, its operating environment, etc.

7.2.1 Fault Diagnosis and Prognosis Requirements

For fault detection, we typically specify the ability to detect mechanical system degradation below required performance levels (incipient or impending failure) owing to physical property changes through detectable phenomena. Requirements are stated as the CBM/PHM system shall detect no less than a specific percentage of all mechanical system failures specifically identified through Reliability Centered Maintenance (RCM) and Failure Modes and Effects Criticality Analysis (FMECA) analyses.

For fault isolation, the CBM/PHM system shall demonstrate a fault isolation rate to one or multiple organizational removable assemblies (ORAs) of x percent (say, 2 percent) for all faults that are detected by CBM/PHM.

For fault identification, requirements may be defined as to the impact of a particular fault on the system's operating characteristics. For false-alarm rates, the CBM/PHM system must have a demonstrated mean operating range (flight hours for aircraft, for example) between false alarms of a specific number of hours.

If the system (avionics, for example) has built-in test (BIT) capability, then the CBM/PHM diagnostic test shall complete an end-to-end test within an x-minute test period. For prognosis, requirements typically are stated as

- The prediction of impending system failures shall be made with a 90 percent (for example) (threshold) and 98 percent (for example) (objective) confidence level.
- The prognostic module of the CBM/PHM system shall predict failures on life-limited components to enable part replacement with 10 percent (for example) remaining life.

An important concept in prognosis relates to the amount of time between when a prognostic system predicts an impending failure, within the required confidence limits, and when the operator or maintainer can effectively take action to address the failure and its consequences. This time interval is defined as the *critical prediction horizon* (CPH). A requirement may be stated as the prognostic system shall function with a maximum false prediction rate of x percent (threshold) and y percent (desireable). The diagnostic/prognostic requirements may be assembled into a spreadsheet/matrix format, as shown in Table 7.1.

7.3 FEATURE-EVALUATION METRICS

Feature selection and extraction constitute the foundation of an effective fault diagnosis and prognosis methodology. Converting raw sensor data into useful information that can be processed accurately and expeditiously is not only important but also a requirement when implementing CBM/PHM architectures for complex dynamic systems. We are addressing in this chapter such questions as: What are good features? and How do we compare various features for a specific system when the latter operates under various conditions? It is useful, therefore, to establish metrics for evaluating feature performance before selecting the "optimal" feature set. The feature-evaluation task presents major challenges because a feature set is most likely a vector function of such control variables or parameters as the load, environmental conditions,

TABLE 7.1   The Diagnostic/Prognostic Requirements

	Threshold	Objective
Fault detection	<75% of mechanical system failures monitored <90% of mechanical systems identified through FMECA	NA
Failure diagnosis	<75% subsystem isolation <5% subsystem incorrect identification	<75% failure mode isolation <5% failure mode incorrect identification
Failure prognosis	>1 day >90% confidence	>7 days > 98% confidence

etc. Our task in feature selection for fault classification and prediction is to maximize the classification accuracy, that is, to distinguish as unequivocally as possible a particular fault condition from others and from the healthy state of the system at hand while providing maximum prediction accuracy and precision.

7.3.1 The Ideal Classifier

We need an ideal classifier to estimate the predictive accuracy of a feature. Adopting Bayesian analysis, the true class of a given object is considered as a random variable C taking values in the set $\{c_1, c_2, \ldots, c_d\}$. The initial uncertainty regarding the true class is expressed by the prior probability $P(c_i)$. The classification criterion is to select the class c_i with the greatest posterior probability for a given pattern of evidence or feature $X = x_k$ such that

$$P(c_i|X = x_k) \geq P(c_j|X = x_k) \qquad j \neq i$$

Since

$$P(c_i|X = x_k) = P(X = x_k|c_i)P(c_i)/P(X = x_k)$$

for a certain feature value, c_i is chosen such that

$$P(X = x_k|c_i)P(c_i) \geq P(X = x_k|c_j)P(c_j) \qquad j \neq i$$

The error for a feature X then is given by

$$P_e(X) = E\{1 - \max[P(c_1|X = x_k), \ldots, P(c_d|X = x_k)]\}$$

where E is the expectation operator with respect to all possible values of X. The error values for different features can be used to rank the features. However, this technique cannot be used to reduce a given feature set to be of minimum number or to exclude redundant features because the predictive accuracy of a Bayesian classifier cannot be improved by eliminating a feature. There are also serious obstacles to the direct application of the Bayesian approach. Finite sample sizes are rarely sufficient to reliably estimate all the probabilities and statistics that the approach requires. Consequently, we need to rely on nonideal classifiers to evaluate features and remove redundant ones.

7.3.2 Information Measures

Information entails the uncertainty of a message or a class of data in the context of classification (Liu and Motoda, 1998). An information measure of uncertainty concerning a true class U is defined so that larger values of U represent higher levels of uncertainty. Given an uncertainty function and the prior probabilities $P(c_i)$, where $i = 1, 2, \ldots, d$, the information gain from a feature X, $IG(X)$, is defined as the difference between the prior uncertainty $\Sigma_i\, U[P(c_i)]$ and the expected posterior uncertainty using X, that is,

$$IG(X) = \Sigma_i\, U[P(c_i)] - E\{\Sigma_i\, U[P(c_i|X)]\}$$

We choose a feature X over another feature Y if

$$IG(X) > IG(Y)$$

That is, a feature is preferred if it reduces further the uncertainty content, and if it cannot reduce it, then it is irrelevant and can be removed from the feature set.

A commonly used uncertainty function is Shannon's entropy, that is,

$$U[P(c_i)] = -P(c_i)\, \log_2[P(c_i)]$$

7.3.3 Distance Measures

These types of measures are also known as *separability, divergence,* or *distinguishability measures.* Typically, they are derived from distances between the class-conditional density functions. For a two-class case, one may choose a feature X over another Y if

$$D(X) > D(Y)$$

where $D(X)$ is the distance between $P(X|c_1)$ and $P(X|c_2)$.

Another way of quantifying a feature's ability to differentiate between various classes is to find the area of the region in which the two classes overlap, as shown in Fig. 7.1. The distinguishability measure of feature X in a two-class case is given by

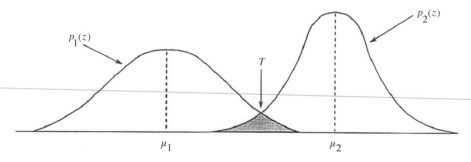

Figure 7.1   The distinguishability measure of feature in a two-class case.

$$d_X = \int_{-\infty}^{T} \frac{1}{\sqrt{2\pi}\sigma_2} \exp\left[-\frac{(z-\mu_2)^2}{2\sigma_2^2}\right] dz + \int_{T}^{\infty} \frac{1}{\sqrt{2\pi}\sigma_1} \exp\left[-\frac{(z-\mu_1)^2}{2\sigma_1^2}\right] dz$$

where μ and σ^2 are the mean and the variance of the feature's distribution (which is assumed to be Gaussian), respectively. Feature X is preferred over Y if $d_Y > d_X$.

7.3.4 Dependence Measures

These types of measures are also known as *association* or *correlation measures*. They are designed to quantify how strongly two variables are associated or correlated with each other so that by knowing the value of one variable one can predict the value of the other. We look at how strongly a feature is associated with a specific class in order to evaluate the feature, and we choose a feature that associates closely with class C. The dependence measures also can be applied to find out the correlation between two features and thereby remove any redundant ones. One example of a dependence measure is the correlation between two features X and Y, which is a measure of linear association between the two features, that is,

$$r_{X,Y} = \frac{\Sigma (X - \bar{X})(Y - \bar{Y})}{(n-1)s_X s_Y}$$

where

$$s_X = \sqrt{\frac{\Sigma (X - \bar{X})^2}{n-1}}$$

is the sample standard deviation for feature X. For example, if $Y = X + 1$, then the correlation measure between the two features is the highest and is 1.

7.3.5 Feature Evaluation Methods: Two-Sample Z-Test Procedures

As defined in the distance measure, we prefer features that exhibit the farthest distance between the distributions of the feature values for a two-class case. The two-sample Z-test assesses whether the means of two groups are statistically different from each other and thereby provides a useful tool to measure the distance between two distributions. The distance in this case is not simply the difference between the two means, but the difference between group means normalized by the variability of the groups. Consider the three situations shown in Fig. 7.2. The three cases have the same difference between the means; however, they tell very different stories with a different variability in each group. We would conclude that the two groups appear most distinct

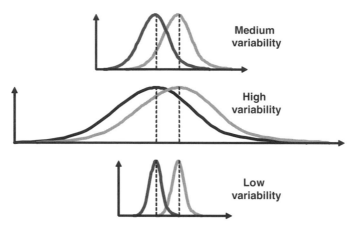

Figure 7.2 Three scenarios for differences between means.

in the bottom or low-variability case because there is relatively little overlap between the two bell-shaped curves. In the high-variability case, the group difference appears the least striking because the two bell-shaped distributions overlap significantly.

The formula for the Z-test can be regarded as the signal-to-noise ratio of the distribution, with the top part being the difference between the two means and the bottom part a measure of the variability of the distributions. The formula is expressed as follows:

$$Z = \frac{\overline{X}_1 - \overline{X}_2}{\sqrt{\dfrac{s_{x_1}^2}{n_1} + \dfrac{s_{x_2}^2}{n_2}}}$$

where s_X and $\overline{X}$ are the standard deviation and mean of a distribution, respectively, and n is the number of incidences for each distribution. The Z value calculated in this way defines the distance between the distributions of a feature for two classes in a classification problem.

7.4 FAULT DIAGNOSIS PERFORMANCE METRICS

7.4.1 Performance Metrics for Diagnostic Systems

Major performance metrics for diagnostic systems include, among others,

- False positives
- False negatives

- Time delay—time span between the initiation and the detection of a fault (failure) event
- Percent isolation to one line-replaceable unit (LRU)

Performance requirements for diagnostic algorithms typically specify the maximum allowable number of false positives and false negatives as a percentage of the total faults present in a particular subsystem over its expected life. For example, a typical requirement may be stated as "no more than 5 percent of false positives and no more than 3 percent of false negatives." A case should be made for the relative significance of these two metrics: False negatives may present major risks to the health of the equipment under test. Missed fault conditions may lead to a catastrophic failure resulting in loss of life or loss of the system. False positives, on the other hand, could be accommodated with an unavoidable loss of confidence on the part of the system operator as to the effectiveness of the diagnostic tools. For this reason, tradeoffs that naturally arise between the two metrics favor a stricter false-negative requirement. The time-delay metric is most significant not only as an early warning to the operator of an impending failure but also in terms of providing a sufficient time window that allows a prognostic algorithm to perform its intended task. We will detail in the next sub-section the receiver operating characteristic (ROC) method used to answer the question: How do we evaluate how well a fault is diagnosed?

Fault-detection events may be evaluated through the decision matrix (Liu and Motoda, 1998). It is based on a hypothesis-testing methodology and represents the possible fault-detection combinations that may occur (Table 7.2).

From this matrix, the detection metrics can be computed readily. The probability of detection (POD) given a fault (a.k.a. *sensitivity*) assesses the detected faults over all potential fault cases:

TABLE 7.2 Decision Matrix for Fault-Detection Evaluation

Outcome	Fault (F_1)	No Fault (F_0)	Total
Positive (D_1) (detected)	a Number of detected faults	b Number of false alarms	$a + b$ Total number of alarms
Negative (D_0) (not detected)	c Number of missed faults	d Number of correct rejections	$c + d$ Total number of non-alarms
	$a + c$ Total number of faults	$b + d$ Total number of fault-free cases	$a + b + c + d$ Total number of cases

$$\text{POD} = P(D_1/F_1) = \frac{a}{a + c}$$

The probability of a false alarm (POFA) considers the proportion of all fault-free cases that trigger a fault detection alarm

$$\text{POFA} = P(D_1/F_0) = \frac{b}{b + d}$$

A metric of accuracy is used to measure the effectiveness of the algorithm in correctly distinguishing between a fault-present and fault-free condition. The metric uses all available data for analysis (both fault and no fault):

$$\text{Accuracy} = P(D_1/F_1 \text{ \& } D_0/F_0) = \frac{a + d}{a + b + c + d}$$

Fault isolation implies the ability to distinguish between possible failing-unit candidates once a fault has been detected and reported. An appropriate metric in this case is the percent isolation to one LRU. Such a metric may be significant for both maintenance and operational reasons. Coupled with fault identification, that is, the severity of the fault impact on system performance, it provides the operator with information required to assess the criticality of the fault event and the corrective action to be taken.

7.4.2 The Receiver Operating Characteristic

The accuracy of diagnostic methods typically is characterized, as noted previously, in terms of such metrics as false positives and false negatives. The two metrics may change if the detection threshold varies and tradeoffs are needed. A question that often arises is stated as: How do we evaluate how well a fault is diagnosed? The *receiver operating characteristic* (ROC) method has been suggested and employed as an effective tool to provide answers to this question. It is recognized, though, that its basic limitation refers to the two-dimensional nature of the ROC curve and that higher-dimensional attributes/metrics cannot be accounted for. We will use an example to illustrate the method. Suppose that a temperature measurement is used to diagnose a particular machine (component/subsystem) fault. A stepwise ROC procedure proceeds as follows.

Step 1: Aggregate a large sample population of test results whose temperature levels and corresponding fault conditions are known. Separate those which are healthy (not faulty) from those with a fault, and plot the number of events against temperature levels, as shown in Fig. 7.3. The graph reveals

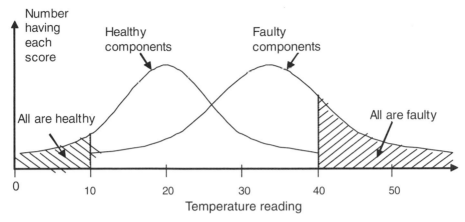

Figure 7.3 The number of events against temperature levels.

that temperature readings in the 10 to 40 range cannot conclusively distinguish healthy components from those with faults.

Step 2: Calculate the probability that a "yes" diagnosis at or above any given threshold would be correct for a new component. Find such probabilities by determining the fraction of components in the sample population that would have been diagnosed properly if that threshold were applied. The area under the curve represents 100 percent of each population. If the threshold were 20, 90 percent of components that truly failed would be diagnosed correctly (true positives), and 50 percent of healthy components would be diagnosed incorrectly as having failed (false positives), as shown in Fig. 7.4.

Step 3: Construct an ROC curve for each potential threshold, the rate of true positives against the rate of false positives, as shown in Fig. 7.5. A straight line signifies that the diagnosis had 50/50 odds of being correct (no better that the flip of a coin). As curves bow more to the left, they indicate greater accuracy (a higher ratio of true positives to false positives). Accuracy *A* is indexed more precisely by the amount of area under the curve, which increases as the curves bend.

Step 4: If the accuracy is acceptable, select a threshold for yes/no diagnoses. Choose a threshold that yields a good rate of true positives without generating an unacceptable rate of false positives. Figure 7.6 shows that each point on the curve represents a specific threshold, moving from the strictest at the bottom left to the most lenient at the top right. Strict thresholds (bottom inset) limit false positives at the cost of missing many affected components; lenient thresholds (top inset) maximize discovery of affected components at a cost of many false positives. Which threshold is optimal for a given population depends on such factors as the criticality of the

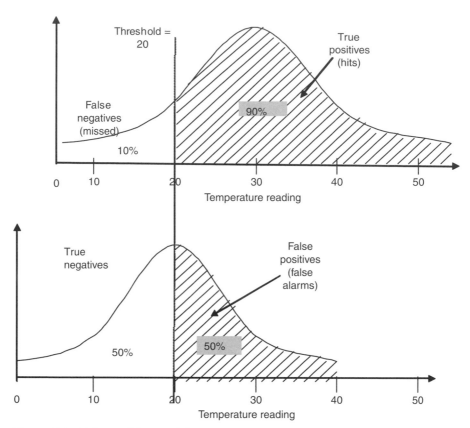

Figure 7.4 False positives and false negatives corresponding to a specific threshold.

failure condition being diagnosed, the prevalence of the condition in a population, the availability of corrective measures, and the financial and other costs of false alarms.

7.4.3 Fault Severity, Detection Threshold/Confidence, and Complexity Metrics

The ability of CBM/PHM systems to detect and isolate faults or to predict failures typically depends on fault severity. Incipient faults and mild forms of performance degradation usually present the greatest challenge to CBM/PHM systems. Consequently, CBM/PHM performance must be considered a function of fault severity. Fault severity must be established by objective and irrefutable measures to ensure that the assessments based on it are accurate and impartial. This measure of severity hereafter will be referred to as the *ground-truth severity level.* Measurements of the fault severity are mapped

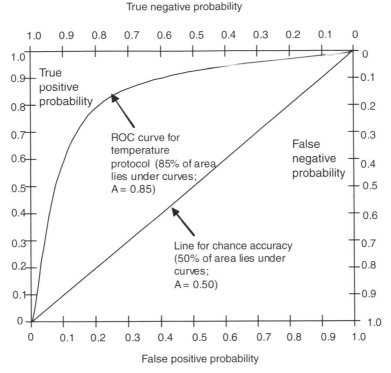

Figure 7.5 The ROC curve.

onto the ground-truth severity scale, where 0 represents a healthy operating condition, and 1 represents an unacceptable level of performance degradation.

Figure 7.7 shows the confidence level reported by a hypothetical CBM/PHM technology and the corresponding fault severity level as a function of time. This could be the confidence that an anomaly exists or the confidence in a particular fault diagnosis. Varying operating conditions or noise can cause fluctuations in this confidence level. The success function of the CBM/PHM technology is defined as the function of the difference between the average confidence and the average severity level. Note that this function may be used to assess either Boolean (0 or 1) confidence levels or continuous confidence levels within the same interval. The success function for the hypothetical CBM/PHM technology is plotted in Fig. 7.8.

A detection threshold metric is required that can measure a CBM/PHM algorithm's ability to identify anomalous operation associated with incipient faults with a specified confidence level. Confidence levels of 67 and 95 percent, corresponding to one and two standard deviations, are used to calculate the detection threshold metric. An assessment of the detection confidence level over the entire severity range for the failure mode also will be achieved

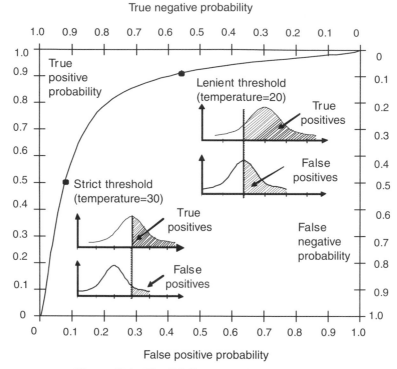

Figure 7.6 The ROC curve and the threshold.

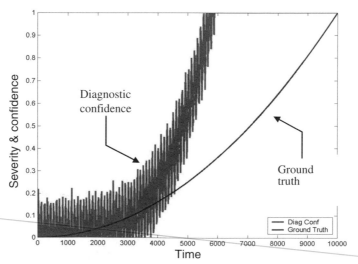

Figure 7.7 Diagnostic and ground truth information.

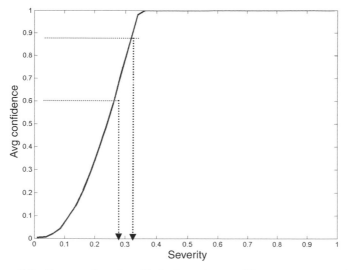

Figure 7.8 Success function (fault detection confidence versus severity).

with an overall detection confidence metric. With such a metric, an algorithm that detects an incipient fault with high confidence will receive a high overall confidence score, whereas an algorithm that does not report a fault until it becomes very severe would receive a low score. The detection threshold is defined as

$$\text{Detection threshold} = 1 - S(s) \qquad \text{(i.e., success function normalized)}$$

And the overall detection confidence is computed as

$$\text{Overall confidence} = \int_0^1 C(s)\, ds$$

where $S(s)$ is the success function, as shown in Fig. 7.8, and $C(s)$ is the detection confidence for a specific severity value.

A CBM/PHM technology whose output fault confidence level fluctuates wildly is difficult to interpret and therefore is undesirable. For example, a diagnostic tool that produces a Boolean result of either no fault or fault may flicker as the fault severity approaches the detection level. A stability metric is developed that measures the range of confidence values that occur over the fault transition by integrating the peak-to-peak difference at each point in the transition, that is,

$$\text{Stability} = 1 - \int_0^1 [C_H(s) - C_L(s)]\, ds$$

where $C_H(s)$ and $C_L(s)$ are the high and low values of the confidence at each point in the transition. In addition, CBM/PHM systems should detect anomalies over the full range of operating (duty) conditions, such as loads, speeds, etc. A detection duty sensitivity metric is developed that measures the difference between the outputs of a CBM/PHM technology under various duty conditions and is formulated as

$$\text{Duty sensitivity} = 1 - \sqrt{\int_I^1 [C_1(s) - C_2(s)]^2\, ds}$$

where $C_1(s)$ and $C_2(s)$ are the outputs of a CBM/PHM technology under two different load or duty conditions.

A CBM/PHM technology that incorrectly reports anomalies is unacceptable because it reduces availability and increases maintenance costs for the associated system. Hence, a false-positive confidence metric is developed that measures the frequency and upper confidence limit associated with false anomaly detection by the CBM/PHM technology. Calculation of the false confidence metric is based on the false-positive function as defined in Orsagh et al. (2000). In an operational environment, sensor data sometimes are contaminated with noise that may interfere with the operation of diagnostic algorithms. The robustness of a CBM/PHM algorithm to noisy data will be measured by a noise sensitivity metric, which is defined as follows:

$$\text{Noise sensitivity} = \left\{ 1 - \sqrt{\int_0^1 [C_1(s) - C_2(s)]^2\, ds} \right\} (\Delta FalsePositive)$$

where $C_1(s)$ is the success function under noise condition 1, $C_2(s)$ is the success function under noise condition 2, and s is the fault severity.

Acquisition and implementation costs of the CBM/PHM algorithm under investigation may have a significant effect on the overall system's cost-effectiveness. An implementation cost metric is proposed that simply will measure the cost of acquiring and implementing a CBM/PHM system on a single application. If the CMB/PHM system is applied to several pieces of equipment, any shared costs are divided among them. Operation and maintenance (O&M) costs also may play a significant role in determining whether a CBM/PHM system is cost-effective. An O&M cost metric is also needed to measure the annual cost incurred to keep the CBM/PHM system running. These costs may include manual data collection, inspections, laboratory testing, data archival, relicensing fees, and repairs.

The ability of the CBM/PHM algorithms or system to be run within a specified time requirement and on traditional aerospace computer platforms with common operating systems is important when considering implementation. Therefore, a metric that takes into account computational effort as well as static and dynamic memory allocation requirements is necessary. A computer resource metric may be developed to compute a score based on the normalized addition of CPU time to run (in terms of floating-point operations), static and dynamic memory requirements for RAM and static source-code space, and static and dynamic hard-disk storage requirements.

Complex systems generally are more susceptible to unexpected behavior owing to unforeseen events. Therefore, a system complexity metric may be used to measure the complexity of the CBM/PHM system under investigation in terms of the number of source lines of code (SLOCs) and the number of inputs required.

7.4.4 Diagnostic System Classifier Performance Metrics

Diagnostic classifier performance metrics are used to evaluate classification algorithms, typically considering multiple fault cases, and are based on the confusion matrix concept. In general, a confusion matrix is intended to highlight in a table format the degree of correlation between several parameters (i.e., features, classes, etc.). It is a square matrix consisting of identical columns and rows where its entries represent the degree or strength of correlation between its ith and jth element. The confusion matrix used as a tool to define the diagnostic classifier metrics is shown in Table 7.3, which illustrates the results of classifying data into several categories. The table shows actual defects (as headings across the top of the table) and how they were classified (as headings down the first column). The shaded diagonal represents the number of correct classifications for each fault in the column, and subsequent

TABLE 7.3   Confusion Matrix Used for Diagnostic Metrics

↓ Classified as Problem →	Collector Gear Crack Prop.	Quill Shaft Crack Prop.	Spiral Bevel Input Pinion Spalling	Input Pinion Bearing Corrosion	Total
Collector Gear Crack Propagation	5	1	1	0	7
Quill Shaft Crack Propagation	0	6	0	1	7
Spiral Bevel Input Pinion Spalling	1	0	5	0	6
Input Pinion Bearing Corrosion	0	0	1	6	7
Total	6	7	7	7	27

numbers in the columns represent incorrect classifications. Ideally, the numbers along the diagonal should dominate. The confusion matrix can be constructed using percentages or actual cases witnessed.

The probability of isolation (fault isolation rate, or FIR) is the percentage of all component failures that the classifier is able to isolate unambiguously. It is calculated using

$$FIR = \frac{A_T}{A_T + C_T} \times 100$$

$$A_T = \sum_i A_i$$

$$C_T = \sum_i C_i$$

where A_i is the number of detected faults in component i that the monitor is able to isolate unambiguously as due to any failure mode (numbers on the diagonal of the confusion matrix), and C_i is the number of detected faults in component i that the monitor is unable to isolate unambiguously as due to any failure mode (number off the diagonal of the confusion matrix).

The confusion matrix is simple and easy to compute and suitable for both binary and multiple classifiers. However, we assume that classification errors for all classes are equal in cost. A more general classifier performance metric is based on the misclassification cost, which is defined as

$$MC = \sum_{i,j} \overline{cm}(i, j)C(i, j)$$

and
$$\overline{cm}(i, j) = CM(i, j)/\Sigma \, CM(i)$$

where $C(i, j)$ and $CM(i, j)$ are cost and confusion matrices, respectively. Its main advantages are that it is a general classifier performance measure, that it can be used for multiclass classifiers, and that it can take care of unequal costs of classes. However, it requires the evaluation of a cost matrix.

These metrics form the basis of top-level detection and diagnostic effectiveness. A challenge for the research community is to generate realistic estimates to these metrics given the currently limited baseline data and faulted data available. The basis for the statistical analysis introduced here is offered next.

Example 7.1 Evaluation of Aircraft Engine Fault Diagnostic Systems. As an example, we evaluate diagnostic classifiers of aircraft engine fault diagnostic systems according to the classification accuracy and misclassification cost metrics. The diagnostic classifiers are part of an overall project that designs and tests a condition-based intelligent maintenance advisor for turbine engines (IMATE) system (Ashby and Scheuren, 2000). For a typical aircraft engine, consider the following six engine hot-gas path faults:

1. FAN = fan damage
2. COM = compressor fault
3. HPT = high-pressure turbine fault
4. LPT = low-pressure turbine fault
5. CDP = customer discharge pressure fault
6. VBV = variable bleed valve leak

For evaluation purposes, we will consider three types of classifiers:

1. Multiple-layer neural network
2. Fuzzy neural network
3. CART

Figure 7.9 depicts the IMATE functional architecture, and the IMATE system configuration is shown in Fig. 7.10. The output of the classifier is fused in order to arrive at better results. The classifier fusion architecture in IMATE is shown in Fig. 7.11.

The faults are ranked according to their criticality, as shown in Table 7.4. The estimated cost matrix is shown in Table 7.5 (with $m = 0.9$ and $S = 100$).

A confusion matrix, as shown in Table 7.6, is intended to depict the coupling or cross-correlation between classifiers in their ability to classify the related fault classes.

Tables 7.7 and 7.8 aggregate the results of the diagnostic evaluation process. Different criticality rankings of the faults (as shown by the R vector above

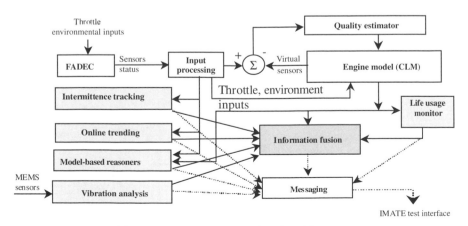

Figure 7.9 IMATE functional architecture.

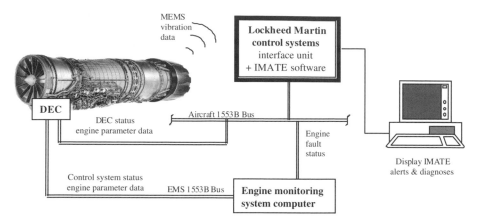

Figure 7.10 IMATE system configuration.

the tables) lead to different estimated cost matrices and thereby different measures of the misclassification cost and the accuracy.

7.4.5 Statistical Detection Concepts

Statistical detection is based on the separability of features between the no-fault and faulted conditions. While a detection probability of 0.9 or higher is desirable, this may not always be possible. In the first place, POFA and POD are correlated because both are measured on the high side of the threshold.

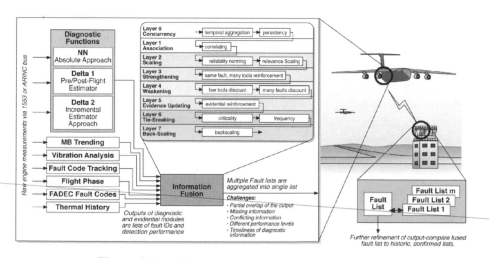

Figure 7.11 Classifier fusion architecture in IMATE.

TABLE 7.4 Fault Criticality Rank

Faults	NF	FAN	COM	HPT	LPT	CDP	VBV
Ranking	0	0.9	0.8	0.8	0.8	0.1	0.1

If the threshold is raised to decrease the probability of false alarms, the probability of detection is decreased accordingly.

This can be seen in Fig. 7.12, in which the signal for no fault is on the left and faulted-feature values are on the right side. The probability of detection can be seen on the right side of the threshold. The lower figure emphasizes the no-fault (detection noise) distribution. POFA is also on the right side of the threshold. To decrease POFA, we raise the threshold (move it to the right), and this has the unfortunate effect of decreasing the probability of detection.

Statistical Analysis. The fault and no-fault conditions can be estimated using probability density functions (PDFs). Obviously, with sufficient data, these distributions can be identified directly. Within the current limited-data environment, we assume the data statistics to be a Rician distribution, which is also known as the *noncentral chi distribution.* If the real and imaginary parts of the data, with mean values α_Q and α_I, respectively, are corrupted by independent Gaussian noise with variance σ^2, it is easy to show that the magnitude data will be Rician distributed (Papoulis, 1984), given by

$$p(x) = \frac{x}{\sigma^2} \exp\left[-\frac{1}{2\sigma^2}(x^2 + \alpha^2)\right] I_0\left(\frac{\alpha x}{\sigma^2}\right) \quad x > 0$$

$$p(x) = 0 \quad x < 0$$

where $\alpha^2 = \alpha_Q^2 + \alpha_I^2$, and $I_0(u)$ is the modified 0th-order Bessel function of the first kind, given by

TABLE 7.5 Estimated Cost Matrix

True Faults	Predicted Faults						
	NF	Fan	Comp	HPT	LPT	CDP	VBV
NF	0	23.1	20.6	20.6	20.6	2.6	2.6
Fan	25.7	0	2.9	2.9	2.9	22.9	22.9
Com	22.9	2.6	0	2.9	2.9	20.0	20.0
HPT	22.9	2.6	2.9	0	2.9	20.0	20.0
LPT	22.9	2.6	2.9	2.9	0	20.0	20.0
CDP	2.9	20.6	18.0	18.0	18.0	0	2.9
VBV	2.9	20.6	18.0	18.0	18.0	2.9	0

TABLE 7.6   The Confusion Matrix

Classifiers		NF	Fan	Comp	HPT	LPT	CDP	VBV
				Predicted Faults				
CART	NF	2074	0	126	106	347	96	56
	Fan	0	2781	24	0	0	0	0
	Comp	112	2	2577	36	55	22	1
	HPT	110	0	93	2419	179	3	1
	LPT	394	0	70	309	1864	146	22
	CDP	77	0	11	1	105	2539	72
	VBV	114	0	22	1	31	90	2547
FNN	NF	1460	0	174	316	463	278	114
	Fan	0	2805	0	0	0	0	0
	Comp	264	170	1995	72	159	134	11
	HPT	155	0	60	2314	241	35	0
	LPT	432	0	75	908	914	459	17
	CDP	69	0	6	0	157	2496	77
	VBV	167	0	32	2	34	78	2492
NN20	NF	2450	11	73	51	131	50	39
	Fan	4	2750	32	0	1	4	14
	Comp	84	56	2582	29	18	15	21
	HPT	60	7	97	2509	121	6	5
	LPT	188	5	34	162	2364	45	7
	CDP	83	5	21	2	27	2595	72
	VBV	47	6	14	7	4	52	2675
NN25	NF	2462	5	52	39	164	71	12
	Fan	0	2718	63	1	1	22	0
	Comp	97	31	2590	21	29	36	1
	HPT	56	11	88	2499	142	8	1
	LPT	151	1	27	141	2450	31	4
	CDP	68	9	12	1	37	2654	24
	VBV	59	2	16	0	9	113	2606
NN30	NF	2483	4	45	37	166	48	22
	Fan	1	2771	29	4	0	0	0
	Comp	96	30	2594	33	19	24	9
	HPT	62	16	93	2493	137	1	3
	LPT	180	6	36	153	2375	47	8
	CDP	81	5	16	0	23	2654	26
	VBV	68	21	12	0	4	91	2609

$$I_\theta(u) = \frac{1}{\pi} \int_0^\pi \exp(u \cos \theta) \, d\theta$$

Given the limited data sets available for the metrics assessment [for the Westland seeded-fault data set (Cameron, 1993), we have five fault and no-

TABLE 7.7   The Results of the Diagnostic Evaluation Process (1)
$R = [0.0, 0.9, 0.8, 0.8, 0.8, 0.1, 0.1]$

Classifiers	M.C. Measure		Accuracy Measure	
	Cost	Ranking	Accuracy	Ranking
CART	13.03	4	85.57	4
FNN	22.54	5	73.73	5
NN20	7.02	3	91.29	3
NN25	6.61	1	91.57	1
NN30	6.72	2	91.57	1

fault data sets for each sensor], an estimate of the statistical variance is evaluated using a chi-squared approximation with user-defined confidence bounds. This approximation of variance is represented by the interval (Papoulis, 1984)

$$\left[s^2 \frac{v}{\chi_v^2\left(1 - \frac{\alpha}{2}\right)}, s^2 \frac{v}{\chi_v^2\left(\frac{\alpha}{2}\right)} \right]$$

where s^2 is the calculated variance from the sample, and σ^2 is the true variance that lies within the interval, that is,

$$s^2 \frac{v}{\chi_v^2\left(1 - \frac{\alpha}{2}\right)} \le \sigma^2 \le s^2 \frac{v}{\chi_v^2\left(\frac{\alpha}{2}\right)}$$

The PDF of a chi-squared (non-central) distribution is given by

TABLE 7.8   The Results of the Diagnostic Evaluation Process (2)
$R = [0.0, 0.95, 0.01, 0.1, 0.01, 0.01, 0.2]$

Classifiers	M.C. Measure		Accuracy Measure	
	Cost	Ranking	Accuracy	Ranking
CART	6.29	2	85.57	4
FNN	14.25	5	73.73	5
NN20	6.68	3	91.29	3
NN25	6.75	4	91.57	1
NN30	5.83	1	91.57	1

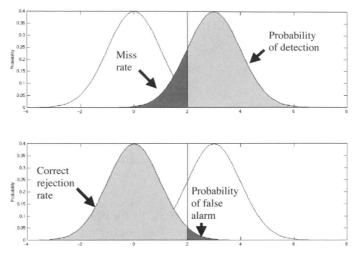

Figure 7.12   Relationship of statistical detection metrics.

$$p(x) = \begin{cases} \dfrac{1}{2}\left(\dfrac{x}{\lambda}\right)^{[(v-2)/4]} \exp[-(1/2)(x+\lambda)]I_{(v/2)-1}\sqrt{\lambda x} & x > 0 \\ 0 & x < 0 \end{cases}$$

where $I_r(u)$ is the rth modified Bessel function of the first kind and is defined as

$$I_r(u) = \frac{\left(\dfrac{1}{2}u\right)^r}{\sqrt{\pi}\Gamma\left(r + \dfrac{1}{2}\right)} \int_0^\pi \exp(u\cos\theta)\sin^{2r}\theta\, d\theta$$

$I_r(u)$ has the series representation:

$$I_r(u) = \sum_{k=0}^{\infty} \frac{\left(\dfrac{1}{2}u\right)^{2k+r}}{k!\Gamma(r + k + 1)}$$

Using a series expansion, the PDF can be restated as

$$p(x) = \frac{x^{(v/2)-1}\exp\left[-\frac{1}{2}(x+\lambda)\right]}{2^{v/2}} \sum_{k=0}^{\infty} \frac{\left(\frac{\lambda x}{4}\right)^k}{k!\,\Gamma\left(\frac{v}{2}+k\right)}$$

Impact Technologies, Inc., implemented the PDFs given above using C++. A numeric approximation of the 0th-order modified Bessel function was adapted from code in "Numerical Recipes in C." In order to simplify computation, the 0th-order Bessel function is used in the computation of the chi-squared PDF. The very small error introduced by this simplifying approximation is less than the expected finite arithmetic errors in the computation. A visualization of variance estimate using the chi-squared confidence interval is shown in Fig. 7.13. This code module is embedded within the metrics evaluation tool developed by Impact Technologies, Inc., which will be introduced in the next subsection.

7.4.6 An Architecture for Diagnostic Metrics Assessment

Impact Technologies, Inc., in collaboration with Sikorsky Aircraft and the Penn State Applied Research Laboratory, has developed a metrics evaluation tool (MET) under Boeing's RITA HUMS Technology project. The tool was motivated by the need to evaluate the performance and effectiveness of algorithms used to detect and diagnose faults. The MET is being demonstrated using a prototype database with seeded-fault data from an H-46 aft transmission exercised with typical diagnostic algorithms for gear, shaft, and bearing faults.

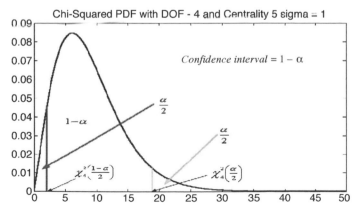

Figure 7.13 Visualization of variance estimate using the chi-squared confidence interval.

Vibration-feature results are analyzed using probability of detection and false-alarm metrics, as well as diagnostic accuracy metrics. In addition, the effects of signal-to-noise ratio and threshold settings on the detection and false-alarm metrics are evaluated within the tool. This allows helicopter manufacturers and HUMS end users to evaluate effectiveness of an algorithm by statistical analysis of seeded-fault test data. Moreover, it directly evaluates the risk associated with changing detection threshold or relying on data containing varying noise levels. The tool, along with a diagnosis database, will provide an enhanced capability to assess and record the performance of vibration-based diagnostic algorithms.

The development of the MET first required the development of a suitable architecture to allow users to gain access, explore, calculate, and evaluate the data and information. We deemed the most useful form to be a client-server model for requests, processing, and viewing of results. The MET needed to allow its users remote access to the RITA HUMS database, which contains amounts of data much too large to be stored locally by the client application. Figure 7.14 illustrates the architecture and interface between the MET client and the database server. In addition to eliminating client-side data storage, users also gain the advantage of access to useful applications such as MATLAB® and the Oracle 8i database, which is the relational database used within RITA HUMS.

Tool Command Language (TCL) was used to create the customized interface required to use the existing algorithms and new algorithms from their storage in the database to their execution in engineering software (MATLAB). The advantages of using TCL include its ability to communicate with a client-side application (such as the current MET) using the TCP/IP protocol, as

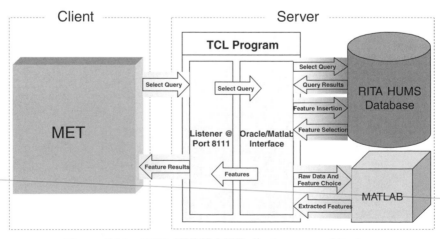

Figure 7.14   RITA HUMS MET client-server architecture.

well as its easily available and highly usable interface to Oracle. TCL also allows the flexibility of interfacing with different algorithm platforms, including MATLAB, C++, and Java. With this client-server design and application program, the complex data manipulations are managed on the server side of the application.

Database Development and Data Evaluated. The RITA HUMS database (Rozak, 2001) provides a common environment to allow an end user (or client) a means to query available test data and algorithms. The database is structured as a client-server relationship between the Oracle database server and the primary Oracle form. The Oracle application is used as the database engine because it can handle and store large binary objects (BLOB), provides an interface to external C++ coding, and provides comprehensive table/user security. Algorithms are stored in the database in a similar fashion as raw data. They are also treated as large binary objects.

To place data or algorithms into the database, the user may use containers embedded into the Oracle form or write Sequential Querying Language (SQL) scripts to automate the process of data upload. For the current MET demonstration, Impact Technologies populated and analyzed data from the Westland seeded-fault data set (Cameron, 1993). Lessons learned from this process are offered at the end of this chapter.

Several conversions were performed on the data before they were added to the RITA HUMS database. The 22-second (100k samples/s) Westland data (Cameron, 1993) was parsed into 4-second segments to allow for additional analysis and to enhance the accuracy of algorithm assessment by the MET. To facilitate automated script upload, the subdivided Westland data were first entered into Microsoft Access using database population scripts written in Visual Basic. The final version of the Westland data resulted in 98,000 individual required field updates to the raw data table alone in the RITA HUMS database. The magnitude of this effort clearly requires automated means to populate the database. Other relational fields in associated sensor and test tables also required updates to maintain both connectivity and database integrity. SQL scripts were written to accomplish the task of entering these data in the RITA HUMS Oracle. A summary of this process is shown in Fig. 7.15.

Within the design of the RITA HUMS database, the user has the option to insert the data into the Oracle database either directly as a BLOB or indirectly as a pointer to its storage location external to the actual RITA HUMS database. This flexibility does not prevent the data from being readily retrieved and used for subsequent analysis, and using pointers to external files was found to be much faster and more robust as the database size increases. Therefore, the current implementation of the RITA HUMS database uses an external file referencing system. To complete the data storage, a file structure was created consistent with the identifying fields. The use of such an approach requires a logical assignment of the stored high-bandwidth data on the hard

			Torque								
			27%	40%	45%	50%	60%	70%	75%	80%	100%
	2	Planetary Bearing Corrosion									99
F	3	Input Pinion Bearing Corrosion	22, 50	27, 55	26, 63	23, 60	37, 56	24, 61	25, 62	38, 57	30, 58
A	4	Spiral Bevel Input Pinion Spalling	17, 89	13, 85	12, 84	81	86	10, 82	83	15, 87	16, 88
U	5	Helical Input Pinion Chipping						85	86	90	91
L	6	Helical Idler Gear Crack Propagation						40	41	45	46
T	7	Collector Gear Crack Propagation	13	92		41	19	42	43	20	95
	8	Quill Shaft Crack Propagation	79	80	81	82	83	84	85	86	87
	9	No Defect	02	07	06	03	08	04	05	00	01

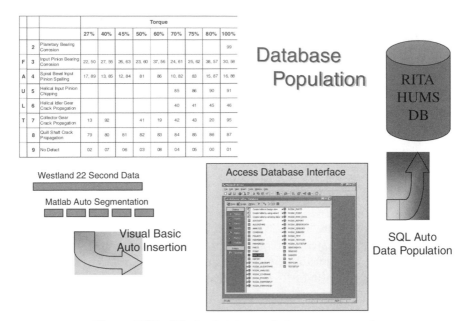

Westland 22 Second Data

Matlab Auto Segmentation

Visual Basic Auto Insertion

Access Database Interface

Database Population

RITA HUMS DB

SQL Auto Data Population

Figure 7.15 Westland data population process used.

drive or like storage media. The current method uses a directory structure that incorporates the platform, subsystem, data, and test ID information stored in the database:

../CH46-Chinook/Aft_Transmission/6_1_1992/Test_ID/Raw

This directory structure is consistent with the RITA HUMS fields in the database. This identification scheme may need to be revisited as new data are organized into the database. The ability to automatically link the appropriate baseline data with faulted data and a nominal threshold with each data file are two additional feature processing capabilities that aid metrics evaluation from the database.

MET Client Application and Function. As described earlier, one of the strengths of the MET is its provision of convenient access to a central repository of data and algorithms to its users. This is accomplished through the use of a native TCP/IP interface to Oracle. The client-side application interfaces with the user and directly computes the detection and diagnostic metrics from the server-returned feature results.

Metric Calculation Window. The current MET client application is in development and incorporates a tabbed GUI with five separate tabs/panels for a user to view. The primary panel to start the analysis is the metrics calculation, as shown in Fig. 7.16. Data-selection choices in this panel include algorithms, aircraft, systems, and components. These selection lists are dynamically populated from Oracle SQL queries to the RITA HUMS database. This screen also provides pictures of the system being analyzed, faults, and full test report when available in the database. Within this tab, users also can select and unselect individual faults, features, loads, and sensors. The hierarchical choices are dynamically generated from available data in the database. The user also chooses detection metrics, diagnostic metrics, or both, according to preference.

MET Detection Window. The MET detection window, as shown in Fig. 7.17, not only displays the feature discriminability plot at normal and faulted condition but also displays results by individual sensor or as composite. Sorts also can be done by load and fault conditions. Within this window, thresholds are determined from false-alarm criteria and the statistical confidence bounds provided by the user. The resulting probability of detection, false alarms, and accuracy are also calculated for the identified threshold.

MET Sensitivity Window. Users can view a receiver operating characteristic (ROC) curve in the MET sensitivity panel shown in Fig. 7.18. The ROC plots the probability of detection versus the probability of false alarm for a range of thresholds, as introduced in Sec. 7.4.2. In the figure, the dark shaded lines represent the theoretical values based on notional Rician distributions for various signal-to-noise ratios. The lighter shaded lines represent the results for the data being analyzed. The results are fictitious in the figure. Within this

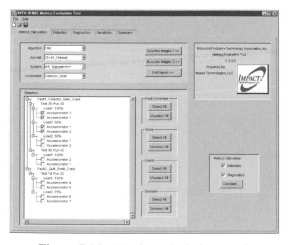

Figure 7.16 Metrics calculation panel.

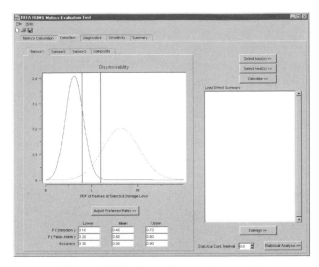

Figure 7.17   Detection results panel.

screen, the user can investigate the effects of additional noise on the data or change threshold settings to better characterize performance. Additionally, the actual predicted POD and POFA are displayed.

MET Summary Window. The summary panel in Fig. 7.19 is the final tab in the user interface. A summary of the metrics evaluation project is displayed.

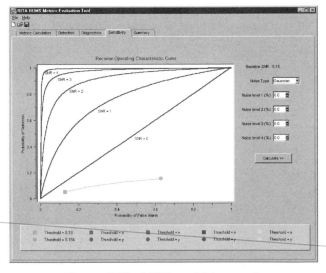

Figure 7.18   MET sensitivity panel.

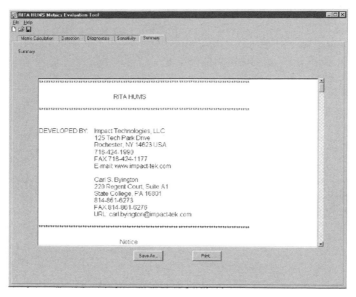

Figure 7.19   Summary panel.

Users are provided with "Save As" and "Print" buttons for their future review of the summary file. A user also can choose to save the entire project by using the icon at the top left in the File menu.

7.5   PROGNOSIS PERFORMANCE METRICS

There is no general agreement as to an appropriate and acceptable set of metrics that can be employed effectively to assess the technical performance of prognostic systems. We define below a possible class of metrics that stem primarily from the main objectives of prognosis and the uncertain (statistical) nature of prognostic algorithms.

A prognostic algorithm should yield two basic results: the predicted time of failure (the expected value based on historical data) and the confidence interval. Quantitative measures of the prediction performance should take both results into account; that is, we should consider the width of the confidence interval as well as the closeness of the predicted value to the actual value. The measures should be based on statistical methods, and many experiments will be carried out to test the algorithm. At one specific moment, during one test, the prognostic algorithm produces a predicted failure time and a confidence interval. We will measure the performance of the prognostic algorithm based on the results obtained from many experiments. The performance measures will evolve with time as more data are available, and the measures are expected to improve over time.

Two performance measures are considered to dominate the prognosis task: accuracy and precision. In general, *accuracy* measures the closeness of the predicted value to the actual value, and *precision* implies how close the predictions are bunched or clustered together. Precision in our context involves the confidence level as well as the distribution of the predictions during experiments.

7.5.1   Accuracy

Accuracy is a measure of how close a point estimate of failure time is to the actual failure time. Assuming that for the ith experiment, the actual and predicted failure times are $t_{af}(i)$ and $t_{pf}(i)$, respectively, then the accuracy of the prognostic algorithm at a specific predicting time t_p is defined as

$$\text{Accuracy}(t_p) = \frac{1}{N} \sum_{i=1}^{N} e^{D_i/D_0}$$

where $D_i = |t_{pf}(i) - t_{af}(i)|$ is the distance between the actual and predicted failure times, and D_0 is a normalizing factor, a constant whose value is based on the magnitude of the actual value in an application. N is the number of experiments. Note that the actual failure times for each experiment are (slightly) different owing to inherent system uncertainty. The exponential function is used here to give a smooth monotonically decreasing curve. The value of e^{D_i/D_0} decreases as D_i increases, and it is 1 when $D_i = 0$ and approaches 0 when D_i approaches infinity. In another words, the accuracy is the highest when the predicted value is the same as the actual value and decreases when the predicted value deviates from the actual value. The exponential function also has higher decreasing rate when D_i is closer to 0, which gives higher measurement sensitivity when $t_{pf}(i)$ is around $t_{af}(i)$, as in normal scenarios. The measurement sensitivity is very low when the predicted value deviates too much from the actual value.

Figure 7.20 illustrates the fault evolution and the prognosis, the actual and predicted failure times, and the prediction accuracy. Three evolution curves split from the predict time and represent three possible evolutions of the fault dimension. The accuracy is the highest (1) when the predicted failure time is equal to the actual failure time.

7.5.2   Precision

Precision is a measure of the narrowness of an interval in which the remaining life falls. Precision is defined on the basis of the variance of the predicted results for many experiments because the precision associated with a prediction algorithm is high if the predicted values are clustered together around the actual value and it is low if the predicted values are scattered over the

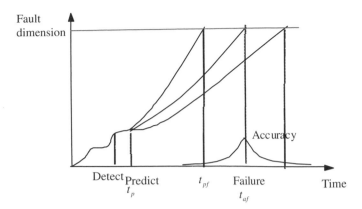

Figure 7.20 Predicted and actual failure times and the associated accuracy.

output range. We also will consider the average of the width of the confidence intervals of the prediction algorithm in the precision definition because a narrower confidence interval gives higher precision. Figure 7.21 shows how the distribution of the prediction is related to the precision. Figure 7.22 shows how the confidence bound of the prediction, R, is counted in the precision.

Defining the prediction error $E_i = t_{pf}(i) - t_{af}(i)$, we have the mean and the variance of the prediction error:

$$\overline{E} = \frac{1}{N} \sum_{i=1}^{N} E_i$$

$$\sigma^2 = \frac{1}{N} \sum_{i=1}^{N} (E_i - \overline{E})^2$$

And the formula for the precision at a specific prediction time t_p is given by

$$\text{Precision}(t_p) = \left(\frac{1}{N} \sum_{i=1}^{N} e^{-R_i/R_0} \right) \times e^{-\sigma^2/\sigma_0}$$

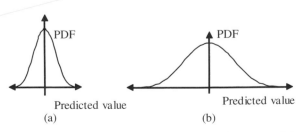

Figure 7.21 (*a*) Higher precision output, and (*b*) lower precision output.

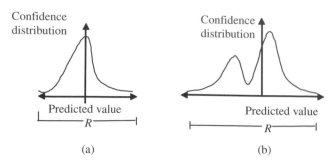

Confidence distribution / Predicted value / R / (a)

Confidence distribution / Predicted value / R / (b)

Figure 7.22   (*a*) Higher precision output, and (*b*) lower precision output.

where σ_0^2 and R_0 are normalizing factors depending on specific applications, and R_i is the confidence interval of the prediction for each experiment. Similarly, an exponential function is used here to define the relations between the prediction variance and the confidence interval and the precision. Precision also has a value ranging between 1 and 0, with 1 designating the higher precision and 0 the lowest. Narrower confidence bounds and more clustered predictions will give a higher precision value.

7.5.3 Timeliness of the Predicted Time to Failure

Performance metrics for prognostic algorithms, such as accuracy and precision, defined above, do not account for the impact of the timeliness of the predicted time to failure (TTF), that is, the PDF's relative position along the time axis with respect to the occurrence of the actual failure event. Consider, for example, the situations shown in Fig. 7.23.

In case (*a*), the failure happens to occur after the predicted time-to-failure PDF, whereas in case (*b*), the failure occurs during the prediction horizon. We must agree that case (*b*) is the desirable outcome if the prognostic algorithm employed is to be declared accurate. Now let us consider the situation shown in case (*c*). Here, the failure occurs much earlier than predicted. The consequences of such an event happening might be catastrophic if immediate action after fault detection is not taken, and indeed it is undesirable (Hannah, 2004). We may conclude that accuracy and precision metrics may need to be augmented with severity requirements as to the timing of the prediction vis-à-vis the actual occurrence of the failure event. We should be seeking, therefore, to achieve an accurate prediction, case (*b*), or at least to avoid a situation such as case (*c*).

7.5.4 Prediction Confidence Metrics

In defining and implementing performance metrics for prognostic models, two possible viewpoints/objectives must be considered:

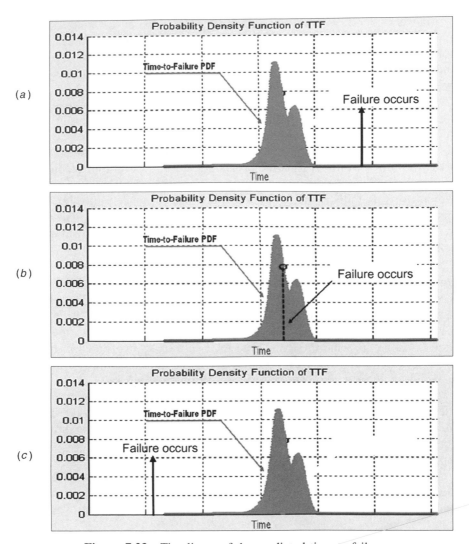

Figure 7.23 Timeliness of the predicted time to failure.

- *The maintainer's viewpoint.* We pose here the questions: When should I perform maintenance? What is the most appropriate time (given such constraints as availability of spares, etc.) to maintain critical equipment?
- *The field commander's viewpoint.* The questions posed here are: Are my assets available to perform the mission? What is the confidence level that critical assets will be available for the duration of the mission? Or, given a certain confidence level, what is the expected time to failure of a particular asset?

Answers to the maintainer's questions will require an estimate of the remaining useful lifetime of a failing component and assignment of uncertainty bounds to the trending curve that will provide him or her with the earliest and the latest (with increasing risk) time to perform maintenance and the associated risk factor when maintenance action is delayed.

Answers to the commander's questions will require an estimate of the confidence level assigned to completion of a mission considering mission duration, expected operating and environmental conditions, other contingencies, etc., or, given a desired confidence level, means to estimate the component's time to failure. Thus the prognostic curve (the output of prognostic routines) must be modulated by a time-dependent distribution (statistical/Weibull or possibilistic/fuzzy) that will provide the uncertainty bounds. An appropriate integral of such a distribution at each point in time will result in the confidence level committed to completing the mission or, given a desired confidence level, the time to failure of a failing component.

Uncertainty and Confidence. Assume that a prognostic algorithm predicts the following progression or evolution of a fault with the associated uncertainty bounds. At current time t_0 a fault has been detected and isolated, and the prognostic routine is predicting a mean time to failure of T_{fm}, the earliest time to failure T_{fe}, and the latest time to failure of T_{fl}, as shown in Fig. 7.24 (Barlas et al., 2003). The hazard line specifies the fault magnitude (dimension) at which the component ceases to be operational (failure).

Let us further assume that failure data are available to superimpose a distribution or distributions, as shown pictorially in Fig. 7.25. Strictly speaking, these "distributions" are either possibilistic (fuzzy) functions or probability density functions. Suppose further that the distribution at T_{fm} (crossing the hazard line) is as shown in Fig. 7.26.

As an example, let us assume that the preplanned mission requires the availability of the asset under consideration for time T'. The integral under the distribution curve from T' to infinity will give us an estimate of the

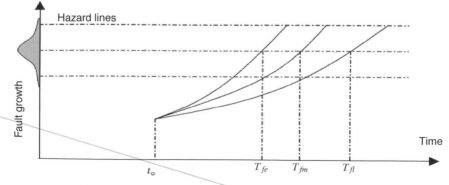

Figure 7.24 Confidence limits and uncertainty bounds.

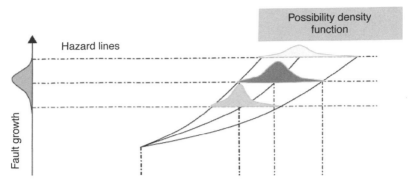

Figure 7.25 Possibility density functions for confidence bounds.

confidence level in terms of probability (or possibility), that is, how confident we are that the asset will not fail before the mission is completed. Consider next the same distribution. Now we specify a certain confidence level, say, 95 percent, and would like to find the time T'' that the component will remain operational before a complete failure. Integrating the distribution curve from T'' to infinity and setting the result equal to the confidence limit (95 percent in our example), we solve for T'', thus arriving at the length of time (starting at t_0) that the asset will be available within the desired confidence limit, as shown in Fig. 7.27. We view this procedure as a dynamic evolution of our estimates. That is, as more data become available and time marches on, new confidence limits are derived, and the uncertainty bounds shrink through appropriate learning routines. The procedure just outlined may lead eventually to specification of performance metrics for prognostic systems. Such additional metrics may refer to the reliability of the confidence limits, risk assessment, safety specs, and others.

Example 7.2 Bearing Failure and Remaining Useful Life. Assume that we have 2300 bearings and that we perform failure analysis for two random variables, crack size for which the bearing fails (hazard line) and the time it

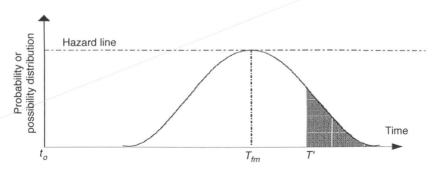

Figure 7.26 Distribution at the hazard line.

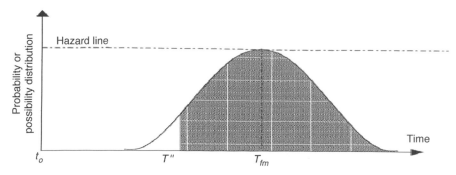

Figure 7.27 Distribution at hazard line.

takes for the bearing to fail. For the sake of simplicity, we assume that the crack size distribution is discrete and triangular in shape. We assume a continuous normal distribution in time. Results of the tests are tabulated in Table 7.9 and shown in Fig. 7.28.

For the second random variable, we will assume linear growth in crack size with time, starting at 0. The three radiating lighter shaded lines in the graph shown in Fig. 7.29 represent three crack-size growth rates. The horizontal dotted lines represent the five discrete distribution points of bearing failures.

From the probability distribution curve shown in Fig. 7.30, we can estimate the confidence level for a time T', that is, how confident we are that the bearing will not fail for, say, 90 minutes. The shaded portion shows the confidence estimate. From Fig. 7.29, say, 2200 of 2300 bearings failed within 90 minutes. Thus our confidence that bearings will not fail is $100/2300 \sim 0.043$, or 4.3 percent, as shown by the shaded portion in Fig. 7.30.

For a given confidence level, say, 90 percent, the estimate of time T'' corresponds to the shaded portion shown in Fig. 7.31. From the numbers above, 90 percent is of $2300 \sim 2100$. Therefore, we can be 90 percent confident that at least 2100 bearings will not fail. As can be seen from the figure, T'' is around 40 minutes.

TABLE 7.9   Test Results of Bearing Failures

No. of Bearings Failed	Crack Size (inches)
300	0.7″
500	0.85″
700	1.0″
500	1.15″
300	1.3″

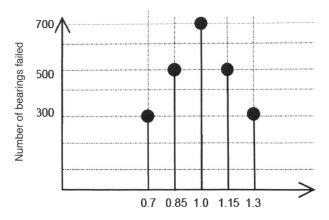

Figure 7.28   Distribution of bearing failures.

7.5.5   Other Prognosis Metrics

Additional performance metrics may be defined that find application in specific cases. For example, *similarity* compares more than one predicted time series against the real series from a starting point to a certain future point in time. This measurement can be computed as

$$
\text{Similarity}(x, y) = \sum_{i=1}^{n} \left(1 - \frac{|x_i - y_i|}{\max_i - \min_i} \right)
$$

where x_i and y_i are two ith elements of two different time series, and $\max_i$ and $\min_i$ are the maximum and minimum of all ith elements (Fig. 7.32).

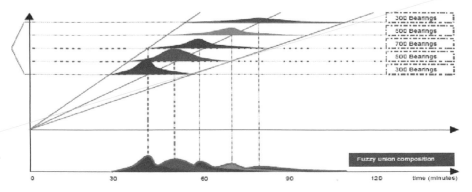

Figure 7.29   Distributions of bearing failures and fuzzy composition.

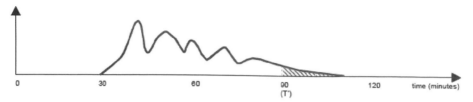

Figure 7.30   Confidence calculations for a given time frame.

Sensitivity.   Sensitivity measures how sensitive the prognostic algorithm is to input changes or external disturbances. It is defined as

$$SN = \frac{\displaystyle\sum_{i=1}^{N} \frac{\Delta_i^{\text{Out}}}{\Delta_i^{\text{In}}}}{N}$$

where Δ^{Out} is the distance measure of two successive outputs, and Δ^{In} is the distance measure of two successive inputs.

Prediction-to-Failure Times.   The mean of the prediction-to-failure time is calculated as

$$E\{t_{pf}\} = \frac{1}{N} \sum_{i=1}^{N} t_{pf}(i)$$

where $t_{pf}(i)$ denotes the prediction-to-failure time for the ith experiment, and N is the number of experiments. The standard deviation of the prediction-to-failure time is calculated as

$$S\{t_{pf}\} = \left[\frac{1}{N} \sum_{i=1}^{N} (t_{pf}(i) - E\{t_{pf}\})^2 \right]^{1/2}$$

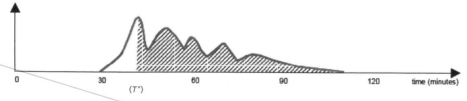

Figure 7.31   Confidence calculation for a given confidence level (90 percent).

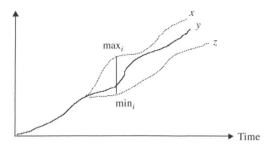

Figure 7.32   Determining similarity.

7.6   DIAGNOSIS AND PROGNOSIS EFFECTIVENESS METRICS

Technology performance and effectiveness metrics associated with CBM/ PHM algorithms for fault diagnosis and prognosis that capture system implementation and cost issues may be grouped as shown in Fig. 7.33.

The benefits achieved through accurate detection, fault isolation, and prediction of critical failure modes are weighed against the costs associated with false alarms, inaccurate diagnoses/prognoses, licensing costs, and resource requirements of implementing and operating specific techniques. The simplified cost function shown below states the technical value provided by a specific diagnostic or prognostic technology for a particular system. The value of a CBM/PHM technology in a particular application is the summation of the benefits it provides over all the failure modes that it can diagnose or

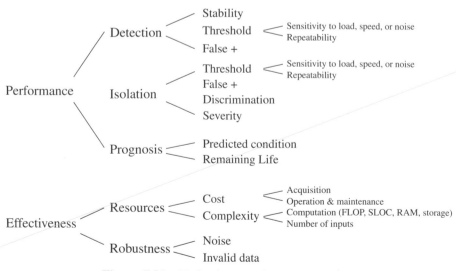

Figure 7.33   Technology performance metrics.

prognose less the implementation cost, operation and maintenance cost, and consequential cost of incorrect assessments as stated in the "total value" equation layout as savings costs.

$$\text{Technical value} = P_f(D\alpha + I\beta) - (1 - P_f)(P_D\phi + P_I\theta)$$

where P_f = probability (time-based) of occurrence for a failure mode
D = overall detection confidence metric score
α = savings realized by detecting a fault prior to failure
I = overall isolation confidence metric score
β = savings realized through automated isolation of a fault
P_D = false positive detection metric score
ϕ = cost associated with a false positive detection
P_I = false-positive isolation metric score
θ = cost associated with a false positive isolation

Thus

$$\text{Total value} = \sum_{\text{Failure modes}} \text{Technical Value}_i - A - O - (1 - P_c)\delta$$

where A = acquisition and implementation cost
O = life-cycle operation and maintenance cost
P_c = computer resource requirement score
δ = cost of a standard computer system

For all the metrics, a low score indicates an undesirable result, and a high score indicates a desirable one. For example, a high computational resource requirement score is awarded to CBM/PHM algorithms that use a small portion of the computer's resources.

7.7   COMPLEXITY/COST-BENEFIT ANALYSIS OF CBM/PHM SYSTEMS

When considering the economic feasibility of prognostic systems, it is necessary to factor in the impact of the integrated CBM/PHM system over the life cycle of the hybrid hardware/software complement that is especially designed to perform the dedicated prognostic task. Unfortunately, data are not yet available from existing systems to permit a thorough analysis of the economic benefits of CBM/PHM. Related technologies (sensing, diagnosis, prognosis, etc.) have been developed only recently, and very few have been implemented in actual systems and at a scale that will result in a historical database. It is reasonable, therefore, that a number of recent studies from the military and industrial sectors are based on concepts of "avoided cost." A comprehensive performance metric must include a complexity component as well as appropriate cost-benefit ones. Thus the overall performance must be a combination of

1. Complexity measure:

$$\text{Complexity} = E\left\{\frac{\text{computation time}}{\text{time to failure}}\right\} = E\left\{\frac{t_p - t_d}{t_{pf}}\right\}$$

where t_p and t_d are the moments when the detection of the fault happens and when the remaining useful lifetime is predicted.
2. Cost-benefit analysis:
   - Frequency of maintenance
   - Downtime for maintenance
   - Dollar cost
   - Etc.
3. Overall performance:

$$\text{Overall performance} = w_1 \cdot \text{accuracy}$$
$$+ w_2 \cdot \text{complexity} + w_3 \cdot \text{cost} + \cdots$$

A possible approach to the cost-benefit component of the overall metric could proceed along the following steps:

1. *Establish baseline condition.* If breakdown maintenance or time-based planned preventive maintenance is followed, estimate cost of such maintenance practices and their impact on operations; frequency of maintenance, downtime for maintenance, dollar cost, etc. should be available from maintenance logs. Provide qualitative estimates of impact on other operations by interviewing key personnel.
2. Condition-based maintenance (CBM) will assist to reduce breakdown maintenance (BM) costs. A good percentage of BM costs may be counted as CBM benefits.
3. If time-based planned maintenance is practiced, estimate how many of these maintenance events may be avoided by thresholding failure probabilities of critical components and computing approximate times to failure. The cost of such avoided maintenance events is counted as benefit to CBM.
4. *Intangible benefits.* The impact of breakdown maintenance or extensive breakdown on system operations can only be assessed via interviewing

affected personnel and assigning an index to the impact (severe, moderate, etc.).

5. Estimate roughly the projected cost of CBM, that is, dollar cost of instrumentation, computing, etc. and cost of personnel training, installation, and maintenance of the monitoring and analysis equipment.

6. Aggregate life-cycle costs and benefits from the information detailed above.

Example 7.3   CINCLANTFLT Complexity/Cost-Benefit Analysis of the Value of Prognosis. A recent study by CINCLANTFLT, the U.S. Atlantic Fleet, is addressing the question: What is the value of prognosis? Its findings are summarized below.

1. Notional development and implementation program for predictive CBM based on CINCLANTFLT I&D maintenance cost savings

2. Assumptions
   - CINCLANTFLT annual $2.6B (FY96$) I&D maintenance cost.
   - Fully integrated CBM yields 30 percent reduction.
   - Full realization occurs in 2017, S&T sunk cost included.
   - Full implementation costs 1 percent of asset acquisition cost.
   - IT 21 or equivalent in place prior to CBM technology.

3. Financial factors
   - Inflation rate: 4 percent
   - Investment hurdle rate: 10 percent
   - Technology maintenance cost: 10 percent of installed cost

4. Financial metrics: 15 year   20 year
   - NPV $337M $1,306M
   - IRR   22 percent   30 percent

7.8   REFERENCES

M. Ashby and W. Scheuren, "Intelligent Maintenance Advisor for Turbine Engines (IMATE)," *Proceedings of the IEEE Aerospace Conference,* 2000, Big Sky, Montana, IEEE.

I. Barlas, G. Zhang, N. Propes, and G. Vachtsevanos, "Confidence Metrics and UnCertainty Management in Prognosis," *MARCON,* Knoxville, TN, May, 2003.

P. Bonissone and K. Goebel, "Soft Computing Techniques for Diagnostics and Prognostics," *AAAI Spring Symposium on Artificial Intelligence in Equipment Maintenance and Support,* 1999, Palo Alto, CA: AAAI Press.

B. Cameron, "Final Report on CH46 AFT Transmission Seeded Fault Testing," Westland Helicopter Mechanical Research Paper RP907, September 1993.

Dowling, Boeing Company, "HUMS METRICS, Monitoring the Monitor," Rotorcraft Industry Technology Association Final Report, January 3, 2000.

B. Hannah, "Embedded Diagnostics and Prognostics Synchronization (EDAPS) Transitioning to a Common Logistics Operating Environment," U.S. Army Logistics Transformation Agency Materiel Logistics Division, 2004.

J. Keller and P. Grabill, "Vibration Monitoring of a UH-60A Main Transmission Planetary Carrier Fault," American Helicopter Society 59th Annual Forum, Phoenix, AZ, May 6–8, 2003.

N. Khiripet and G. Vachtsevanos, "Machine Failure Prognosis Using a Confidence Prediction Neural Network," *Proceedings of 8th International Conference on Advances in Communications and Control,* Crete, Greece, 2001, Optimization Software, Inc.

H. Liu and H. Motoda, *Feature Selection for Knowledge Discovery and Data Mining.* Boston: Kluwer Academic, 1998.

P. D. McFadden and J. D. Smith, "An Explanation for the Asymmetry of the Modulation Sidebands about the Tooth Meshing Frequency in Epicyclic Gear Vibration," *Proceedings of the Institution of Mechanical Engineers, Part C: Mechanical Engineering Science* **199**(1): 65–70, 1985.

R. F. Orsagh, M. J. Roemer, C. Savage, and K. McClintic, "Development of Metrics for Mechanical Diagnostic Technique Qualification and Validation", COMADEM Conference, Houston, TX, December 2000.

A. Papoulis, *Random Variables and Stochastic Processes,* 2d ed. New York: McGraw-Hill, 1984.

M. J. Roemer, G. J. Kacprzynski, and R. F. Orsagh, "Assessment of Data and Knowledge Fusion Strategies for Prognostics and Health Management," IEEE Aerospace Conference, Big Sky, MT, March 2001.

Rozak, RITA HUMS Diagnostic Database, RHDDB 00-06-1.1(4), Draft 4.0, Sikorsky Aircraft Corporation, March 2001.

CHAPTER 8

LOGISTICS: SUPPORT OF THE SYSTEM IN OPERATION

GARY O'NEILL
Co-Director of the Logistics and Maintenance Applied Research Center, Georgia Tech Research Institute

JOHN REIMAN
Georgia Institute of Technology

8.1 INTRODUCTION

Product support and *logistics* are terms used to describe all the processes, personnel, equipment, and facilities necessary to support a system during its intended life. In the commercial sector, the support processes, personnel, and resources used to provide sustaining support for a system are known as *customer support* or *product support*. In the U.S. defense industry and the military services, these are commonly referred to as *logistics*. We use the terms *product support* and *logistics* interchangeably.

Every system has a purpose, such as producing power, transportation, or the means to manufacture goods. Each system has design objectives of the amount of useful time it should operate, commonly known as *up-time,* and cost targets that define economic operation. Traditionally, systems have been built with established periods of preventive (scheduled) maintenance that require the system to be either shut down or operated in "standby." This period of *downtime* affects the system economics or profitability of a commercial organization, and for a military organization, it represents time "not ready" for its mission. For both the private and public sectors, there are increasing pressures to minimize downtime and operate as cost-effectively as possible.

To do so, the operator must have a continuous and accurate assessment of system performance and awareness as early as possible of the presence of faults that will degrade system performance. This knowledge makes possible the use of condition-based maintenance (CBM), or maintenance tasks that sustain the system and its components without arbitrary constraints of calendar time or operating cycles. This approach holds the promise of increased up-time and lower cost because unnecessary preventive maintenance tasks are eliminated.

8.2 PRODUCT-SUPPORT ARCHITECTURE, KNOWLEDGE BASE, AND METHODS FOR CBM

Product support is a system of systems that delivers the necessary information, material, and skilled labor to sustain the operations of a system and its constituent components. In most cases, the system being supported is also a system of systems. Therefore, for effective product support with CBM as its basis, an architecture is needed that develops data from the components and subsystems of the supported system (i.e., an aircraft or industrial process), and that data must be shared across the disciplines that provide support: engineering, material distribution, maintenance, and operations (Fig. 8.1). A notional architecture is shown in Fig. 8.2, and has three zones of action: at the system/equipment, at the support base for the equipment, and at the basic design engineering/product support center (either the OEM or the cognizant technical organization charged with its life-cycle support). The architecture is meant to support decision processes at each of the three zones of action, typically with a deeper and broader capability for system maintenance and modification as you go further out in the zones. Systems analysis and comprehensive failure modes effects criticality analysis (FMECA) determine the amount and extent of actions appropriate in each zone, with the intent to maximize the ability to cost-effectively sustain the system and provide the optimal level of up time.

The knowledge base (Fig. 8.3) that supports the architecture comes from three key sources:

- *Design engineering.* The initial data and information regarding the system and its components, including failure modes, failure models

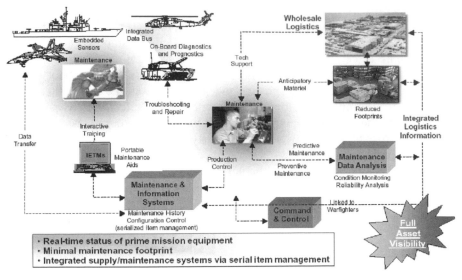

Figure 8.1   Maintenance-centric logistics support for the future.

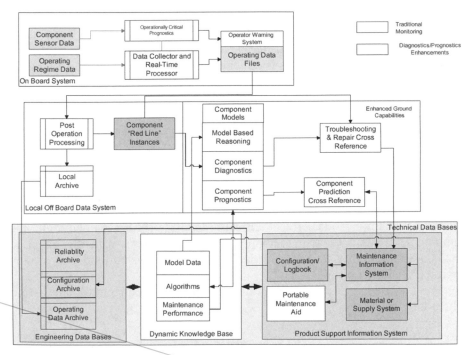

Figure 8.2   Notional data architecture for CBM.

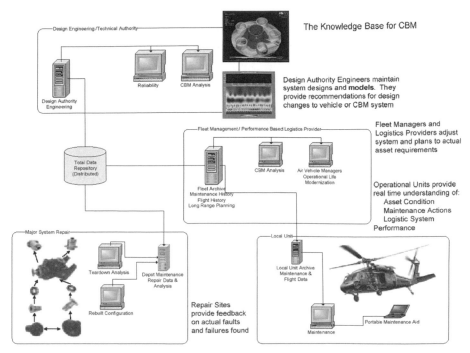

Figure 8.3 The CBM knowledge base.

(behavior of the faults over time), operating mode behavior, and data from development and operational testing.
- *Operations.* The data gathered as the system is operated, including operational data recorded (sensor data and operating modes, such as speed, rpm, etc), maintenance history, and material consumption history.
- *Teardown analysis.* Detailed observations of failed equipment that is undergoing repair at service centers or repair providers.

Developing the ability to identify and characterize faults (diagnosis) and predict the system or component's useful life (prognosis) is only useful if those facts are turned into action. The first objective of any diagnostic or prognostic system is to provide the operator or maintenance technician with the facts: There is a fault in a specific component, and within a given period of time, the system will degrade beyond useful limits. But there is much more to consider because those facts are important to the full spectrum of activities in logistics or product support: getting the right information, material, equipment, and skilled personnel to the right place at the right time to sustain the system.

We have described the elements needed to develop the infrastructure and data management architecture to effectively assess the condition of a complex system and turn that assessment into action by the operator and maintenance staff. That information is also extremely important to the rest of the enterprise that operates the system. For example, having accurate predictions of system condition affects

- *The operational planning and scheduling of the system being operated.* Can it operate long enough to complete its next scheduled mission? Can the industrial equipment continue until all the orders planned for the week are complete?
- *The procurement of spare parts and material needed to restore the system to normal operations.* Is the notification of predicted useful life remaining sufficient to allow order and receipt of parts? If not, how much inventory, and what parts should be held in local inventory to guarantee system operation?
- *Training of system operators.* Is the cause of a recent number of failures the result of improper operation or servicing? What new procedures should be in place to eliminate a specific failure mode?
- *Design of the system.* Has the expected life of recently failed components followed the expectations of design engineers? Has the operation of the system followed their anticipated usage parameters? Does the component or system require redesign?
- *Financial performance.* Is the system and its expected failures and out-of-service time following the estimates for system operation? Is the system operating economically?

Without a structured and effective means to assess the condition of a component or system, the support processes associated with sustaining the system are largely incapable of predicting future requirements. The current mode for product-support teams for major systems is to gather as much information as possible about what has happened and any available information about what is scheduled to happen in the near future. Using these data from various kinds of enterprise or maintenance information systems, the support team uses statistical analysis and data mining to discern any patterns regarding reported failures and maintenance actions. They use this information as the primary means to drive them to potential root causes for system degradation. Then they attempt to isolate and mitigate the root causes through a variety of means: new processes, training, more spares inventory, and as a last resort, design change. All these changes require time, money, and action by a multidisciplined group of people across multiple organizations so that improvements are incremental and generally slow moving.

8.3   PRODUCT SUPPORT WITHOUT CBM

The best product-support systems in current practice often lack effective diagnosis and prognosis and therefore emphasize the use of reliability-centered maintenance (RCM) (Moubray, 2001; Narayan, 2004). RCM uses extensive statistical analysis of the system components (or ones similar in design) to establish a set of maintenance processes and inspections that minimize unscheduled, or corrective, maintenance to restore system operation. RCM uses a well-established set of procedures to define optimal inspection and maintenance intervals based on an assumed set of operating parameters. Because of its reliance on average life expectancies and "average" or static estimates of operating profiles, RCM can perform maintenance in some cases before it is needed. This loss of system operation in a conservative, scheduled maintenance environment has wide acceptance and practice because it does reduce unscheduled, emergency repairs and is beneficial to planning and scheduling. RCM is dependent on data collection and analysis to remain in synchronization with the system in operation as it ages and potentially as its usage changes over time.

The current state of the practice of product support has several major impediments associated with collection of data. First, there is often a significant lag between data collection (which is becoming more automated but is often manual) and analysis, which reduces the support team's effectiveness at tracing root causes. Second, much of the data collected are flawed because accuracy in data collection is nearly always secondary to repairing the system and restoring normal operation. Third, system support is often a fragmented network that combines the system operators, its maintainers (often a third-party provider of labor, spare parts, or both), and the system's original design and support team. This fragmentation tends to degrade information sharing across the boundaries, if any sharing occurs at all. Thus the data used to base decisions regarding product support (especially within the framework of RCM) often are flawed and can lead to ineffective action.

In summary, RCM and similar maintenance programs use a calendar- or event-based series of tasks to define the maintenance plan for the system being operated. The tasks are scheduled and executed based on the number of operating hours or cycles accomplished by the system, regardless of the system's condition. Scheduling the tasks is normally done to optimize downtime and cost (Helm et al., 2002). This approach works well when the system being operated has relatively few operating modes and its failure mechanisms are well understood. It is also necessary for systems that do not have effective sensors or data collection that enable the system's condition to be well understood.

8.4 PRODUCT SUPPORT WITH CBM

Establishing a reliable and robust network of sensors, data collection, and data analysis can create the means to change from a calendar- or event-based maintenance program to a maintenance approach that continuously assesses the condition of the system and makes decisions related to maintenance and system shutdown based on an optimization of the system's condition and its operational requirements. This approach is often referred to as *condition-based maintenance* (CBM).

CBM can be an effective maintenance approach, but it must be tested for effectiveness before it is adopted as the established method for a specific system. The cost and complexity of sensors, signal processing, and data management architecture must be compared against the benefits in system operation (or up-time) and the specific response times allowed by the diagnosis and prognosis of the various system components and their failure modes. Some systems may be stable and durable enough that only a handful of sensors is needed to adequately assess the useful life of the system and its components, and therefore, CBM may not demand many resources. Other systems, such as fighter aircraft, may be so highly complex with many interrelated failure modes that the cost of CBM is large and only makes economic sense through its reduction in costly inventory or its ability to have a much smaller-sized fleet of aircraft. In any case, selection of CBM as the product-support approach should be based on positive economic analysis and ability to meet customer requirements.

8.4.1   How Condition-Based Maintenance Changes System Operation and Product Support

Establishing CBM as the maintenance approach for a system has several immediate effects on operation and maintenance of the system. Because CBM is affected by the failure modes of the system and the way that different operating modes affect those failure modes, CBM can influence the operational scheduling of a system. For example, if a vehicle has a flaw in a transmission that is accelerated by high power settings or rapid power cycling, it is possible to extend the useful life remaining by assigning it to missions that minimize those accelerants. This gives the operator an option of extending the useful life by changing the way the system is employed over the short run until the useful life has expended to a point that demands corrective action. Having accurate diagnosis/prognosis also can be used to modify operator behavior. If rapid power changes accelerate damage accumulation, mandating smooth acceleration and minimal power changes during system use can help to extend useful life, and the data collected by the diagnostic/prognostic system can verify correct operation of the vehicle.

Providing maintenance technicians with accurate predictions of useful life remaining and associated maintenance actions transforms maintenance execution into a forward-looking optimization of resources and material. Instead of fixing things after finding out they are defective and racing to complete the corrective tasks with whatever resources are at hand, CBM allows maintenance tasks to be scheduled for the optimal time that matches operational requirements with maintenance personnel, parts, and tool availability. "Maintenance control" becomes less a reactive process and more of an optimization of resources, including the resources of time and money. This transformation is an important point to consider because current maintenance information systems have limited optimization capability. The good news is that such routines are not hard to find (see Web sites: *www.cogz.com/*, *www. maintstar.com/*, and www.keeptrak.com/).

CBM changes the supply-chain management processes that support system operation in significant ways. Accurate and reliable predictions of useful life remaining (ULR) can be used as inputs to forecast algorithms in enterprise requirements planning (ERP) systems by inserting the forecast ULR as the lead time required for the degrading component and its dependent material needs. Coupled with serial number tracking and other data, supply-chain forecast algorithms can break away from forecasts based on average monthly or quarterly demand and instead focus on targeted material requirements by operating location or unit and roll these requirements up to a forecast of total requirements. Having more specific data on how components fail and how failure is related to operational tempo and use can help SCM algorithms to reduce the amount of inventory in the pipeline without sacrificing material availability. For components that are repaired and returned to inventory (a large portion of defense-related material needs), linking the specific failure mode data to the serial-numbered component also benefits the repair process and can contribute to faster repair turnaround times and more effective piece-part consumption at the designated repair point.

CBM transforms engineering and technical support of the system by providing design and support engineers with far more accurate data regarding the failure modes of the system. Accurate data and systemic sharing of information is made possible by the CBM data management architecture. This environment of improved and nearly real-time engineering analysis of the system can shorten the time span for maintenance process evaluation, material substitution, or system redesign.

Example 8.1   Cost Analysis of Different Maintenance Schemes for the Intermediate Gearbox on a Military Aircraft. Table 8.1 shows an example for cost analysis and a comparison of maintenance and inspection methods for a typical component for a military aircraft. The component, the intermediate gearbox, has a maintenance replacement cost of approximately $39,800 dollars based on its reorder cost (assumes that the defective gearbox it replaces is returned for a carcass credit), maintenance labor, and system checks necessary. To eliminate premature failures, there are three possible methods: (A) disassembly and physical/visual checks, (B) period vibration sampling and analysis during ground operations, and (C) use of installed HUMS to collect data and analyze the condition of the aircraft. Table 8.2 shows a comparison of the costs and readiness impact of the various inspection methods, based on inspection frequencies of 50-, 100-, 200-, or 300-hour inspection intervals. It is clear that HUMS can provide substantially better up-time and lower cost to perform the inspections.

TABLE 8.1   Example of a Component Cost Model

H60 intermediate gearbox
Replacement cost
   Material $37,500
   Labor $800 (40 mhrs at $20/hr both grnd crew + flt crew)
   Other $1,500 ground turn costs (at nominal $1500/flt hr)
   Total $39,800
   Elapsed time   12 Hour average
Inspection method
   A Disassembly/physical check
           Material $200
           Labor $400 (20 hours at $20/hr)
           Other $1,500 ground turn costs (at nominal $1500/flt hr)
           Total $2,100
           Time   8 Hour average
   B Special ground turn for vibration analysis
           Material $20
           Labor $120 (6 mhrs: ground crew + flt crew)
           Other $1,550 ground turn costs (at nominal $1500/flt hr) + cost
                        of .5 hr of AED/AMC support
           Total $1,690
           Time   3 hours
   C HUMS analysis
           Material $10 Incidentals for data system
           Labor $10 (.5 maint mhr)
           Other $30 Amortized cost of infrastructure for UH-60 fleet per
                        acft flt hr
           Total $50
           Time   1 hour

TABLE 8.2   Comparison of Inspection Methods

Inspection	Frequency Flight Hours	Annual Cost	Annual Downtime	Availability	Cost Ratio
A	50	25,200	96	0.98904	0.633
	100	12,600	48	0.99452	0.317
	200	6,300	24	0.99726	0.158
	300	4,200	16	0.99817	0.106
B	50	20,280	36	0.99589	0.510
	100	10,140	18	0.99795	0.255
	200	5,070	9	0.99897	0.127
	300	3,380	6	0.99932	0.085
C	50	600	12	0.99863	0.015
	100	300	6	0.99932	0.008
	200	150	3	0.99966	0.004
	300	100	2	0.99977	0.003

8.5   MAINTENANCE SCHEDULING STRATEGIES

In logistics support operations, maximizing equipment up-time, minimizing time to repair (MTTR), and optimizing maintenance costs are the major objectives. Other objectives and constraints, such as minimizing logistics footprint or transportation, may be included in the optimization scheme if dictated by specific system requirements and the logistics network. Logistics support as currently practiced by military and commercial enterprises does not fully implement CBM-related concepts. For example, many organizations do not allow ordering parts prior to removal of defective ones as a means to control cost and maintain control over the retrograde, or defective parts, that can be repaired. Those that do allow ordering in advance of replacement typically place such orders at a much lower priority, effectively impeding the delivery at the time of need. With CBM entering the maintenance picture, both the maintainer's and the logistician's objective are altered significantly. The logistician now must use the prediction of asset condition to base the logistic planning to support maintenance rather than using updated historical demands for material. The maintainer must now look for optimal ways to combine maintenance tasks to support system restoration and look beyond the immediate task to restore a failing component and include other tasks that should be done while the system is inactive for repair. This shift requires new and innovative scheduling algorithms that optimize the use of resources and time to restore the right number of assets needed for operations. In this subsection we describe some emerging methods and algorithms designed to enable maintenance scheduling to optimize maintenance execution by taking the new capabilities of prediction of remaining useful life from the prognostic module of the CBM architecture.

The logistician must be able to support the specific maintenance and repair tasks that will restore normal operation of the system after a fault has been detected, isolated, and the remaining useful life predicted. At the same time, the maintainer must be able to schedule the task for execution by technicians who are properly trained and ready for the task. The prognostic algorithm provides a PDF, or a representation of the probabilistic distribution of the time to failure. In the CBM data architecture, additional data are available regarding the availability of parts, technical data, maintenance personnel, and special equipment and facilities that are required to support the maintenance task. The objective, within the CBM framework, is to optimize equipment up-time and availability, minimize maintenance costs, and preserving the system's operational integrity. The maintenance scheduling problem is viewed, therefore, as a constrained optimization problem. These types of problems have been researched widely, and a number of effective optimization tools may be called on to provide a viable and robust solution for the emerging need of maintenance optimization. Potential challenges include the probabilistic/statistical nature of the input parameters (PDF of time to failure, probability of parts availability, and labor), the multiobjective and possibly conflicting criteria, and the dynamic and evolving nature of the components entering into the optimization scheme. The resulting objective function may not be smooth, differentiable, or unimodal. Multiple constraints must be accounted for and quantified, which is often difficult.

The logistician's objectives can be described as hybrid, consisting of easily quantitative terms, for example, MTTR or cost, and some more qualitative, for example, repair quality, which can be more difficult to express quantitatively. This hybrid nature of the maintenance task scheduling problem can call for unconventional methods that are capable of addressing the complexities of this problem. Optimization routines for fairly simple optimizations abound in the literature, and a large number are available commercially. We provide a simple example below.

For the problem at higher levels of complexity (i.e., more constraints and more complex input relationships) an emerging class of algorithms known as *genetic algorithms* (GAs) has shown capability to handle multimodal objective functions that may not be differentiable and are viable optimizers to address the scheduling problem. Although they are effective for complex optimization problems because they search for a global maximum or minimum, they are also known for the considerable length of time required to arrive at a solution. They differ from brute-force search techniques by selectively rewarding the most promising solution candidates, thus speeding the algorithm's convergence toward a global optimum. In the search process, the quality of a member of the populations is evaluated in terms of a fitness function. An intelligent selection process then is used to decide which members of the current population will survive to the next generation. Such genetic operators as crossover and mutation are employed to combine or change members of the population and generate new ones (see *http://lancet.mit.edu/ga/* for a library of GA components; see *http://en.wikipedia.org/wiki/Genetic_algorithm* for a concise, understandable discussion of GAs). They are not, therefore, suitable for real-time applications where the speed of response is an essential requirement.

8.6 A SIMPLE EXAMPLE

To illustrate the issues related to optimizing maintenance, let us consider the example of a helicopter requiring the replacement of a line-replaceable unit (LRU) or component of the helicopter drive system that the maintenance personnel at the flight line can replace. To accomplish the repair, the flight line needs the correct replacement parts, technical data (providing the repair steps), tools and special equipment, trained and available personnel, and a designated place or maintenance "spot" that facilitates the repair.

Without condition-based maintenance, the logistician and maintenance supervisor are reacting to an unplanned requirement to replace the component, and they typically patch together a suboptimized set of activities that prepares the helicopter for maintenance; gather the necessary replacement parts; and use the available personnel, equipment, and maintenance spot to accomplish the repair. Often the repair task steps are interrupted while some other resource is obtained, lengthening the time to repair and providing greater opportunity for error

With CBM, the logistician and maintenance supervisor are lining up the resources to choose the most opportune time for maintenance, ensuring that all the resources are there in synchronization with the maintenance task execution. This leads to more effective repair and shorter repair times.

In this example, we assume the existence of a CBM data management architecture that can provide a high-confidence prediction of remaining useful life and that the prediction time span enables the logistics and maintenance portions of the CBM data system to respond to the prediction of remaining useful life. We add several realistic constraints to the scheduling problem to be optimized:

- Maintenance must be scheduled such that we can achieve a 90 percent certainty that maintenance will be completed successfully in order to have the helicopter available at a specified time (for operational scheduling).
- There must be at least two available maintenance "spots" or facilities to ensure that the maintenance can be performed at the specified time.
- Personnel availability places a constraint on how many people can be assigned a maintenance task at any given point in time.

The factors incorporated into the optimization process are shown in Figs. 8.4 through Fig. 8.8. For this simple case we use a generic time scale that does not differentiate between operating hours and calendar time, but obvi-

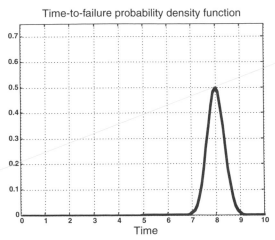

Figure 8.4   Estimated failure probability density function.

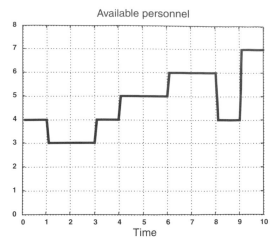

Figure 8.5 Current estimate of the available personnel.

ously, for the real-world case, the two different timelines need to be correlated and synchronized. The time-to-failure probability density function shown in Fig. 8.4 is provided by the diagnostic and prognostic routines and shows that failure is predicted to occur between $t = 7$ and $t = 9$. The estimated time to failure is in this example set to the mean at $t = 8$. The hard schedule requirement for the helicopter up-time is set to $t = 10$ for the example. The available maintenance personnel and workbenches are shown in Figs. 8.5 and 8.7, respectively. The graphs depict the current availability; hence it is likely that the curve changes in the future depending on the demand for other mainte-

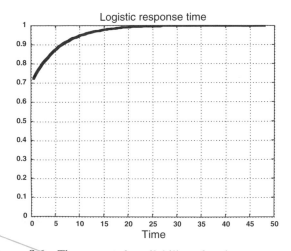

Figure 8.6 The expected availability of maintenance parts.

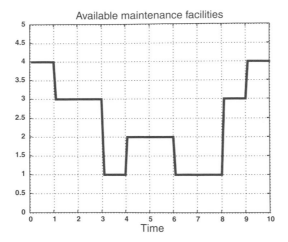

Figure 8.7 The expected number of available workbenches.

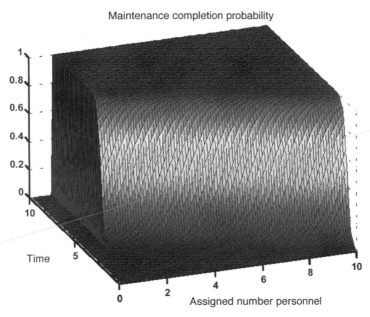

Figure 8.8 The probability that a task is completed as a function of time and the assigned personnel.

nance on this helicopter or others that might be assigned to that unit. The logistic response time curve shown in Fig. 8.6 expresses the probability that the parts needed to complete the maintenance task are available at planeside when maintenance is performed. This distribution shows a 70 percent chance that all parts are available on demand (without prior notice) and that 85 percent are available with 5 time units of lead time, increasing to 95 percent at $t = 10$.

In this example we are optimizing cost of repair. We simplify the cost function to contain three terms: the first related to the availability of parts, the second related to the number of personnel to accomplish the repair, and the last dealing with the timing of the repair action and its relationship to the probability that failure will occur. This conforms to the well-known mantra of the maintainer: "Give me the parts, and I can fix the problem." In this day of multiple modes of transportation, the cost of parts is governed by their availability, in that when they are not available, premium shipping is used to bring them to the point of need.

For this example we can control or vary two elements: the number of people assigned to the task and the time at which the maintenance is to be performed. We assume that the time to accomplish the repair is defined by the constraint (90 percent confidence that it is completed in time for helicopter assignment at $t = 10$).

The cost functional to be optimized is expressed by

$$J(n, t_f) = c_1 \cdot (1 - \text{logistic response time}) + c_2 \cdot n$$

$$+ \int_0^{t_f} (c_3 \cdot \text{PDF}_{\text{time to failure}} - c_4)\, dt$$

where c_1, c_2, c_3, and c_4 are constants used to weigh the importance of each term. For instance, c_2 must be increased if one would like to reduce the number of people assigned to a certain task. The controllable quantities are the time at which the maintenance is performed (t_f) and the personnel assigned to the maintenance task (n).

For this example we chose arbitrary values of the constants, as shown in Table 8.3. It should be noted that c_3 is much larger than c_4 because the product

TABLE 8.3   Arbitrary Values of the Constants

Variable	Value
c_1	5
c_2	1
c_3	25
c_4	.25

of the vanishing tail of the time-to-failure PDF and c_3 otherwise would be insignificant.

A graphic representation of the cost is shown in Fig. 8.9. If maintenance is not possible, the cost is set to infinity. Since the solution is determined numerically, we define infinity to be a large value, or in this case 25. The best possible time to perform maintenance was found to be at $t = 5.7$. At that time, the maintenance task requires at least three mechanics because the task must be completed before the hard-up time with a probability above 90 percent.

If the maintenance is scheduled as early as possible, the maintenance cost is significantly higher, as shown in Table 8.4.

For the example we use a simple simulated annealing algorithm to determine the optimal maintenance schedule (Ravi et al., 1997). This algorithm provides better results because of the constraint on the number of available maintenance "spots" for the repair, which had the effect of dividing the solution space into two parts. A simple gradient-descent algorithm would not produce satisfactory results. If the gradient-descent algorithm's starting point is not in the part of the solution space containing the minimum cost, the algorithm will only find a locally optimal solution. The simulated annealing

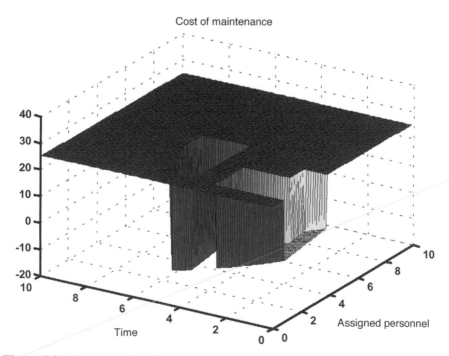

Figure 8.9   The cost of performing maintenance as a function of the scheduled maintenance time and the number of people assigned the maintenance task.

TABLE 8.4   Maintenance Cost Comparison

Time	Personnel	Cost
0	1	≈ 2
5.7	3	≈ -11

algorithm avoids this situation by introducing random effects into the optimization scheme; consequently, the search for the optimal solution does not "get stuck" at a local minimum.

Several software packages provide efficient and customizable optimization routines that can be readily applied to a wide variety of scheduling problems. Two such packages are Matlab by MathWorks (*www.mathworks.com*) and CPLEX by ILOG (*www.ilog.com*).

8.7   REFERENCES

T. M. Helm, S. W. Painter, and W. R. Oakes, "A Comparison of Three Optimization Methods for Scheduling Maintenance of High Cost, Long-Lived Capital Assets," *Proceedings of the Winter Simulation Conference,* San Diego, CA: IEEE Press, 2002, pp. 1880–1884.

J. Moubray, *Reliability-Centered Maintenance.* New York: Industrial Press, 2001.

V. Narayan, *Effective Maintenance Management: Risk and Reliability Strategies for Optimizing Performance.* New York: Industrial Press, April 2004.

Y. Ravi, B. Murty, and P. Reddy, "Nonequillibrium Simulated Annealing Algorithm Applied to Reliability Optimization of Complex Systems," *IEEE Transactions on Reliability* **46**(2):233–239, 1997.

APPENDIX

SOFTWARE FOR DIAGNOSIS

Kalman Filter MATLAB Code

```
A=[0 1; -0.9425 1.9]; B=[0; 1]; H=[1 0]; G=[1 0;0 1]; R=0.1; Q=[0.1 0;0 0.1];

%Initialization
x(:,1)=[0;0];
X_hat(:,1)=[0; 0];
X_hat_bar(:,1)=[0 ; 0];
p(:,:,1)=[0 0; 0 0];
p_bar(:,:,1)=[0 0;0 0];
var = sqrt(0.1);

for i=1:200
 u(i)=1;
 w(:,i)=[randn*var ; randn*var];
 %Discrete Time System
 x(:,i+1)=A*x(:,i)+B*u(i)+G*w(:,i);
 z(i)=H*x(:,i)+ randn*var;
 %error covariance - Time update
 p_bar(:,:,i+1)=A*p(:,:,i)*A'+G*Q*G';
 %estimate - time update
 X_hat_bar(:,i+1)=A*X_hat(:,i)+B*u(i);
 %error covariance - Measurement update
 p(:,:,i+1)=((p_bar(:,:,i+1))^-1+H'*R^-1*H)^-1;
 %estimate -Measurement update
 X_hat(:,i+1)=X_hat_bar(:,i+1)+p(:,:,i)*H'*R^-1*(z(i)-H*X_hat_bar(:,i+1));
end
```

RLS MATLAB Code

```
%Load data from text files
load u_data.txt;
U=u_data;
m=length(U);
load y_data.txt;
```

```
Y=y_data;
n=length(Y);

var =1;
mean =0;
gama=500;

I=eye(3,3);
P=gama*I;
%Initialize first two theta_hat
theta_hat(:,1)=[0;0;0];
theta_hat(:,2)=[0;0;0];
for k=3:n

%RLS Algorithm
h=[-Y(k-1); -Y(k-2); U(k-2)];
K=P*h*(h'*P*h+var)∧-1;
P = (I-K*h')*P;
theta_hat(:,k)=theta_hat(:,k-1)+K*(Y(k)-h'*theta_hat(:,k-1));
end

%System parameters
a1=theta_hat(1,n)
a2=theta_hat(2,n)
b0=theta_hat(3,n)
```

TSA in the Frequency Domain Code

```
%%%%%%%%%%%%%%%%%%%%%%%%%%%%%%%%%%%%%%%%%%%%%%%%%%%%
%%% Generated by Biqing (Becky) Wu, Sept. 2005 %%%
%%%%%%%%%%%%%%%%%%%%%%%%%%%%%%%%%%%%%%%%%%%%%%%%%%%%

clear;
clc;

nofSegments = 19;      % about 19 revolutions for each shapshot
NMR_THRESHOLD = -5000; % tachometer signal threshold

load torqueIndex1352-.mat Runidx Torque % corresponding run number
                                        % and the torque condition

n = 0;
for k = 1:1:length(Runidx)
 file_name = ['G:\Accels\PaxRun0' num2str(Runidx(k)) '.mat'];
 load(file_name, 'VMEP2', 'NMR');
 n = n+1;
 Torque2(n) = Torque(k);

 data = VMEP2;
 data2 = NMR;
 clear VMEP2 NMR;

 % Calculate the starting/ending point of each revolution
 i = 1;
 j = 1;
 idx = [];
 while(i<length(data2))
  if data2(i) < NMR_THRESHOLD
```

```
   idx(j) = i;
   i = i + 15000;
   j = j + 1;
  else
   i = i + 1;
  end
end


minRevlength = min(diff(idx));

for i=1:nofSegments
 % get a segment of data
 seg = data(idx(j):idx(j+1)-1);

 % fft each segment Data
 SpecSeg = fft(seg);
 Spec1(i,:) = SpecSeg(1:(floor(minRevlength/2)-1))/(length(SpecSeg)/2);
end

% averaging the thre frequency domain, discarding the extra points
% at the end of the axis
TSAinFDdata = mean(Spec1(1:nofSegments,:));

file_name = ['C:\VibData\TSAinFD_VMEP20' num2str(Runidx(k)) '.mat'];
save(file_name, 'TSAinFDdata');

clear Spec1 TSAinFDdata;

%Reconstruct data in the time domain
% FDdata = [TSAinFDdata 0 conj(fliplr(TSAinFDdata(2:end)))];
% TDdata = ifft(FDdata*(length(FDdata)/2));
% figure; plot(TDdata);
end
```

Note: IFFT of a conjugate-symmetric vector is real; that is, mirroring gives each block conjugate symmetry in mirroring; each block is flipped around, conjugated, and tacked onto the end of the original block. There are a few finer points, however. If the block is length M before mirroring, the $(M + 1)$th element of the mirrored block takes any real value (we set it to zero), whereas the $(M + 2)$th through $2M$th elements take the first through $(M - 1)$th elements (flipped) of the original block. Also, the first element in any block cannot be complex valued; it must be artificially set to some real value. This is so because the first point in any block corresponds to a frequency of zero after after the IFFT; if this element is complex, the resulting IFFT vector will have a complex DC offset and thus also will be complex.

WEB LINKS

Labs and Organizations

Intelligent Control Systems Laboratory, Georgia Institute of Technology
  http://icsl.marc.gatech.edu/

Impact Technologies, LLC
www.impact-tek.com/

Condition-Based Maintenance Laboratory, University of Toronto
www.mie.utoronto.ca/labs/cbm/

Advanced Diagnosis, Automation, and Control (ADAC) Laboratory, North Carolina State University (NCSU).
www.adac.ncsu.edu/

Maintenance and Reliability Center, University of Tennessee
www.engr.utk.edu/mrc/

Society for Machinery Failure Prevention Technology
www.mfpt.org/

Machinery Information Management Open Systems Alliance (MIMOSA)
www.mimosa.org

Open System Alliance for Condition Based Maintenance
www.osacbm.org

OSD CBM+
www.acq.osd.mil/log/mppr/

Reference Sites

AskaChe, a Chemical Engineering Help desk
www.askache.com

Condition Monitoring and Diagnostic Engineering Management (COMADEM)
www.comadem.com

The Consumer Review Site for Plant Reliability Community
www.reliabilityweb.com

DARPA Joint Strike Fighter Program
www.jsf.mil

NASA Facilities Engineering Division (JX Home Page)
www.hq.nasa.gov/office/codej/codejx/codejx.html

Reliability Center, Inc.
www.reliability.com

Plant Maintenance Resource Center
www.plant-maintenance.com

Links to Predictive Maintenance Resources
http://steelynx.net/PdM.html

Integrated Product and Process Development (IPPD)
www.goldpractices.com/practices/ippd/index.php

Department of the Navy, IPPD
www.abm.rda.hq.navy.mil/navyaos/acquisition_topics/program_management/ippd

Bollinger Bands
www.bollingerbands.com

Articles on Bearings and Vibration Analysis
www.plant-maintenance.com/maintenance_articles_vibration.shtml

Sensor OEM

Mechanical/Structural Sensor System
www.sensorland.com/TorqPage001.html,
www.oceanasensor.com

Accelerometers and Vibration Sensors
www.bksv.dk/
www.wilcoxon.com/
www.sensotec.com/accelerometers.asp

Temperature Senosrs/Thermography OEM
www.watlow.com/products/sensors/se_rtd.cfm
www.sensortecinc.com/
www.peaksensors.co.uk/
www.rdfcorp.com/products/industrial/r-21_01.shtml
content.honeywell.com/sensing/prodinfo/thermistors/isocurves.asp
www.thermistor.com/ptctherms.cfm
www.nanmac.com/handbook/a33.htm

Eddy Current Proximity Probes
http://proximity-sensors.globalspec.com/SpecSearch/Suppliers/
*Sensors_Transducers_Detectors/Proximity_Presence_Sensing/Eddy_*
Current_Proximity_Sensors

Panasonic Electric Works/Formerly Aromat
http://pewa.panasonic.com/acsd/

Lion Precision
www.lionprecision.com/

Metrix Instrument Co.
www.metrix1.com/

Micro-Electromechanical System (MEMS) Sensors
www.analog.com/en/subCat/
0,2879,764%255F800%255F0%255F%255F0%255F,00.html
http://content.honeywell.com/dses/products/tactical/hg1900.htm

Fiber Optic Sensors OEM

Luna Innovations
www.lunainnovations.com/r&d/fiber_sensing.asp

KEYENCE
http://world.keyence.com/products/fiber_optic_sensors/fs.html

Techno-Sciences
www.technosci.com/capabilities/fiberoptic.php

Wireless Sensor Networks
Crossbow
www.xbow.com
Microstrain's X-Link Measurement System
www.microstrain.com/

Software Tools

ANSYS
www.ansys.com
ABAQUS
www.abaqus.com
CPLEX
www.ilog.com/products/cplex/

INDEX